The World Re
by Artificial Int

## Also of Interest and from McFarland

*Keeping Schools Safe: Case Studies and Insights,*
edited by Joaquin Jay Gonzalez III and Roger L. Kemp (2023)

*Climate Change and Disaster Resilience: Challenges, Actions and Innovations in Urban Planning,*
edited by Joaquin Jay Gonzalez III, Roger L. Kemp and Alan R. Roper (2022)

*Syringe Exchange Programs and the Opioid Epidemic: Government and Nonprofit Practices and Policies,*
edited by Joaquin Jay Gonzalez III and Mickey P. McGee (2022)

*Gun Safety and America's Cities: Current Perspectives and Practices,*
edited by Joaquin Jay Gonzalez III, Roger L. Kemp (2022)

*Brownfields Redevelopment: Case Studies and Concepts in Community Revitalization,*
edited by Joaquin Jay Gonzalez III, Tad McGalliard and Ignacio Dayrit (2021)

*Cities and Homelessness: Essays and Case Studies on Practices, Innovations and Challenges,*
edited by Joaquin Jay Gonzalez III and Mickey P. McGee (2021)

*Senior Care and Services: Essays and Case Studies on Practices, Innovations and Challenges,*
edited by Joaquin Jay Gonzalez III, Roger L. Kemp and Willie Lee Brit (2020)

*Veteran Care and Services: Essays and Case Studies on Practices, Innovations and Challenges,*
edited by Joaquin Jay Gonzalez III, Mickey P. McGee and Roger L. Kemp (2020)

*Legal Marijuana: Perspectives on Public Benefits, Risks and Policy Approaches,*
edited by Joaquin Jay Gonzalez III and Mickey P. McGee (2019)

*Cybersecurity: Current Writings on Threats and Protection,*
edited by Joaquin Jay Gonzalez III and Roger L. Kemp (2019)

# The World Remade by Artificial Intelligence

*Perspectives on Applications, Impacts and Ethics*

*Edited by*
SHALINI S. GOPALKRISHNAN *and*
JOAQUIN JAY GONZALEZ III

ADVANCES IN PUBLIC PLANNING

McFarland & Company, Inc., Publishers
*Jefferson, North Carolina*

**ISBN (print) 978-1-4766-9753-6**
**ISBN (ebook) 978-1-4766-5645-8**

Library of Congress cataloging data are available



Printed in the United States of America

*McFarland & Company, Inc., Publishers*
*Box 611, Jefferson, North Carolina 28640*
*www.mcfarlandpub.com*

Shalini dedicates this book to her daughters Trisha
and Leka and to the future of humanity

* * *

Jay dedicates this book to his daughters
Elise and Coral and grandchildren Aiden and Alena

# Acknowledgments

We are grateful for the support of the Mayor George Christopher Professorship at Golden Gate University. We appreciate the encouragement from President David Fike, Provost Bruce Magid, Dr. Mick McGee, Dr. Siamak Zadeh, Dr. Nate Hinerman, Dr. Rocco Lamanna, Professor Robert Shoffner, our wonderful colleagues at GGU Worldwide and the Edward S. Ageno School of Business and our ever-supportive spouses, Gopal Krishnan and Michelle Hong-Gonzalez.

It took a kind and generous village to write and compile this unique, accessible, one-of-a-kind book. Our sincere, heartfelt thanks goes to the contributors listed in About the Contributors section and the individuals, organizations, and publishers listed below for granting their permission to publish or reprint the material in this volume. We also thank others listed below for their guidance, promotion, support, meals, and inspiration. Authors waived fees as an expression of their support for practical research, creative commons, and information sharing that benefits our communities and global community.

Adrian Hopgood
Aihua Yang-Zelinski
Alex Ignatov
Andrew Leonard
Beacon Education (China)
Bruce Schneier
City and County of San Francisco
*The Conversation*
Courtlin Holt-Nguyen
Dang Thi Thanh Tam
Daniel N.Y. Tan
Delaram Farzaneh
Doan Ngoc Duy
Emmanuel C. Lallana
Emmanuel G. Zara, Jr.
Ethan Zuckerman
European Commission
Francis Wang
Gautam Roy
Gopal Krishnan
Gopi Kallayil
Goutham Balasubramanian
Hannah Norman
Jagadish Gona
Jody C. Salas
Kannan Santharaman
Kevyn Eng
*KFF Health News*
Lakshmi R. Pillai
Larry Ebert
Lichao Zhang
May Bulanon
McKenzie Funk
Michael Wang
Michelle Andrews
Mike Horne
Nicole C. Jackson
Paula Andalo
Pearly Ee
Phalgun Kompali
Phil Galewitz

Praneet Singh
Pratik N. Shah
*ProPublica*
R. "Doc" Vaidhyanatha
Ratheesh Venugopal
Richard R. Khan
Ronald Powell
Ryan McGrady
Sachin Dole
Sameeksha Sahni
Sandeep Arora
Saroja M. N.
Seetha Anitha
Severo C. Madrona
Smrite Goutham
Sonny Panesar
State of Washington
Sulbha Shantwan
Sunil Manikani
Tachanun Rattanasiriwilai
Toscano and Salt Restaurants (India)
Tran Minh Viet
Umar Farooq
Upal Roy
upGrad Education
Vandinika Shukla
Veejay Madhavan
Vinay Singh
Virgil Labrador
White House Office of Science and Technology Policy

# Table of Contents

### • J. AI and Public Trust •



## Part III. The Future



## Part IV. Appendices

# Preface

Artificial intelligence (AI) has become a transformative force, shaping industries, societies, and individual lives in ways we are only beginning to comprehend. This edited volume brings together a diverse collection of perspectives to explore AI's promises, perils, and profound implications across sectors and disciplines. From its roots in theoretical computation to its modern applications in industries like healthcare, finance, and education, AI represents both an opportunity and a challenge—demanding critical reflection, ethical consideration, and innovative application.

As this compilation will demonstrate, AI offers tremendous potential benefits, enhancing human capabilities across various sectors such as healthcare, government, education, marketing and sales, arts, financial services, and data analysis, among others. AI promises more accurate medical diagnoses, groundbreaking scientific discoveries, efficient public services, and deep insights from vast data pools. However, AI also presents significant risks. The development of superintelligent AI systems that exceed human cognitive abilities poses significant risks, including the displacement of jobs and the potential for catastrophes caused by misaligned superintelligence with goals that do not align with human ethics, laws, and values.

With lucid contributions from global scholars, journalists, practitioners, and influencers, this book delves into both the remarkable opportunities and the challenges posed by AI across fields, industries, and geographies. We are all united in arguing for the adoption of this powerful technology while advocating for strong safeguards, ethical standards, social justice measures, and regulatory oversight to prevent the promising future of AI from being compromised by its perils.

We've been doing research on business, technology, and society for years, and we feel that it is important for us to always start any conversation by stressing important milestones from when we started to the present:

1936: Invention of the computer
1950: Alan Turing's paper on "Can machines think?" and the "Turing Test"
1955: The term "artificial intelligence" is coined by John McCarthy
1956: The field of Artificial Intelligence is born
1969: UNIMATE replaces humans at GM assembly line
1966: ELIZA, the first chatbot from MIT
1972: SHAKEY, the first robot from Stanford
1979: American Association of Artificial Intelligence
1996: DEEP BLUE, chess-playing computer
2002: ROOMBA, first mass produced autonomous robotic vacuum cleaner

2011: SIRI, Apple's virtual assistant, is integrated into the iPhone
2014: ALEXA, Amazon's virtual assistant, is launched
2016: Google Assistant initial release
2017: Machine Learning paper "Attention is all you need" is published
2022: ChatGPT is released by OpenAI
2023: Microsoft Copilot is born

The book is divided into four distinct yet interconnected sections, each providing a lens through which to understand the multi-faceted impact of AI. In Part I, Introduction, we set the stage with an exploration of AI's historical trajectory and its societal and economic implications. Contributions examine the ethical dilemmas surrounding generative AI, its effects on employment, and the potential to address global challenges, offering readers a foundational understanding of where AI has been and where it is headed.

Part II, Perils and Promises, delves into the intersection of AI with specific industries, showcasing its transformative role in marketing and sales, finance and financial services, human resources and training, hospitality, business consulting, education and the arts, industrial ecosystems, data infrastructure, and healthcare. We have balanced this section with important cases and perspectives on government, politics, and the public trust. These contributions highlight AI's capacity to enhance operational efficiency and innovation while also navigating its inherent risks—such as bias, ethical dilemmas, and legal and privacy concerns. The case studies included provide real-world insights into how organizations, companies, entrepreneurs, and governments are leveraging AI to drive value and adapt to a rapidly changing technological landscape.

In Part III, The Future, the book shifts focus to the future of AI. As emerging, evolving, and disruptive technologies push boundaries, we do in-depth explorations of pressing questions about trust, regulation, and societal impact. Contributions address existential concerns about AI, the black-box nature of machine learning, and the role of responsible governance. These discussions are particularly crucial as humanity seeks to balance crucial technological advancement with transparency, care, and accountability.

Finally, Part IV, Appendices, offers practical tools and frameworks through a series of appendices. These include glossaries, policy guidelines, and regulatory blueprints from governments and organizations worldwide, providing a roadmap for navigating the ethical and practical complexities of AI deployment.

This collection is the result of contributions from experts across academia, industry, and public policy, representing a global view of AI's perils and promises. By bringing together such a rich array of perspectives, the book aims to foster critical thinking, inspire innovative approaches, and encourage collaborative efforts in shaping AI's future. It is our hope that this volume serves as a valuable resource for citizens, consumers, researchers, practitioners, policymakers, and anyone interested in understanding and leveraging the power of artificial intelligence equitably and responsibly.

We hope that you have a "inspiring read" on these important and timely narratives and actions on AI.

Part I

# Introduction

# 1. A Brief History of AI*

*How We Got Here and Where We Are Going*

Adrian Hopgood

With the current buzz around artificial intelligence (AI), it would be easy to assume that it is a recent innovation. In fact, AI has been around in one form or another for more than 70 years. To understand the current generation of AI tools and where they might lead, it is helpful to understand how we got here.

Each generation of AI tools can be seen as an improvement on those that went before, but none of the tools are headed toward consciousness.

The mathematician and computing pioneer Alan Turing published an article in 1950 with the opening sentence: "I propose to consider the question, 'Can machines think?.'" He goes on to propose something called the imitation game, now commonly called the Turing test, in which a machine is considered intelligent if it cannot be distinguished from a human in a blind conversation.

Five years later, came the first published use of the phrase "artificial intelligence" in a proposal for the Dartmouth Summer Research Project on Artificial Intelligence.

Do experts have something to add to public debate?

From those early beginnings, a branch of AI that became known as expert systems was developed from the 1960s onward. Those systems were designed to capture human expertise in specialized domains. They used explicit representations of knowledge and are, therefore, an example of what's called symbolic AI.

There were many well-publicized early successes, including systems for identifying organic molecules, diagnosing blood infections, and prospecting for minerals. One of the most eye-catching examples was a system called R1 that, in 1982, was reportedly saving the Digital Equipment Corporation US$25m per annum by designing efficient configurations of its minicomputer systems.

The key benefit of expert systems was that a subject specialist without any coding expertise could, in principle, build and maintain the computer's knowledge base. A software component known as the inference engine then applied that knowledge to solve new problems within the subject domain, with a trail of evidence providing a form of explanation.

---

*Originally published as Adrian Hopgood, "A Brief History of AI: How We Got Here and Where We Are Going," *The Conversation*, https://theconversation.com/a-brief-history-of-ai-how-we-got-here-and-where-we-are-going-233482 (June 28, 2024). Reprinted with permission of the publisher.

These were all the rage in the 1980s, with organizations clamoring to build their own expert systems, and they remain a useful part of AI today.

## *Enter Machine Learning*

The human brain contains around 100 billion nerve cells, or neurons, interconnected by a dendritic (branching) structure. So, while expert systems aimed to model human knowledge, a separate field known as connectionism was also emerging that aimed to model the human brain in a more literal way. In 1943, two researchers called Warren McCulloch and Walter Pitts had produced a mathematical model for neurons, whereby each one would produce a binary output depending on its inputs.

One of the earliest computer implementations of connected neurons was developed by Bernard Widrow and Ted Hoff in 1960. Such developments were interesting, but they were of limited practical use until the development of a learning algorithm for a software model called the multi-layered perceptron (MLP) in 1986.

The MLP is an arrangement of typically three or four layers of simple simulated neurons, where each layer is fully interconnected with the next. The learning algorithm for the MLP was a breakthrough. It enabled the first practical tool that could learn from a set of examples (the training data) and then generalize so that it could classify previously unseen input data (the testing data).

It achieved this feat by attaching numerical weightings on the connections between neurons and adjusting them to get the best classification with the training data, before being deployed to classify previously unseen examples.

The MLP could handle a wide range of practical applications, provided the data was presented in a format that it could use. A classic example was the recognition of handwritten characters, but only if the images were pre-processed to pick out the key features.

## *Newer AI Models*

Following the success of the MLP, numerous alternative forms of neural network began to emerge. An important one was the convolutional neural network (CNN) in 1998, which was similar to an MLP apart from its additional layers of neurons for identifying the key features of an image, thereby removing the need for pre-processing.

Both the MLP and the CNN were discriminative models, meaning that they could make a decision, typically classifying their inputs to produce an interpretation, diagnosis, prediction, or recommendation. Meanwhile, other neural network models were being developed that were generative, meaning that they could create something new, after being trained on large numbers of prior examples.

Generative neural networks could produce text, images, or music, as well as generate new sequences to assist in scientific discoveries.

Two models of generative neural network have stood out: generative-adversarial networks (GANs) and transformer networks. GANs achieve good results because they are partly "adversarial," which can be thought of as a built-in critic that demands improved quality from the "generative" component.

Transformer networks have come to prominence through models such as GPT4 (Generative Pre-trained Transformer 4) and its text-based version, ChatGPT. These large-language models (LLMs) have been trained on enormous datasets, drawn from the Internet. Human feedback improves their performance further still through so-called reinforcement learning.

As well as producing an impressive generative capability, the vast training set has meant that such networks are no longer limited to specialized narrow domains like their predecessors, but they are now generalized to cover any topic.

## *Where Is AI Going?*

The capabilities of LLMs have led to dire predictions of AI taking over the world. Such scaremongering is unjustified, in my view. Although current models are evidently more powerful than their predecessors, the trajectory remains firmly toward greater capacity, reliability and accuracy, rather than toward any form of consciousness.

As Professor Michael Wooldridge remarked in his evidence to the UK Parliament's House of Lords in 2017, "the Hollywood dream of conscious machines is not imminent, and indeed I see no path taking us there." Seven years later, his assessment still holds true.

There are many positive and exciting potential applications for AI, but a look at the history shows that machine learning is not the only tool. Symbolic AI still has a role, as it allows known facts, understanding, and human perspectives to be incorporated.

A driverless car, for example, can be provided with the rules of the road rather than learning them by example. A medical diagnosis system can be checked against medical knowledge to provide verification and explanation of the outputs from a machine learning system.

Societal knowledge can be applied to filter out offensive or biased outputs. The future is bright, and it will involve the use of a range of AI techniques, including some that have been around for many years.

# 2. The Zero World

## *Envisioning a Future Where AI Eliminates Global Challenges*

SHALINI GOPALKRISHNAN

In recent years, the concept of a "Zero World" has emerged as a compelling vision of humanity's future—a world where artificial intelligence helps eliminate fundamental challenges like hunger, disease, poverty, and war. While this might sound like science fiction, rapid advances in AI technology are making such a future increasingly plausible. This piece explores how AI is and could help achieve these ambitious "zero" goals and examines both the opportunities and challenges in realizing this vision.

### *Zero Hunger: AI-Driven Food Security*

The elimination of global hunger represents one of the most promising applications of AI technology. According to the UN Food and Agriculture Organization, nearly 735 million people suffered from hunger in 2023 which is around 8 percent of the world, a large number (WHO, July 2024). However, AI solutions are already showing remarkable potential in addressing this challenge.

Artificial Intelligence offers unprecedented opportunities for agriculture. From enhancing crop yield and quality to optimizing resource usage, AI's impact is far-reaching. Whether analyzing land use with high-precision satellite imagery or predicting crop diseases through real-time monitoring, AI applications are gradually taking root globally. Some areas include:

a. Precision agriculture, powered by machine learning algorithms, is revolutionizing food production. These systems analyze satellite imagery, soil data, and weather patterns to optimize crop yields while minimizing resource usage. For example, Microsoft's FarmBeats project has demonstrated yield increases of up to 30 percent while reducing water consumption by 30 percent (Smith & Johnson, 2023).

b. Vertical farming, enhanced by AI-controlled environments, is another promising development. Studies suggest that AI-optimized vertical farms could produce up to 100 times more food per square meter than traditional farming methods (Hopkin 2023). Companies like AeroFarms are already implementing these technologies, showing how urban areas could become self-sufficient in food production.

## *Zero Disease: AI-Powered Healthcare Revolution*

The vision of a world without disease might seem utopian, but AI is already transforming healthcare in remarkable ways. Deep learning systems are now matching or exceeding human experts in disease diagnosis across multiple specialties (Razzaki et al. 2018).

Recent breakthroughs in protein folding prediction by systems like AlphaFold have accelerated drug discovery dramatically. The ability to accurately predict protein structures has reduced the time needed to develop new treatments from years to months. DeepMind's research suggests that AI could eventually help develop treatments for all known diseases. In another study "ML algorithms play an important role in testing systems and can predict important aspects such as the pharmacokinetics and toxicity of drug candidates" (Dhudum et al. 2024).

Preventive medicine, powered by AI analysis of genetic and lifestyle data, is enabling increasingly personalized healthcare approaches. The aim of precision medicine is to design and improve diagnosis, therapeutics and prognostication through the use of large complex datasets that incorporate individual gene, function, and environmental variations.

The implementation of high-performance computing (HPC) and artificial intelligence (AI) can predict risks with greater accuracy based on available multidimensional clinical and biological datasets. AI-powered precision medicine provides clinicians with an opportunity to specifically tailor early interventions to each individual. Here we discuss the strengths and limitations of existing and evolving recent, data-driven technologies, such as AI, in preventing, treating and reversing lifestyle-related diseases (Subramanian et al. 2020).

## *Zero Poverty: Economic Transformation Through AI*

While critics argue that AI might exacerbate economic inequality, proponents of the Zero World vision suggest that properly implemented AI systems could actually eliminate poverty. The World Bank has identified several promising applications of AI in poverty reduction.

Financial Inclusion: AI-powered mobile banking and microlending systems are already providing financial services to previously unbanked populations. M-PESA's success in Kenya demonstrates how digital financial services can lift millions out of poverty. M-Pesa in Kenya revolutionized financial inclusion by enabling the unbanked poor to participate in the digital economy through their mobile phones. Users can store, send, and receive money electronically at low cost. Through a network of local agents, such as shopkeepers, they can convert between cash and digital money ("Cash In, Cash Out" or CICO)—essential for participating in the predominantly cash-based economies of developing nations. The system's success has spread globally, with over 850 million registered mobile money accounts across 90 countries now processing USD $1.3 billion in daily transactions (Pasti, 2018).

Access to financial services plays a crucial role in expanding opportunities and building resilience for those in poverty, with particularly significant impact on women. However, in the world's poorest economies, 65 percent of adults lack even a basic transaction account—a fundamental tool for safe and efficient payments. These basic accounts serve as an entry point to broader financial services including savings, insurance, and credit. Currently, only 20 percent of adults in developing economies save through formal financial institutions, while the majority resort to informal and more expensive saving methods (Pazarbasioglu et al. 2020).

Skills Development: AI-driven educational platforms are democratizing access to high-quality education and job training. Studies show that AI-powered adaptive learning systems can accelerate skill acquisition. In one study, Komasawa and Yokohira (2023) applied AI for experiential learning for medical education. In an increasingly AI-driven healthcare environment, experiential-based medical education (EXPBME) is crucial for preparing medical practitioners. Through clinical settings and simulations, EXPBME enables healthcare professionals to understand and respond to potential AI system limitations and failures. As artificial intelligence becomes more prevalent in medicine, it's essential to adapt educational frameworks to develop both technical and non-technical skills. The future of medical education must therefore combine learner-centered experiential training with comprehensive AI literacy to create competent, adaptable healthcare professionals (Komasawa and Yokohira, 2023).

Resource Optimization: AI systems can help optimize resource distribution and public services, ensuring more efficient allocation of limited resources to those most in need.

## *Zero War: AI as a Peacekeeper*

Perhaps the most ambitious aspect of the Zero World vision is the elimination of war. While this goal might seem unrealistic, AI could contribute to peace by using data to predict and thus control it before the event.

Conflict Prediction: Advanced AI systems can analyze vast amounts of data to identify potential conflicts before they escalate. The Peace Technology Lab has demonstrated 85 percent accuracy in predicting regional conflicts six months in advance (Owolabi et al, 2024).

## *Challenges and Considerations*

While the Zero World vision is compelling, several significant challenges must be addressed:

### Ethical Considerations

The implementation of AI systems at this scale raises important ethical questions. Issues of privacy, autonomy, and decision-making authority must be carefully considered. The IEEE Global Initiative on Ethics of Autonomous and Intelligent Systems provides important guidelines for addressing these concerns (IEEE, 2024).

### Technical Challenges

Despite rapid advances, significant technical challenges remain. These include:

- Ensuring AI system reliability and safety
- Managing computational resources sustainably
- Developing robust security measures
- Maintaining privacy in data-driven systems

### Social and Political Obstacles

The transition to a Zero World requires significant social and political changes. Questions of governance, equality, and cultural adaptation must be addressed. The UN's AI for Good initiative provides a framework for managing these transitions (United Nations, 2024).

## *Implementation Roadmap*

Achieving the Zero World vision requires a coordinated global effort. Experts suggest a phased approach:

Foundation Phase (2025–2030)

1. Establish global AI governance frameworks
2. Develop core technologies and infrastructure
3. Build public-private partnerships

Acceleration Phase (2030–2035)

1. Scale successful pilot programs
2. Implement global AI systems
3. Address emerging challenges

Integration Phase (2035–2040)

1. Achieve full integration of AI solutions
2. Monitor and adjust systems
3. Ensure equitable access

## *Conclusion*

The Zero World vision represents an ambitious but increasingly achievable goal. While significant challenges remain, the convergence of AI technologies with global development efforts offers unprecedented opportunities to eliminate humanity's most persistent challenges.

Success will require careful attention to ethical considerations, robust technical solutions, and strong international cooperation. However, the potential benefits—a world free from hunger, disease, poverty, and war—make this effort worthwhile.

## References

Dhudum, R., Ganeshpurkar, A., & Pawar, A. (2024). Revolutionizing Drug Discovery: A Comprehensive Review of AI Applications. *Drugs and Drug Candidates*, 3(1), 148–171.

Hopkin 2023: https://aimagazine.com/articles/ai-and-vertical-farms-to-keep-world-on-top-of-food-demand.

IEEE. (2024). "Ethical Guidelines for Autonomous Systems." *IEEE Global Initiative*.

Komasawa, N., & Yokohira, M. (2023). Learner-centered experience-based medical education in an AI-driven society: a literature review. *Cureus*, 15(10).

Owolabi, H.A., Oyedele, A.A., Oyedele, L., Alaka, H., Olawale, O., Aju, O., & Ganiyu, S. (2024). Big data innovation and implementation in projects teams: towards a SEM approach to conflict prevention. *Information Technology & People*.

Pasti, Francesco. State of the Industry Report on Mobile Money 2018. GSMA, 2019. https://www.gsma.com/r/wp-content/uploads/2019/05/GSMA-State-of-the-Industry-Report-on-Mobile-Money-2018-1.pdf.

Pazarbasioglu, C., Mora, A.G., Uttamchandani, M., Natarajan, H., Feyen, E., & Saal, M. (2020). Digital financial services. World Bank, 54, 1–54.

Razzaki, S., Baker, A., Perov, Y., Middleton, K., Baxter, J., Mullarkey, D., ... & Johri, S. (2018). A comparative study of artificial intelligence and human doctors for the purpose of triage and diagnosis. *arXiv preprint arXiv:1806.10698*.

Sharma, K.K., Pawar, S.D., & Bali, B. (2020). Proactive preventive and evidence-based artificial intelligene models: future healthcare. In International Conference on Intelligent Computing and Smart Communication 2019: *Proceedings of ICSC 2019* (pp. 463–472). Springer Singapore.

Subramanian, M., Wojtusciszyn, A., Favre, L., Boughorbel, S., Shan, J., Letaief, K.B., ... & Chouchane, L. (2020). Precision medicine in the era of artificial intelligence: implications in chronic disease management. *Journal of translational medicine*, 18, 1–12.

United Nations. (2024). "AI for Good: Global Summit Report." UN Publications.

Vasisht, D., Kapetanovic, Z., Won, J., Jin, X., Chandra, R., Sinha, S., ... & Stratman, S. (2017). {FarmBeats}: an {IoT} platform for {Data-Driven} agriculture. In 14th USENIX Symposium on Networked Systems Design and Implementation (NSDI 17) (pp. 515–529).

WHO (2024). Hunger numbers stubbornly high for three consecutive years as global crises deepen: UN report: https://www.who.int/news/item/24-07-2024-hunger-numbers-stubbornly-high-for-three-consecutive-years-as-global-crises-deepen—un-report.

# 3. AI Is Changing the World of Business

Gopi Kallayil

We live in an incredible time, the time of artificial intelligence (AI), the most recent of the three greatest shifts we've experienced during our lives. First came the consumer internet in the 1990s, then the mobile revolution of the 2000s, both shifting how we live, shop, and do business. And now we're witnessing the platform shift of AI. Artificial intelligence is also the most profound technology created by human beings, ever, on par with or even more significant than those once-in-a-century technology shifts such as the printing press in 1414, electricity in 1762, and the automobile in 1886. And now, here we are, in 2024.

In December 2023, Google announced Gemini, our most versatile and powerful AI model to date. While the world has seen models built around single modalities, Gemini is built from the ground to be multimodal, which means it can generalize and seamlessly understand, operate across, and combine different types of information, including text, audio, image, video, and even code. This makes our Gemini models the most capable and general AI model.

"We're excited by the amazing possibilities of a world responsibly empowered by AI," stated Google Deepmind CEO Demis Hassibis—"a future of innovation that will enhance creativity, extend knowledge, advance science, and transform the way billions of people live and work around the world."

## *AI Deconstructed*

Before I go any further, let me demystify artificial intelligence for you. To do that, I need to first discuss a unique human capability, NI, natural intelligence, the type of intelligence that created today's ballpoint pen. It's a simple instrument, with elements that have taken thousands of years to develop. The forging process, used to melt and shape metal, began in the Middle East between 4000 and 7000 BCE. The ancient Egyptians and Chinese graduated from communicating thoughts on a tablet with a chisel to painting them with rudimentary ink. Then came the engineering of the compression spring and the invention of plastic. Finally, the function: easily convey readily available ideas for others to read. Add branding, pricing, production, and distribution. All these elements are built into this simple product.

But NI natural intelligence is a limited resource. If your leadership team consists of five people, then that team possesses the intelligence of five people. What if, with this same team of five people, you could tap into the intelligence of fifty people? This is what artificial intelligence is about and what computer scientists have been trying to do: mimic our natural intelligence in computer systems and *amplify* that capability, emphasizing the word *amplify* rather than *replace*.

Now, AI mimics the human brain and amplifies it. It also mimics the two aspects, or two sides, the left and right hemispheres, of the brain. Conventional wisdom holds that people tend to have two ways of interacting with the world that is either right-brained (more intuitive and creative) or left-brained (more quantitative and analytical). Artificial intelligence systems mimic different types of cognitive functions. Systems that mimic the right side of the brain are what we broadly refer to as generative AI, while AI systems that mimic the left side of the brain are referred to as analytical AI.

In recent years, we've come to understand that the brain is far more complex and interconnected and plastic than this earlier division suggests. Just as the human brain integrates both hemispheres to function effectively, advanced AI systems are being developed to combine generative and analytical capabilities, allowing them to manage a wider range of tasks with greater sophistication.

## *Google's AI Strategy*

Google has been working on AI for a long time. In fact, we declared ourselves to be an AI-first company in 2016. So what is Google's AI strategy? We're looking at three audiences—consumers, communities, and companies—and helping each of these audiences reach their highest potential with AI. Let me explain this one step at a time.

## *Consumers*

There are fifteen Google products that are used by half billion people or more. In fact, six products are used by two billion people or more.

We're looking at each of these products and asking, "How can we enhance the experience for you as a consumer by using the power of AI across all of these products"? Let me give you one example: search. Historically, when you wanted to search for something on Google, you've been forced to go to a rectangular box on your desktop, mobile device, or tablet and ask a question.

But the natural way we search is different. We look at something, point to it, and ask, "What is that"? Today, you can look, point, and ask on your device. Let's say you're on your phone, maybe looking at a picture of a friend who's wearing an amazing pair of shoes, so you point to the shoes and circle them, initiating a valid search experience and conversation around it. This capability is called Circle to Search.

## *Communities*

We're facing a number of societal problems right now: violence, mental health, inequality, and climate change, to name a few. At Google, we've been asking, "Are there any of these that we can use AI to solve for?" For example, floods affect more people than any other environmental hazard. And today, due to climate change and vulnerable populations being forced to live in these areas, half the world's population lives in a flood-prone area. When you have an unpredictable flooding situation, which we have, there's a lot of damage to property. It hinders sustainable building. So we asked the question, "Can we use AI to predict flooding?"

1. We start by collecting thousands of satellite images to build a digital model of the terrain.
2. Based on these maps, we generate hundreds of thousands of simulations of how the river could possibly behave.
3. We receive the measurements from the government and cross those measurements with our simulations.
4. We can then send those forecasts to individuals using search maps and natural notifications.

## *Companies*

How we can help companies grow with the power of AI is best illustrated in the case study of Omoda, an innovative Dutch fashion retailer.

They've been going through a period of extreme change because of the acceleration of online shopping, the variety of payment methods that are available, and the introduction of fast fashion. And what all of these have meant for Omoda, means more returns. Up to 50 percent of the items are returned. And it's not simply a business challenge, but very specifically, they have to address two things: financial so the counting is done correctly when items are returned and environmentally, you have to dispose of these items in a responsible manner.

Omoda came to us and asked, "Can we build an AI-based system that will predict whether an order will be returned? And if so, how much of the order?" So we built a model looking at both historical data where we know the items with a higher probability of being returned and new real-time data, which included variables like payment methods or order size. The model then gave a prediction, and Omoda used the prediction to change its operating process model. As a result, the profit margin increased by 40 percent.

## *Conclusion*

To close this off, I'd like you to think about this simple question as you start off with your business strategy: What is your business strategy for growth this year? Then, the important question you need to ask yourself is, *How can I use AI to enable the strategy?*

# 4. AI and Employment: Bridging Skill Gaps or Expanding Inequality

PEARLY EE

Recent development of artificial intelligence has impacted the global employment sector intensively, which raised crucial questions about the future of work. As prospects for AI become less abstract, researchers and policymakers have been faced with the increasingly complex and nuanced nature of the impacts of AI on the labor market.

A major issue is that AI can change the purpose of skills and affect jobs. Even though many basic tasks at the workplace will be automated by AI, such skill discrepancy of labor force might be polarized by jobs that create new demand for certain skills and jobs that do not (Bian et al., 2022; Chen et al., 2022; Braiki et al., 2020). AI productivity gains, on the one hand, might specifically displace human work, aggravating density of employment discrimination. Moreover, AI has the potential to drive new jobs and employment opportunities, which may balance the jobs being lost to automation due to innovations like rapid scaling of industrial processes and structural upgrades made solely due to AI technologies (Bian et al., 2022; Braiki et al., 2020).

Coupled with this AI impact, the changing nature of skills and the education and training landscape that goes along with this are tremendous. Researchers found that although the growth of the labor supply persists, the impact of technological progress is still considerable on the laborers through employment and wage premiums, a finding known as skill-biased technological change (Frank et al, 2019; Chen et al, 2022). This presents the necessity to prepare for the displacement risk from AI by developing digital skills as well as reforming technical and vocational education.

The key question is whether AI will close skill gaps or create inequality in society. To do so, this contribution will analyze the interrelationship between displacement risk, new competency requirements, and labor market outcomes, with the help of recent empirical evidence.

## *The Promise of AI in Employment*

Supporters of AI claim that the potential of this technology can revolutionize productivity, generate new jobs, and contribute to economic growth (Ernst et al., 2019). AI will revolutionize many such industries including education, defense, business, law and science, by automating menial tasks, improving decision-making, and offering new modes of service-delivery.

The potential for AI to boost productivity could be particularly large in labor-intensive sectors; in particular, it can be highly impactful in developing countries that have greatly lowered the price of capital (Ernst et al., 2019). An example of a manual task that can be automated by AI includes processing data periodically, inventory management, and in some manufacturing processes such as assembly and quality control (Furman & Seamans, 2018).

Moreover, productivity growth driven by AI could result in industrial scale expansion and structural upgrades, creating new jobs and employment. The demand for emerging industry repairers, conductors, managers, and financiers could compensate for replacement in other professions through AI automation, for example.

Finally, the lower capital costs that come with certain AI use cases can help developing countries, which may also translate into higher productivity and economic growth (Ernst et al., 2019). AI is also credited with making personalized learning or re-skilling easier by understanding employee needs and characteristics to easily determine the most effective way of re-skilling for each employee (Georgieff & Hyee, 2022; Frank et al., 2019).

Discourse surrounding the potential benefits of AI in employment are common yet cannot overshadow the economic displacement risks associated with high-technology economic transformations.

## *The Perils of AI in Employment*

The concern regarding AI in the labor market for jobs is job displacement, either through full automation or through change in the demand for skill. On the one hand, the adoption of AI may directly automate many human jobs and activities, which would cause massive unemployment, especially for routine and monotonous jobs. Thus, while AI will change the demand for skills, it will also contribute to skill misalignment, the mismatch between people's skills and jobs, making it harder for workers to adjust (and thereby remain competitive) on the labor market.

Further, the skills needed in the workforce are changing at a fast pace and the gap between the skills of the workers and the skills needed for the jobs of today and tomorrow are growing larger. Skill-biased technological change can contribute to rising income inequality and social stratification, as demand for skilled workers explode, and wage premium for skilled workers expand while low-skilled workers face shrinking labor-market opportunities and stagnant wages. This is exacerbated by skill-biased technological change as the creation of job opportunities supported directly by re-skilling benefits in the context of their employment is always available to highly skilled workers but not to lower skilled workers (Deming & Noray, 2018).

Additionally, in some sectors, such as healthcare, it is feared that AI systems may be used to rationalize the hiring of lower grades of personnel who may not be able to maintain the quality-of-service delivery (Hazarika, 2020). Inadequate training affiliated with AI systems leads to incompetency among staff, and when this happens it also reduces the likelihood of identifying errors or even the ability to perform essential tasks without the help of the computing machine. AI could be a source of employment discrimination, with algorithmic bias in AI systems being a potential contributor to biased hiring and promotion. They are complicated, multi-faceted, and situational; and they tend to not target communities experiencing marginalization equally.

One form of algorithm bias in hiring is the process of picking candidates based on certain demographic attributes of the candidate (e.g., school attended), while discarding other candidates that are devoid of said demographic characteristics (e.g., color of the applicant), or AI-based resume autopilot showing bias against disabled applicants (Buyl et al., 2022). These are often the result of data that organizations have compiled on their internal performance and attrition trends, which are then manipulated to introduce bias in their AI resume screening process (Rosenblat et al., 2014).

## *Ethical and Operational Challenges in AI-Driven Employment*

While the value creation of AI is greatly acknowledged, the widespread use of this technology in the job market creates several ethical and operational issues. One major issue is the possible bias in the training data for AIs. Training data refers to the data that the AI model is trained on. When this training data is unrepresentative, skewed or biased, it means that formal, systematic bias will be reproduced (and bluntly amplified) in the AI system's decision-making algorithms.

In addition, many of the algorithms used in AI are black boxes, which makes it challenging to comprehend and articulate the rationale behind its decisions. It makes accountability arduous by undermining it, making it difficult to find and fix the biases and errors in these algorithms. Organizations tend to buy these kinds of "black box" AI systems for hiring and performance management without needing to understand how they work (Goldenfein, 2019), which compound accountability problems. To enable responsible AI, the providers of AI systems must be transparent, and so must the end users (the organizations).

Moreover, organizations can use AI widely as a tool for surveillance to measure the performance and productivity of the workforce. Such practices cause serious issues concerning privacy, autonomy, and dignity of individuals, because the surveillance of workers powered by AI and ML can create an oppressive environment and turn workers into distrustful humans. AI-based monitoring software is used by organizations to collect metrics including the volume of keystrokes or mouse clicks, break times, web history and screenshot tracking, and tally the productivity of each individual worker (Hickok & Maslej, 2023).

AI poses many ethical and operational challenges to organizations but by prioritizing transparency, accountability, and worker's consent in the use of an AI, organizations can overcome these challenges. It will be through partnerships between employees, employers, policymakers, and AI specialists to create strong governance and ethical frameworks. Nevertheless, since AI remains a relatively untapped field, particularly across various domains, there are currently no significant governmental regulations governing it (Khan et al., 2023; Payton, 2023). Closing this gap is crucial to addressing the risks and effects of AI on employment. To ensure the transparent use of AI, organizations can introduce "AI Champions" in departments who can oversee and interface between the workforce and AI systems. Government regulations should also compel organizations to be answerable for the impacts and merits of AI (Han et al., 2023; Golbin et al., 2020).

## *Inclusive AI Integration*

In addition to the ethical and operational challenges, the integration of AI in the workplace must also consider the need for inclusive and equitable access to the benefits of this technology. AI can exacerbate existing inequalities if its implementation is not carefully designed to address the needs of diverse stakeholders, including marginalized communities and underrepresented groups.

Key principles for organizations to better integrate inclusive AI are:

- Investment in Re-skilling program: Organizations must invest in large-scale upskilling programs and re-skilling programs to help workers, especially low-skill and vulnerable workers navigate these adaptations, thereby preparing them for future jobs with relevant and essential skills.
- Diversity and representation in AI development: The people building and deploying AI systems should be diverse and representative of the larger population—it does not help to have a homogenous group building and deploying these systems as the concerns of those affected may not be adequately addressed in the algorithms.
- Regulatory standards: Governments and policymakers should work on the creation of regulatory standards that can be applied globally, that govern the usage of AI in the workplace, taking care to protect workers' rights and avoid discrimination.
- Hybrid Human-AI work models: organizations must prioritize hybrid work models that leverage the best of AI and human workers, instead of trying to replace human labor fully.

Different government organizations and industry interest groups must work together and help smaller businesses who cannot financially afford to put some of these measures in place like re-skilling existing employees. These may be in the form of government tax cuts or financial grants, or corporate loans that support smaller businesses undertaking such initiatives (Balbina, 2023). For workers who may be displaced by AI automation either in the economy or in their workplace, re-training programs are necessary to help them transition into other roles, industries, or sectors of the economy. Those affected might disproportionately include lower-skilled workers, hence government agencies could also provide help such as subsidizing training allowance for those who need to take a leave from work to learn new skills (Lee & Hsin, 2004).

Industry groups may come forward with a regulatory framework together with the government on ethical deployment of AI-based technologies across sectors of business, employment and workforce management. The framework of regulations should provide a set of specific parameters around the gathering and use of data on workers by AI systems, as well as organizational obligations and enforcement provisions (Fernández-Martínez & Fernández, 2020). The framework should be reviewed regularly to update it to the fast pace of the AI technology evolution and constructive comments received from the stakeholders. This is why the diversity of those drafting such policies and who gets heard in the process should also be representative enough to consider all segments of the workforce such as disability advocates (Morris, 2020).

## *Conclusion*

AI technology is advancing at such a pace now, that businesses must tread a fine line as to how they can use it and remain on the right side of the moral and operational lines moving forward in the future of work. AI will complement and enhance work, creating a value-added opportunity for organizations to harness to remain competitive. AI is a great equalizer for advancing inclusivity, productivity and social goods, if used carefully. It leads to more equitable access to resources and services and opportunities (Chen, 2020; Ernst et al., 2019). AI assumes a leading role in the future of work and is also seen as the engine of Industry 5.0.

AI is the future, and society should not be silent about its effects on generations to come. All of society and all parties involved must proactively play a role to harmoniously dictate where AI use will be heading in the future. Work ethic models need to be forward thinking and foster collaboration and inclusion that will not leave segments of society (Hernández, 2024; Dautil et al., 2021). Only with such collaborative efforts might we guarantee the ethical and inclusive potential of AI at work, ensuring society maximizes the benefits of AI and minimizes dangers of expanding inequality.

### References

Balbina. (2023). Firm Investments in Artificial Intelligence Technologies and Changes in Workforce Composition. https://www.nber.org/papers/w31325.

Bian, Y., Lu, Y., & Li, J. (2022). *Research on an Artificial Intelligence-Based Professional Ability Evaluation System from the Perspective of Industry-Education Integration*. Hindawi Publishing Corporation, 2022, 1–20. https://doi.org/10.1155/2022/4478115.

Braiki, B A., Harous, S., Zaki, N., & Alnajjar, F. (2020). Artificial intelligence in education and assessment methods. Institute of Advanced Engineering and Science (IAES), 9(5), 1998–2007. https://doi.org/10.11591/eei.v9i5.1984.

Buyl, M., Cociancig, C., Frattone, C., & Roekens, N. (2022). Tackling Algorithmic Disability Discrimination in the Hiring Process: An Ethical, Legal and Technical Analysis. https://doi.org/10.1145/3531146.3533169.

Chen, L Y. (2020). A conceptual framework for AI system development and sustainable social equality. https://doi.org/10.1109/ai4g50087.2020.9310984.

Chen, N., Li, Z., & Tang, B. (2022). Can digital skill protect against job displacement risk caused by artificial intelligence? Empirical evidence from 701 detailed occupations. *Public Library of Science*, 17(11), e0277280-e0277280. https://doi.org/10.1371/journal.pone.0277280.

Dautil, F., Fregin, M., Hofmann, M., Luckin, R., Majchrzak, T A., Broecke, S., Lane, M., Salvi, A., Schleicher, A., & Van, R. (2021). Opportunities and drawbacks of using artificial intelligence for training. https://doi.org/10.1787/22729bd6-en.

Deming, D., & Noray, K. (2018). STEM Careers and the Changing Skill Requirements of Work. https://doi.org/10.3386/w25065.

Ernst, E., Merola, R., & Samaan, D. (2019). Economics of Artificial Intelligence: Implications for the Future of Work. *Springer Science+Business Media*, 9(1). https://doi.org/10.2478/izajolp-2019-0004.

Fernández-Martínez, C., & Fernández, A. (2020). AI and recruiting software: Ethical and legal implications. De Gruyter Open, 11(1), 199–216. https://doi.org/10.1515/pjbr-2020-0030.

Frank, M R., Autor, D., Bessen, J., Brynjolfsson, E., Cebrián, M., Deming, D., Feldman, M P., Groh, M., Lobo, J., Moro, E., Wang, D., Youn, H., & Rahwan, I. (2019). Toward understanding the impact of artificial intelligence on labor. National Academy of Sciences, 116(14), 6531–6539. https://doi.org/10.1073/pnas.1900949116.

Furman, J., & Seamans, R. (2018). *AI and the Economy*. University of Chicago Press, 19, 161–191. https://doi.org/10.1086/699936.

Georgieff, A., & Hyee, R. (2022). Artificial Intelligence and Employment: New Cross-Country Evidence. Frontiers Media, 5. https://doi.org/10.3389/frai.2022.832736.
Golbin, I., Rao, A S., Hadjarian, A., & Krittman, D. (2020). Responsible AI: A Primer for the Legal Community. https://doi.org/10.1109/bigdata50022.2020.9377738.
Goldenfein, J. (2019). Algorithmic Transparency and Decision-Making Accountability: Thoughts for Buying Machine Learning Algorithms. RELX Group (Netherlands). https://papers.ssrn.com/sol3/papers.cfm?abstract_id=3445873.
Han, Y., Chen, J., Dou, M., Wang, J., & Feng, K. (2023). The Impact of Artificial Intelligence on the Financial Services Industry., 2(3), 83–85. https://doi.org/10.54097/ajmss.v2i3.8741.
Hazarika, I. (2020). *Artificial intelligence: opportunities and implications for the health workforce.* Oxford University Press, 12(4), 241–245. https://doi.org/10.1093/inthealth/ihaa007.
Hernández, E. (2024). Towards an Ethical and Inclusive Implementation of Artificial Intelligence in Organizations: A Multidimensional Framework. Cornell University. https://doi.org/10.48550/arxiv.2405.01697.
Hickok, M., & Maslej, N. (2023). A policy primer and roadmap on AI worker surveillance and productivity scoring tools. *Springer Nature*, 3(3), 673–687. https://doi.org/10.1007/s43681-023-00275-8.
Khan, A A., Akbar, M A., Fahmideh, M., Liang, P., Waseem, M., Ahmad, A., Niazi, M., & Abrahamsson, P. (2023). AI Ethics: An Empirical Study on the Views of Practitioners and Lawmakers. *Institute of Electrical and Electronics Engineers*, 10(6), 2971–2984. https://doi.org/10.1109/tcss.2023.3251729.
Lee, J S., & Hsin, P. (2004). Employee training and human capital in Taiwan. *Elsevier BV*, 39(4), 362–376. https://doi.org/10.1016/j.jwb.2004.08.004.
Li, L. (2020). Reskilling and Upskilling the Future-ready Workforce for Industry 4.0 and Beyond. https://www.ncbi.nlm.nih.gov/pmc/articles/PMC9278314/.
Morris, M R. (2020). AI and accessibility. Association for Computing Machinery, 63(6), 35–37. https://doi.org/10.1145/3356727.
Payton, M J. (2023). *Impact of Artificial Intelligence Regulations on Organizational Risks.* George A. Smathers Libraries, 36. https://doi.org/10.32473/flairs.36.133263.
Rosenblat, A., Kneese, T., & Boyd, D. (2014). Networked Employment Discrimination. RELX Group (Netherlands). https://doi.org/10.2139/ssrn.2543507.

# 5. The Perils of Generative AI

## *Normalizing Implicit Bias*

R. "Doc" Vaidhyanathan

The development of Generative AI (Gen AI) marks a significant milestone in technological advancement, enabling machines to generate content that was once the exclusive domain of human creativity. However, with this innovation comes a subtle yet profound danger: the potential for implicit bias to be perpetuated and amplified by these systems. As we explore the implications of Gen AI, we must consider not only the immediate impacts but also the long-term consequences of relying on these systems as sources of information and authority.

## *The Evolution of AI and the Rise of Gen AI*

Artificial Intelligence (AI) has evolved over the past several decades, from simple expert systems to complex machine learning models. Early AI was designed to perform specific tasks, such as predicting chess moves or identifying fraudulent transactions. However, the advent of Gen AI, such as ChatGPT, has introduced a new dimension to AI—one that can create new content, including text, images, and music, on demand. This capability distinguishes Gen AI from its predecessors, as it is no longer limited to processing and analyzing data but can generate entirely new information.

Gen AI models are trained on vast datasets, encompassing virtually all available digital knowledge. These models can produce natural language responses that mimic human-like understanding, offering a seemingly authoritative voice on a wide range of topics. The ease of use and the versatility of these systems have made them increasingly popular, leading to widespread adoption across various domains.

## *The Threat of Implicit Bias*

While the capabilities of Gen AI are undoubtedly impressive, they also present significant risks. One of the most insidious dangers is the potential for implicit bias to be embedded within these systems. Unlike explicit biases, which are relatively easy to identify and address, implicit biases are often subtle and difficult to detect. They can emerge from the underlying data used to train Gen AI models, which may reflect historical and cultural biases present in society.

Consider the analogy of a frog placed in boiling water versus one placed in cold water that is slowly heated. The frog in boiling water will instinctively jump out, sensing immediate danger. However, the frog in gradually warming water remains unaware of the creeping threat until it's too late. Similarly, when we encounter explicit bias that targets a specific demographic, we tend to respond swiftly, seeking corrective action. Yet, subtle, incremental shifts often go unnoticed, and over time, these small deviations can accumulate, leaving us in a perilous situation that is much harder to address.

For instance, consider the practice of naming schools in the United States, scientific institutions, or prominent cosmic discoveries. Schools are often named after U.S. Presidents. Other institutions and prominent physical formations like craters on moons are named after Nobel Laureates. While this may seem like an unbiased practice, it inadvertently reinforces certain stereotypes—all U.S. Presidents to date are white men, and the majority of Nobel Laureates in the twentieth century were from Western Europe or the United States. This reflects a broader issue: the historical dominance of certain groups in shaping knowledge and cultural norms. While these may change over time, making references to them repeatedly in common usage legitimizes happenstance into a pattern, and maybe even imply causality.

## *The Skewed Nature of Knowledge Formation*

The capture and dissemination of knowledge have hardly been uniform. They have always been influenced by those in power, and by those wanting to make themselves heard. In ancient times, only the wealthy could afford to inscribe knowledge on stone tablets, and the victors of wars often controlled the narrative by preserving their records while destroying those of the defeated. As knowledge evolved from oral traditions to written documents, and eventually to digital media the disparities in knowledge production continue to persist. While writing blogs or publishing articles on the web is freely available independent of financial means, all communities and cultures do not feel the same urge to tell their stories. Today, the digitization of knowledge is still skewed, with certain cultures and regions contributing disproportionately to the global knowledge base.

Consider the activity on free social media platforms like Facebook, where certain demographics are more likely to share photos of their meals than others. Similarly, on a free knowledge base like Wikipedia, individuals in the U.S. are more inclined to create pages about themselves compared to equally accomplished individuals in other countries. These cultural differences contribute to a skewed representation of information in these public repositories, leading to imbalances that reflect the interests and behaviors of specific groups rather than a truly global perspective.

A clear example of how one group can inadvertently influence global culture is found in management literature. For instance, the term "ballpark," used to indicate an approximation, is widely recognized among business executives worldwide. However, the term originates from baseball, a sport primarily associated with the U.S. Its widespread use in management and business books—many of which are authored by Americans—has led to its global adoption. While the exchange of ideas and idioms across cultures is valuable, it becomes problematic when one group, even without ill intent, dominates a field and shapes the core knowledge within it. This dominance can overshadow regional nuances and limit the diversity of perspectives within that discipline.

This imbalance is further exacerbated by the way Gen AI systems are trained. The data used to train these models is drawn from digital sources, which are represented as being free, open, and equally accessible everywhere globally. As a result, the knowledge embedded within Gen AI is not neutral but reflects the perspectives of those who have historically dominated the production of knowledge.

## *The Dangers of Accepting AI-Generated Knowledge as Truth*

Many technologies have transformed our behaviors, but they haven't fundamentally altered our understanding of the world or our knowledge of these innovations. For example, the simple calculator has made math calculations and graphing more accessible, yet the underlying equations and theorems of mathematics remain unchanged. Similarly, music players allow us to enjoy music on demand, but the essence of the music—what the artist intended—remains intact. In the realm of books and libraries, editors have long influenced which works get published, determining what was deemed worthy of public consumption. However, these mediums still required a certain level of financial and time investment from readers, who exercised discretion in their choices and provided feedback through their purchasing decisions. Authors, musicians, and publishers all had to respond to this feedback, adapting to the preferences and demands of an audience that, in effect, voted with their wallets

One of the most significant dangers of Gen AI is the perception of its outputs as objective truth. The language generated by these systems is often well-formed, persuasive, and delivered with a tone of authority. This can create the illusion that Gen AI is a neutral and reliable source of information, even when the underlying data is biased.

Moreover, the ease with which users can access information through Gen AI systems reduces the need for critical thinking and discernment. In the past, accessing knowledge required effort—whether it was reading books, attending lectures, or conducting research. This process encouraged individuals to engage with the material, question it, and form their own opinions. However, with Gen AI, information is delivered instantly and effortlessly, which may discourage users from critically evaluating the content.

## *Amplifying Bias Through the Feedback Loop*

As Gen AI systems become more integrated into our daily lives, there is a risk that they will perpetuate and even amplify existing biases. When users interact with Gen AI, they often accept the responses provided without question, assuming that the information is accurate and authoritative. This can create a feedback loop in which biased information is continuously reinforced and disseminated, further entrenching existing disparities.

When a Gen AI model is trained on a dataset that overrepresents certain perspectives or cultural norms, it may produce responses that reflect these biases. Users who rely on Gen AI for information may then internalize these biases, shaping their understanding of the world. Over time, this can lead to a distorted view of reality, where certain voices and perspectives are marginalized or excluded altogether. This could lead to a homogenization of knowledge, where diversity of thought is diminished, and only certain narratives are perpetuated.

For example, if Gen AI continues to draw primarily from Western sources of knowledge, it may reinforce the dominance of Western cultural norms and values. This could result in the erasure or devaluation of non–Western perspectives, contributing to a narrow and incomplete understanding of global issues. These can be clearly evidenced when asking Gen AI systems like ChatGPT to name the greatest philosophers, artists, or religious reformers in the world. The answers are all Euro centric and based on the thoughts and ideas that have flourished there in the past 100 years.

## *Addressing the Perils of Bias in Gen AI*

To mitigate the risks of bias in Gen AI, it is essential to take a proactive approach to the development and deployment of these systems. This includes ensuring that the data used to train Gen AI models is diverse and representative of different cultures, perspectives, and experiences. It also involves implementing mechanisms for detecting and correcting bias in AI-generated content.

A common approach to addressing the ethical challenges of generative AI is to develop guidelines and standards that emphasize transparency, accountability, and fairness in AI design and deployment. However, these measures alone may prove insufficient. The core issue lies in the fact that the very benchmarks for fairness are often inherently biased. Even the most impartial judge can make errors if the laws they uphold are fundamentally unjust. Historical examples from the U.S., such as laws predating the abolition of slavery or women's suffrage, illustrate this point. Those laws, while legal at the time, merely reflected the prevailing views of a bygone era, underscoring the need for continuous scrutiny and evolution in our standards.

General-purpose AI systems need to be trained on a genuinely global body of knowledge. To achieve this, creators of such systems must collaborate with governments and knowledge custodians across various regions to capture, collate, digitize, and incorporate diverse perspectives. Additionally, AI architects should make a concerted effort to amplify underrepresented viewpoints, ensuring they are not overshadowed by more popular or extensively documented perspectives. Technologies like search engines often rely on feedback loops that prioritize answers based on user engagement. However, if generative AI systems aim to provide a neutral and balanced view of the world, they must resist the temptation to favor popular responses, instead striving to present a more comprehensive and inclusive understanding.

Additionally, efforts should be made to educate users about the potential biases in Gen AI and encourage critical engagement with AI-generated content. This requires a collaboration between educators, creators of Gen AI systems, and those in mass communications to emphasize that Gen AI systems are not the source of truth.

## *Conclusion*

Generative AI represents a powerful tool with the potential to revolutionize many aspects of our lives. However, it also carries significant risks, particularly in terms of perpetuating and amplifying bias. As we continue to integrate these systems into our daily lives, it is crucial to remain vigilant and take steps to ensure that they do not reinforce existing disparities. By acknowledging the potential dangers of bias in Gen AI and taking proactive measures to address them, we can harness the benefits of this technology while minimizing its risks.

PART II

# Perils and Promises

## • *A. AI in Marketing and Sales* •

# 6. AI-Driven Training and Gen Z Retail Sales Employees in Vietnam

Tran Minh Viet

The consumer goods industry in Vietnam, particularly within the home appliances sector, has experienced significant growth over the past decade, benefiting from rising incomes, increased urbanization, and a growing middle class (Tran & Le, 2023). The retail industry generates many opportunities, but it also brings aggressive competition to the market (Tien et al., 2022). Therefore, constantly enhancing a company's salesforce is crucial (Kumar et al., 2022). Retail companies are pressured to enhance their sales capabilities, reduce costs, and maintain profitability (Grewal et al., 2018). Hence, given expansion needs, organizations need to be innovative in their training delivery, i.e., equipping an ever expanding retail employee team with the knowledge and skills to engage customers effectively, especially in a market where high-value products like home appliances require deep product knowledge and consultative sales approaches (Grewal et al., 2018).

Furthermore, in recent years, e-commerce has been booming rapidly in Vietnam (Luu & Phan, 2024). Despite this growth, brick and mortar retail remains a dominant channel for home appliances in Vietnam (Tran & Nguyen, 2022). With high expense and complicated features, home appliances often require demonstrations and personalized consultations, underscoring the importance of in-store sales employees who effectively convey product benefits to customers (Tran & Nguyen, 2022).

The average national age in Vietnam is 32 (Singh et al., 2020). Thus, most consumers are Generation Z (Gen Z) buyers, who were born between 1996 and 2012 (Seemiller & Grace, 2018). Most people working in Vietnam's retail sector also belong to Gen Z. Typically, they are full of enthusiasm and highly familiar with technology (Nguyen & Nguyen, 2020). These characteristics bring advantages to working in home appliance sales, such as a love of tech, quick learning, and swift adaptation. But rapidly building a solid retail sales team with Gen Z members requires effective training programs. However, building a strong retail sales team composed of Gen Z members requires the implementation of effective and targeted training programs (Nguyen & Nguyen, 2020).

Thus, there is good potential for the use of artificial intelligence (AI) and other technology driven approaches. These solutions could address many of the limitations of traditional training and support their development in customer-facing roles (Upadhyay & Khandelwal, 2019). This piece explores how AI-driven training contributes to enhancing training effectiveness in Vietnam's home appliance retail. The discussion below draws from a larger primary dataset gathered using a mix-methods approach.

## *Retail Sales in Vietnam*

The retail environment for consumer goods, specifically home appliances, involves high customer interaction and technical expertise (Kacen et al., 2013). Modern trade channels for these products include structured environments such as electronics stores, department stores, and specialty retail outlets (Alba et al., 1997). Employees must possess detailed product knowledge, conduct technical demonstrations, and address complex customer questions (Grewal et al., 2017). The sales process in this environment is consultative (Grewal et al., 2018). It requires employees to understand and effectively communicate the benefits of sophisticated technologies in simple language and be familiar with customer preferences, such as high-definition displays, innovative connectivity, advanced sound systems or specifications of engines, or the benefits of different cooling technologies. Consequently, continuous training to help retail sales update product knowledge and selling skills is critical for peak sales performance and market product education (Peesker et al., 2022).

Gen Z sale team recruits in Vietnam's home appliance sector are ideally positioned to thrive in this modern retail environment since it requires adaptability, technological proficiency, and comfort with digital tools, making them well-suited for engaging with customers on high-tech products (Nguyen & Nguyen, 2020; Seemiller & Grace, 2018). Their familiarity with digital media makes them relate easily to a customer base that increasingly relies on online information but still values in-store consultations for high-value purchases (Seemiller & Grace, 2018). Thus, offering an appropriate training approach will greatly enhance their job performance and satisfaction.

## *Role of Training in Improving Sales Capability for Gen Z Retail Sales*

Training is critical in the consumer goods industry, particularly for high-value home appliance products, where effective sales performance depends on an employee's ability to provide in-depth information, demonstrate features, and build customer trust (Rapp et al., 2015). Comprehensive training helps retail employees develop the confidence to recommend products, handle objections, and tailor their approach based on customer needs. When employees are well-trained in product features and customer engagement techniques, sales performance improves, leading to higher customer satisfaction and repeat purchases (Owens, 2023). Gen Z's enthusiasm for technology-driven learning solutions aligns will facilitate their quick acquisition of the advanced product knowledge required in the home appliances sector, making them an ideal participant in AI-driven training (Nguyen & Nguyen, 2020).

## *Limitations of Traditional Training Approaches*

Traditional training in the retail sector is often limited by its rigidity and lack of engagement, which can be a poor fit for Gen Z's preferences (Chillakuri, 2020). Classroom-style lectures and standardized content fail to capture the interactive, flexible, and feedback-driven learning experience that Gen Z employees value (Seemiller & Grace, 2016). They may not sufficiently cover the specific technical features of complex products like TVs and audio systems, refrigerators, washing machines, or air conditioners. As a result, traditional training may not adequately prepare Gen Z employees for the high demands of customer service in the home appliances industry.

## *Potential Impact of AI-Driven Training*

AI-driven training can provide Gen Z retail employees with a highly personalized, flexible, and interactive learning experience that enhances their performance in the consumer goods sector (Chan & Lee, 2023). Key benefits include:

- **Personalized Learning and Adaptive Content:** AI-powered systems can analyze employee performance data and create customized training pathways. These pathways address individual knowledge gaps and provide focused training on product features like high-resolution displays, smart TV functions, and advanced audio technologies. This tailored approach makes learning more relevant for each employee, enhancing engagement and knowledge retention (Lan et al., 2024).
- **Simulation and Hands-On Practice with Real-Time Feedback:** AI-driven training platforms often incorporate simulations and interactive modules, enabling Gen Z employees to practice sales scenarios, product demonstrations, and customer interactions in a virtual and game-like setting. Immediate feedback allows employees to improve their approach continuously, building confidence and competence for in-store applications (Feritas & Oliver, 2006; Chan & Lee, 2023).
- **Flexibility and On-Demand Access:** AI-driven training supports convenient access through digital devices anytime, giving Gen Z employees the flexibility to learn at their own pace. This accessibility is especially beneficial in retail, where employees can complete training modules during downtime or refresh their knowledge as new products are introduced (Chan & Lee, 2023).

Through these benefits, AI-driven training can provide a comprehensive solution that supports Gen Z's preferred learning style and equip them with the necessary skills to excel in the modern trade channel for home appliances.

## *Limitations of AI-Driven Training*

While AI-driven training offers numerous advantages, it also presents certain limitations that are particularly relevant in the context of Vietnam's Consumer Goods industry:

- **High Initial Costs and Technical Requirements:** Implementing AI-driven training systems requires significant upfront investment in software, infrastructure, and ongoing updates, which may be challenging for smaller retail organizations. The cost of regularly updating AI-driven content to reflect the latest home appliance technologies may further impact resource allocation (Shang et al., 2023).
- **Varied Digital Literacy and Resistance to AI:** While Gen Z is generally tech-savvy, not all retail employees may be equally comfortable with AI systems, especially in a country where digital literacy varies. Technical issues, such as software malfunctions or system connectivity challenges, could disrupt training and impact employee satisfaction (Hangl et al., 2023). Additionally, some employees may resist AI-driven training, preferring face-to-face interactions with mentors and peers (Hangl et al., 2023).

In summary, while AI-driven training presents a powerful alternative to traditional approaches, it must be carefully integrated into a balanced learning strategy considering its strengths and limitations.

## *Conclusion and Call to Action*

The primary and secondary research results discussed in this piece highlights the potential of AI-driven training to significantly improve the sales capabilities of Gen Z employees in Vietnam's Consumer Goods industry, specifically within the home appliances sector. In sum, these are:

- **Impact of AI on Training Efficiency:** AI-driven training enhances learning engagement and allows for continuous improvement, resulting in better customer service and increased sales effectiveness. Potentially, the feedback reported improved confidence and faster adaptation to new product lines, positively impacting their sales performance.
- **Gen Z's Adaptability to AI Training:** Gen Z employees strongly preferred AI-driven training methods, particularly real-time feedback and interactive learning features (Chan & Lee, 2023). Their familiarity with digital technology and preference for flexible learning schedules make AI-based training suitable for their professional development needs.
- **Implementation Challenges:** Despite the advantages, implementing AI-driven training presents challenges for organizations. Costs associated with AI integration were retailers' primary concerns, along with technical maintenance and data security (Shang et al., 2023). The organization should consider solutions such as phased AI deployment, employee orientation programs, and continuous support systems to mitigate the challenges.

By providing a more personalized, flexible, and interactive learning environment, AI-driven training aligns with the preferences of Gen Z employees, empowering them to perform effectively in a competitive retail landscape. However, the successful implementation of AI-driven training will require addressing limitations related to cost, data privacy, and the balance between technology and human interaction. This study contributes valuable insights for organizations aiming to enhance their retail sales performance through innovative training solutions.

## References


Alba, J., Lynch, J., Weitz, B., Janiszewski, C., Lutz, R., Sawyer, A., & Wood, S. (1997). Interactive Home Shopping: Consumer, Retailer, And Manufacturer Incentives To Participate In Electronic Marketplaces. *Journal of Marketing, 61*(3), 38–53.

Andrea Marshall, & Murray J. Fisher. (2008). Understanding Descriptive Statistics. *Australian Critical Care, Official Journal of the Confederation of Australian Critical Care Nurses, 22*(2).

Chan, C.K.Y., & Lee, K.K. (2023). The Ai Generation Gap: Are Gen Z Students More Interested In Adopting Generative Ai Such As Chat GPT in Teaching And Learning Than Their Gen X And Millennial Generation Teachers?. *Smart Learning Environments, 10*(1), 60.

Chillakuri, B. (2020). Understanding Generation Z Expectations for Effective Onboarding. *Journal of Organizational Change Management, 33*(7), 1277–1296.

De Freitas, S., & Oliver, M. (2006). How Can Exploratory Learning With Games and Simulations Within The Curriculum Be Most Effectively Evaluated?. *Computers & Education, 46*(3), 249–264.

Grewal, D., Motyka, S., & Levy, M. (2018). The Evolution and Future of Retailing and Retailing Education. *Journal of Marketing Education, 40*(1), 85–93.

Grewal, D., Roggeveen, A.L., & Nordfält, J. (2017). The Future of Retailing. *Journal of Retailing, 93*(1), 1–6. https://doi.org/10.1016/j.jretai.2016.12.008.

Hangl, J., Krause, S., & Behrens, V.J. (2023). Drivers, Barriers and Social Considerations for AI Adoption In Scm. *Technology in Society, 74*, 102299.

Kacen, J.J., Hess, J.D., & Kevin Chiang, W.-Y. (2013). Bricks Or Clicks? Consumer Attitudes Toward Traditional Stores and Online Stores. *Global Economics and Management Review, 18*(1), 12–21. https://doi.org/10.1016/S2340-1540(13)70003-3.

Kumar, N.S., Kapoor, S., & Gupta, S.K. (2021). Is Employee Gratification the Same as Employee Engagement? An In-Depth Theory Perspective. Ad Alta: *Journal of Interdisciplinary Research, 11*(2), 151–156.

Lan, D.H., Tung, T.M., Oanh, V.T.K., & Cuc, T.T.K. (2024). Exploring Gen Z'S Consumption and Perception of Educational Content on Tiktok Platform: A Qualitative Study in Vietnam Context. *Kurdish Studies, 12*(1), 3243–3270.

Lu Phi Nga, & Phan Thanh Tam. (2024). Key Factors Affecting Online Shopping Attitude and Intention: A Case Study of Consumers in Vietnam. *Journal of Eastern European & Central Asian Research, 11*(1), 66–78. https://doi.org/10.15549/jeecar.v11i1.1547.

Nguyen, L.H., & Nguyen, H.P. (2020). Generation Z in Vietnam: The Quest for Authenticity. In The New Generation Z in Asia: Dynamics, Differences, Digitalisation (pp. 135–148). *Emerald Publishing Limited.*

Owens, C.L. (2023). Effective strategies to reduce voluntary employee turnover intentions in the retail sector [ProQuest Information & Learning]. *In Dissertation Abstracts International: Section B: The Sciences and Engineering* (Vol. 84, Issue 6–B).

Peesker, K.M., Kerr, P.D., Bolander, W., Ryals, L.J., Lister, J.A., & Dover, H.F. (2022). Hiring for sales success: The emerging importance of salesperson analytical skills. *Journal of Business Research, 144*, 17–30.

Rapp, A., Baker, T.L., Bachrach, D.G., Ogilvie, J., & Beitelspacher, L.S. (2015). Perceived customer showrooming behavior and the effect on retail salesperson self-efficacy and performance. *Journal of Retailing, 91*(2), 358–369.

Seemiller, C., & Grace, M. (2016). Generation Z goes to college. *John Wiley & Sons.*

Seemiller, C., & Grace, M. (2018). Generation Z: A century in the making. *Routledge.*

Shang, G., Low, S.P., & Lim, X.Y.V. (2023). Prospects, drivers of and barriers to artificial intelligence adoption in project management. *Built Environment Project and Asset Management, 13*(5), 629–645.

Singh, S., Mondal, S., Singh, L.B., Sahoo, K.K., & Das, S. (2020). An Empirical Evidence Study Of Consumer Perception And Socioeconomic Profiles For Digital Stores In Vietnam. *Sustainability, 12*(5), 1716.

Tien, N.H., Van Dung, H., Jose, R.J.S., Bien, B.X., Oanh, N.T.H., & Vu, N.T. (2022). Analysis Of Aeon's Market Penetration Strategy In Vietnam FMCG Industry. *International Journal of Advanced Educational Research, 5*(4), 1–6.

Tran, Q.T., & Nguyen, T.B.H. (2022). Driving Factors Of Online Consumers Of Household Appliances: An Empirical Study Of Consumers In Ho Chi Minh City, Vietnam.

Upadhyay, A.K., & Khandelwal, K. (2019). Artificial Intelligence-Based Training Learning From Application. *Development and Learning in Organizations: An International Journal, 33*(2), 20–23.

# 7. Go-to-Market Strategies and AI in the UAE Technology Services Industry

Ronald Powell

In the dynamic environment of the United Arab Emirates (UAE) technology services sector, the need for innovative and adaptive Go-to-Market (GTM) strategies has become critical. The UAE, characterized by its rapid technological advancement and an economy supported by government-led initiatives to foster innovation, presents unique opportunities and challenges for technology service companies. Against this backdrop, developing GTM strategies that not only align with consumer demands but also anticipate and shape them, and leverage emerging technologies, particularly artificial intelligence (AI), is essential for sustaining a competitive edge. However, companies in the UAE often grapple with the intricacies of integrating innovation into their GTM strategies, facing hurdles such as regulatory compliance, evolving consumer expectations, and intense market competition.

## *Background*

The UAE's ambitious pursuit of digital transformation, supported by initiatives like the UAE National Innovation Strategy, has accelerated the demand for technology-driven solutions that meet high personalization standards, agility, and security. However, while this rapid development presents a fertile ground for business growth, it also challenges companies to innovate continuously. The necessity of continuous innovation cannot be overstated, as without effectively embedding innovation into their GTM strategies, technology services firms may struggle to adapt to the fast-changing landscape, missing growth opportunities, and, ultimately, risking their competitive positioning. The challenge lies in creating GTM strategies that are responsive to consumer behaviors, flexible in navigating regulatory demands, and capable of delivering sustainable business outcomes.

## *Innovation and Go-to-Market Strategies*

The relationship between innovation, mainly through artificial intelligence (AI), and Go-to-Market (GTM) strategies is increasingly relevant in the UAE's technology services sector. Previous research underscores the necessity of leveraging innovation as a core component in strategic market approaches in this rapidly evolving landscape. The UAE's National Innovation Strategy has catalyzed digital transformation, fostering a competitive environment where technology service companies must continuously evolve to meet rising consumer expectations and comply with stringent regulatory standards (UAE Government, 2014). As the UAE positions itself as a leader in technology and innovation, understanding the role of AI in enhancing GTM strategies becomes crucial for sustaining a competitive advantage.

Studies on AI's impact on GTM strategies suggest that data-driven insights enable companies to achieve better market segmentation, customer targeting, and personalization, thereby improving engagement and retention (Andreas et al., 2022). Machine learning algorithms, for instance, offer predictive capabilities that allow firms to anticipate consumer needs, adapt service offerings in real time, and create more personalized user experiences. This shift is particularly relevant in the UAE, where consumers increasingly favor digital-first, customized solutions. Integrating AI-driven analytics in GTM strategies has proven essential for maintaining relevance and growth (Galina et al., 2023).

However, literature also highlights significant challenges. The regulatory landscape in the UAE, particularly in data privacy and cybersecurity, demands that technology services companies balance innovation with compliance. Previous studies emphasize the role of regulatory compliance as both a constraint and a catalyst for innovation. Companies are urged to innovate within the confines of regulatory frameworks, integrating privacy and security considerations into AI-driven GTM strategies (Oihab et al., 2023). This regulatory pressure, coupled with the need for constant innovation, creates a complex environment where companies must strategically position themselves to meet both consumer and governmental expectations. This aligns with Schumpeter's theory of "creative destruction," which posits that innovation drives economic growth by displacing outdated practices, necessitating an adaptive approach to strategy that continually integrates new technologies and processes.

Furthermore, research on consumer behavior in the UAE indicates a heightened preference for personalized and agile service models, which demand more sophisticated AI-enabled GTM strategies. The convergence of AI and consumer behavior analysis allows companies to dynamically adjust their strategies to meet evolving customer demands (Ziyad et al., 2021). The integration of AI in GTM strategies thus becomes a means of enhancing market competitiveness and a fundamental requirement for sustaining customer engagement and loyalty in a highly competitive market landscape.

In the broader context of the UAE's national economic vision, scholars argue that government initiatives, such as the Dubai Future Foundation and Hub71, have fostered an environment conducive to innovation in the technology services sector (Amal et al., 2023). These initiatives underscore the government's commitment to creating an innovative-driven economy, especially in technology and AI, aligning with Kotler's generic marketing framework, emphasizing the need for market adaptability and consumer-centered innovation (Kotler, 1972). However, despite the supportive ecosystem, there remains a gap in empirical evidence on how AI-driven GTM strategies directly impact competitive positioning and market share in the UAE technology sector.

The influence of AI on Go-to-Market (GTM) strategies in the UAE technology sector has been significantly shaped by regulatory frameworks that mandate robust cybersecurity and compliance practices. As the UAE positions itself as a technology-driven economy, it has become necessary for technology service companies to align their AI-powered GTM strategies with evolving compliance requirements. Policies from the UAE Cyber Security Council, such as the Critical Information Infrastructure Protection (CIIP) Policy, Cyber Incident Response Framework (CIRF), and National Cloud Security Policy, provide foundational guidelines for cybersecurity and data protection. These frameworks are pivotal in shaping GTM strategies by embedding security considerations into every aspect of business operations. The CIIP Policy, for instance, mandates sector-wide governance models and resilience-building, prioritizing critical information infrastructures and ensuring alignment with cybersecurity best practices.

Furthermore, the National Cloud Security Policy provides specific guidelines on data governance, cloud interoperability, and risk management. Requiring compliance with data sovereignty and security mandates aligns with the literature on AI's role in creating secure, scalable GTM strategies. This policy's data location awareness and resilience requirements also echo the academic consensus that effective GTM strategies must address innovation and compliance (UAE Cyber Security Council, 2023). Integrating these security frameworks into GTM strategies allows UAE technology firms to align closely with national cybersecurity objectives, ensuring resilience and consumer trust while navigating market demands.

The accelerating influence of artificial intelligence (AI) on Go-to-Market (GTM) strategies in the UAE technology sector reveals how AI-based innovations enhance competitive positioning and compliance with regulatory standards. Existing literature illustrates AI's dual role in promoting operational agility and compliance, positioning it as a key driver of strategic GTM development. Haefner et al. (2021) provides an essential framework that highlights AI's transformative potential within innovation management, illustrating how AI can augment traditional human decision-making processes, overcome information processing constraints, and streamline complex decision-making tasks within regulated environments. Their framework, rooted in the Behavioral Theory of the Firm, emphasizes that AI's capacity for data-driven insights and predictive analytics is instrumental in supporting GTM strategy development. Haefner et al. (2021) thus contextualizes AI as both an operational asset and a compliance facilitator, making their research highly relevant to understanding GTM strategies in the UAE's regulatory landscape.

By leveraging the insights of Haefner et al. (2021), technology service firms in the UAE can better understand AI's role in addressing the evolving regulatory frameworks set forth by the UAE Cyber Security Council. For instance, the Critical Information Infrastructure Protection (CIIP) Policy and the Cyber Incident Response Framework (CIRF) require technology firms to integrate data privacy and cybersecurity measures into their operations. Haefner et al.'s model suggests that AI can alleviate compliance-related pressures by optimizing the management of vast data sets and enhancing real-time risk detection capabilities, effectively aligning operational needs with regulatory demands (Haefner et al., 2021). This model provides a roadmap for UAE technology companies to employ AI for regulatory alignment, demonstrating how AI enables GTM strategies to adapt rapidly to evolving market demands and compliance requirements.

In addition, Haefner et al. (2021) conceptualize AI as a critical component in overcoming the "local search" limitations often seen in traditional GTM strategy formulation, where firms tend to rely on existing knowledge domains. Their findings indicate that AI's advanced information processing capabilities allow for a broader exploration of innovative solutions, moving beyond incremental improvements to generate breakthrough insights and strategies that respond dynamically to market and regulatory conditions. Applying this concept to the UAE's technology sector, AI-based GTM strategies can utilize predictive analytics and machine learning models to foresee and adapt to regulatory shifts, anticipating changes in compliance requirements while continuing to meet evolving consumer demands.

The evolving role of artificial intelligence (AI) in the UAE's technology services sector highlights the need for Go-to-Market (GTM) strategies that are agile, data-driven, and compliant with local regulations. Donthireddy (2024) provides a crucial perspective on this transformation by examining how advanced data analytics, when integrated with AI, can optimize GTM strategies. According to Donthireddy, AI techniques such as regression analysis, clustering, and natural language processing (NLP) empower companies to refine market segmentation, enhance consumer targeting, and improve competitive analysis. This approach directly supports the UAE's technology sector, where rapid technological advancements and data-centric regulatory demands shape business operations and GTM planning.

Donthireddy's work explores the application of predictive models within GTM strategies, showing how these models enable firms to anticipate market shifts, allocate resources more effectively, and adjust strategies proactively. For instance, NLP allows firms to analyze social media sentiment and customer feedback, providing deeper insights into consumer behavior and preferences. Such capabilities align with the UAE's strict regulatory frameworks, including policies like the Cyber Incident Response Framework (CIRF) and the National Cloud Security Policy, which prioritize data privacy, security, and compliance. By leveraging these data-driven techniques, UAE technology firms can adapt their GTM strategies to meet regulatory demands while enhancing their responsiveness to market dynamics (Donthireddy, 2024).

Furthermore, Donthireddy (2024) argues that integrating AI with data analytics creates a competitive edge by fostering GTM strategies that are adaptable, consumer-centric, and compliant. The predictive capabilities of AI allow UAE firms to go beyond traditional, reactive GTM strategies by building anticipatory models that align with both consumer expectations and regulatory frameworks. For example, predictive analytics enables technology companies to anticipate regulatory shifts, adjusting their GTM approaches in real-time to remain compliant. This approach also supports precision in consumer targeting, thus addressing the UAE technology sector's dual need for agility and regulatory alignment.

The dynamic nature of the UAE's technology services industry calls for Go-to-Market (GTM) strategies that are both structured and adaptable. Within this context, Debruyne and Lestiani (2009) offer a critical perspective by framing GTM strategy development as an intersection of art and craft. According to their model, GTM strategies require a balanced approach that combines systematic planning with flexibility to respond to unforeseen market shifts. This view supports the need for GTM strategies in the UAE that can simultaneously leverage the precision of AI for structured decision-making and adapt creatively to the regulatory and competitive complexities of the market.

Debruyne and Lestiani (2009) argue that successful GTM strategies hinge on carefully aligning with foundational business models while allowing room for innovation and responsiveness. Their work emphasizes that GTM strategies should incorporate core elements like market timing, product positioning, and customer alignment, all essential for meeting regulatory demands and fostering market competitiveness in the UAE. This dual focus on structure and adaptability complements the role of AI in enhancing GTM strategies. By enabling predictive analytics, AI allows companies to anticipate regulatory shifts, adapt to consumer behaviors, and refine their market approaches in real-time capabilities that are invaluable in a compliance-heavy environment like the UAE. Debruyne and Lestiani's framework thus provides a foundational understanding of how AI-powered GTM strategies can navigate regulatory constraints while maintaining the flexibility required for sustainable growth in the UAE technology sector.

In summary, the literature establishes a compelling case for the role of AI-driven innovation in the UAE's technology services industry, particularly in GTM strategies. It reveals a consensus that AI enables more precise market targeting, enhanced consumer engagement, and efficient compliance with regulatory frameworks. However, it also identifies a research gap concerning the direct, quantifiable impact of AI on GTM outcomes, highlighting an opportunity for further empirical investigation. Consequently, this study aims to bridge this gap by examining how AI-driven innovation can be strategically implemented to optimize GTM strategies in the UAE, providing both theoretical insights and practical applications for technology services companies within this unique and fast-paced market.

## *Assumptions*

This analysis assumes that technology services companies in the UAE face significant challenges due to rapid technological changes, shifting consumer expectations, and a regulatory environment emphasizing compliance with data privacy and security. Moreover, it presumes that these companies have access to the necessary resources and technological capabilities to adopt AI and other innovations as part of their GTM strategies. Another assumption is that key stakeholders, including decision-makers and government regulators, are not just involved but deeply committed to fostering an environment where innovation-driven GTM strategies can flourish. Finally, this analysis anticipates that integrating AI will contribute significantly to these companies' competitiveness and market responsiveness, enabling them to cater to a digitally driven and increasingly selective consumer base.

## *Recommendations*

To address these challenges, it is recommended that UAE technology services firms prioritize innovation in their GTM strategies, specifically by investing in AI and data analytics capabilities. Companies should focus on creating agile frameworks for GTM strategies that can be swiftly adapted to meet emerging regulatory requirements and evolving consumer needs. Fostering cross-functional teams, including data scientists, marketers, and compliance experts, can enable a cohesive approach to implementing innovative GTM models. Establishing strong partnerships with local government bodies, such as the Dubai Future Foundation and the UAE National Innovation Strategy initiatives, can further align corporate innovation efforts with national objectives, helping firms remain competitive in a highly regulated market. In essence, these recommendations advocate for a holistic approach to GTM strategy, wherein AI and other emerging technologies serve as pillars for sustainable growth, regulatory compliance, and a stronger connection to consumer demands within the UAE's vibrant technology sector.

## References


Amal, E., Sakir, G., & Murat, G. (2023). The GCC's regional roller coaster: Do regional factors affect stock market dynamics in the GCC Region? Evidence from non-parametric quantile regression. *Borsa Istanbul Review*, 23(2), 473–494. https://doi.org/10.1016/j.bir.2022.11.018.

Andreas, J.R., Maximilian, K.D., & Adnane, M. (2022, May). Digital transformation during a pandemic: Stretching the organizational elasticity. *Journal of Business Research*. 144, 1320–1332. https://doi.org/10.1016/j.jbusres.2022.01.088.

Debruyne, M., & Lestiani, F. (2009). Developing a go-to-market strategy: art or craft. Flanders District of Creativity. ttps://repository.vlerick.com/bitstream/handle/20.500.12127/3483/Debruyne_M_FDC_RR_DevelopingAGoToMarketStrategy.pdf?sequence=1&isAllowed=y.

Donthireddy, T.K. (2024). Optimizing Go-to-Market Strategies with Advanced Data Analytics and AI Techniques. *IRE Journals*, 8(2), 537–543. https://www.irejournals.com/formatedpaper/1706176.pdf.

Galina, R. & Inga, L. (2023). Digital transformation as a catalyst for sustainability and open innovation, 9(1). https://doi.org/10.1016/j.joitmc.2023.100017.

Haefner, N., Wincent, J., Parida, V., & Gassmann, O. (2021). Artificial intelligence and innovation management: A review, framework, and research agenda. *Technological Forecasting & Social Change*, 162, 120392. https://doi.org/10.1016/j.techfore.2020.120392.

Kotler, P. (1972). A Generic Concept of Marketing. *Marketing Management*, 36(2), 46–54. https:doi.org/10.2307/1250977.

Oihab, A., Juan, C.C. & Klaus, J.U.B. (2023). Born to be sustainable: How to combine strategic disruption, open innovation, and process digitization to create a sustainable business. *Journal of Business Research*, 154. https://doi.org/10.1016/j.jbusres.2022.113379.

UAE Cyber Security Council. (2023). Cyber Security Council. https://csc.gov.ae/en.

UAE Government. (2014, October). National Innovation Strategy. https://u.ae/en/about-the-uae/strategies-initiatives-and-awards/strategies-plans-and-visions/strategies-plans-and-visions-untill-2021/national-innovation-strategy.

Ziyad, R.A., Mohammed, A., Manmeet, M.S., Yu-Beng, L., Zaher, A.A. & Sami, A.A. (2021). Impact of coronavirus pandemic crisis on technologies and cloud computing applications. *Journal of Electronic Science and Technology*, 19(1). https://doi.org/10.1016/j.jnlest.2020.100059.

## • B. AI in Finance and Financial Services •

# 8. The Transformative Power of Generative AI in Banking

Sonny Panesar

The financial sector is subject to extensive regulation, with compliance requirements mandated by law and additional operational restrictions imposed by regulatory bodies. These constraints differ across countries, each having a unique set of laws and regulatory authorities, resulting in significant complexity and compliance costs. Non-compliance can lead to substantial fines, reputational damage and potential revocation of banking licenses.

Banks have had to innovate to remain competitive, whether due to competition from other banks or fintech companies. They have adopted ATMs, implemented credit cards, online banking, mobile banking and payment networks such as SWIFT to keep up with the demands of a changing financial landscape.

Banks hold vast amounts of non-public data, like transaction histories and behavioral analytics, ideal for leveraging AI due to advances in big data and computing power. However, they face challenges such as outdated applications, siloed data and internal barriers. Fintech companies, unencumbered by old architectures and high overhead costs, push banks to adapt but they themselves face high entry barriers.

Since the 1980s, banks have used AI and expert systems for trading, fraud detection and risk management. In the 1990s, neural networks were introduced for price prediction, trading strategies (Grudnitski & Osburn, 1993), fraud detection, risk management and loan creditworthiness. By the 2000s, machine learning had become popular for credit risk assessment and customer segmentation (Smigel, 2023).

AI made significant advancements in finance during the 2010s, notably with the introduction of robo-advisors and real-time fraud detection systems. In 2016, Bank of America launched Erica (Taylor, H. 2016), one of the first chatbots to offer personalized financial advice through predictive analytics. The emergence of Generative AI has presented new opportunities for productivity gains, efficiency improvements and revenue growth. Nevertheless, it also introduced new challenges that require careful consideration and management.

## *Current Application and Promise of AI in Banking*

GenAI has begun to revolutionize the banking sector, and many boards have recognized the potential revenue opportunities and have begun to invest heavily. A study from Juniper Research (Maynard 2024) has found that banks' spending on GenAI will reach $85 billion in 2030, up from $6 billion globally in 2024. The McKinsey Global Institute (Buehler et al, 2024) estimates that across the global banking sector, generative AI could contribute between $200 billion and $340 billion annually.

In 2024, numerous banks and financial institutions are conducting proof of concepts (POCs) or pilot studies involving Generative AI, with a few already in production. Considering the current levels of investment, a significant increase in applications powered by Generative AI is anticipated soon. Below are some of the applications of Generative AI in Banking.

*Fraud Detection.* Generative AI can enhance fraud detection by examining historical patterns alongside real-time data. In February 2024, Mastercard (Ivanov, 2024) introduced Decision Intelligence Pro to improve its ability to identify suspicious transactions. This has led to increased detection rates by up to 20 percent, with some institutions reporting improvements of up to 300 percent. The model analyzes around 125 billion transactions annually.

*Customer Service.* Several banks have implemented chatbots or virtual assistants powered by large language models to enhance customer interaction and reduce operational costs. These chatbots aim to provide a natural user experience and can deliver accurate responses when combined with methods such as retrieval augmented generation (RAG), parameter-efficient fine-tuning (PEFT) and techniques like low-rank adaptation of original model weights (LoRA).

ING, a global Dutch bank, typically receives 85,000 customer service requests per week, either by phone or online, from its customers in the Netherlands. Around 45 percent are handled by a traditional chatbot, leaving approximately 16,500 to be managed by live agents who are only available during working hours, leading to potentially long waiting times for customers. When ING deployed their generative AI-powered chatbot, using retrieval-augmented generation (RAG) for accessing internal databases and implementing appropriate guardrails, they were able to assist 20 percent more customers in the first seven weeks (*Banking on innovation: How ING uses generative AI to put people first*, 2024). Deutsche expects this to scale up to reach 37 million customers across ten markets as the chatbot is available 24/7 and provides accurate, detailed responses, resulting in a vastly improved customer experience.

*Credit Risk Assessment.* Improved accuracy of risk assessments using additional data points beyond traditional metrics, which typically rely on credit bureaus. Economic indicators, utility payments, employment data, and even social media activity are additional points that can be considered. This may allow the assessment of previously underserved populations.

*Bloomberg and Riskthinking.* AI introduced (*Bloomberg—Are you a robot?, 2023*) an artificial intelligence data tool in 2023 designed to assist businesses in predicting their financial exposure to the physical risks of climate change under various scenarios outlined by the Intergovernmental Panel on Climate Change. The tool will offer data insights that comply with the European Union's climate disclosure rule, the Task Force on Climate-Related Financial Disclosures recommendations and the International Finance Reporting Standards Foundation's sustainability standards.

*Hyper-Personalization.* Hyper-personalization is a concept that will gain increasing prominence. Traditional personalization relies on demographics or customer segmentation, whereas Generative AI will enable personalization through real-time individual behavioral data, thereby ensuring a genuinely personal experience.

Banks can provide custom product recommendations through hyper-personalization. Bank of Ireland is taking this approach to become the "Netflix" of finance by offering truly personalized recommendations via its digital banking app. A truly personal profile will allow for the identification of cross-selling opportunities and help build long-term customer loyalty.

Dutch Bank ABN AMRO, in collaboration with Subaio (Hamann, 2023), a fintech company, is offering personalized loans to its customers. These loans are tailored based on the customers' recurring payments, existing loans, and other financial products within their portfolios. The interest rates for these loans are highly competitive, as the risk profile is individualized rather than relying on a broad market segmentation approach.

*Investment Strategies.* Many financial institutions are using generative AI to perform complex financial analyses and produce research documents more quickly than human analysts. These outputs are then used to develop models that adapt to market changes for trading.

JP Morgan launched Quest IndexGPT, which uses ChatGPT-4 to generate theme-related keywords (JP Morgan, 2024). These keywords help identify relevant news articles, which are then input into the firm's Quest framework to create thematic indices for institutional clients. Areas include AI, cloud computing, e-sports and renewable energy.

*Loan Underwriting.* In partnership with Numerated, a fintech specializing in lending technology, Citibank revolutionized its commercial underwriting process by reducing analysis time from days to hours (Miller, 2024). These gains were achieved by combining technologies such as optical character recognition (OCR), NLP, large neural networks, and logistic regression models with continual feedback loops, allowing the system to improve over time. Plans include using generative AI for predictive analysis, automation and third-party datasets for benchmarking, allowing for comparisons against industry trends and standards.

*Compliance and Regulatory Reporting.* In experiments conducted by Deutsche Bank using a GenAI system (Aggie) in collaboration with Kodex AI, Aggie significantly reduced the time needed to summarize and write complex regulatory text from two hours to just minutes, all while maintaining maker-checker governance (*Adopting generative AI in banking—Corporates and Institutions.* 2024). The use of techniques such as Retrieval Augmented Generation (RAG) and Parameter Efficient Fine Tuning (PEFT) significantly improved the accuracy of the content, reducing the likelihood of hallucinations.

*Future Innovations.* Autonomous agents will interact with humans and other agents capable of tool use, including API calls to existing applications. As a consequence, we will see the rise of hybrid workforces consisting of humans and agentic systems. Workflows that were previously managed by humans will either be fully or partially delegated to autonomous agentic systems (Van Rijmenam, 2024).

As generative AI is embedded into more workflows within Banks, we will begin to see true hyper-personalization of client services at scale. These products could include dynamic investment strategies and personalized insurance policies that adapt in real time to changes in the customer's life and financial status.

The latest iterations of agents can now operate and autonomously use your computer and apps (*Introducing computer use, a new Claude 3.5 Sonnet, and Claude 3.5 Haiku*, 2024). Open AI, Google DeepMind, Anthropic, and several other startups will launch their agent-like systems in 2025.

## *Challenges and Risks of Generative AI in Banking*

Banks are subject to strict regulations for several important reasons. One key reason is that they handle enormous amounts of money, making them attractive targets for both outside criminals and insiders who might be tempted to commit fraud.

A recent case highlights this: An employee of a UK multinational firm in Hong Kong was tricked into releasing payments of HK$200m ($25m USD) after being invited to a video call with the Chief Financial Officer and some other colleagues whom they recognized. The deepfake video call mirrored the scammer's behavior, as the scammer listened, talked, and nodded during the meeting, convincing the employee they were conversing with real colleagues (Kong, 2024). These highly convincing deepfakes and phishing emails are becoming increasingly difficult to distinguish from legitimate communications.

The evolving regulatory landscape remains fragmented and often misaligned as governments and regulators race to define control environments for generative AI. While some countries are taking a more relaxed approach to legislation to prevent stifling the growth of AI, others, like the EU are taking a much more prescriptive approach.

This creates significant uncertainty for organizations navigating these regulations and deploying applications globally. Additionally, banks incur extra costs and resources to stay updated with new regulations and maintain compliance, which slows the pace of adoption.

One key challenge is balancing the potential of generative AI while addressing its inherent risks, including issues related to data privacy and algorithmic bias. Financial institutions must ensure that their AI models are free from biases that could lead to unfair or discriminatory outcomes, not only from an ethical perspective but also to adhere to legal requirements.

The explainability of models is a critical requirement. For instance, if a loan application is denied, the model must clearly articulate the rationale behind the rejection. However, with generative AI, this poses a challenge since these models are built using neural networks, which inherently lack transparency. This area remains an active field of research with innovative techniques such as the use of external "explainer" models (*Uncovering the Enigma: Delving into the Explainability of Large Language Models [LLMs]*, 2024). Additionally, foundation models are trained on large datasets sourced from the internet, which results in the incorporation of human biases present in the training data.

Banks prioritize consistent outcomes, and generative AI is probabilistic by nature. In some use cases, even 98 percent accuracy may not be sufficient. Hence, banks are cautious about deploying client-facing generative AI applications, with most implementations currently focused internally. An inaccurate chatbot can quickly lose trust. Selecting appropriate use cases and determining whether a probabilistic approach is acceptable is crucial. For example, it might be suitable for a support assistant but requires careful consideration for a loan approval process.

In 2022, a precedent was set when Air Canada was held liable for incorrect information provided by their GenAI chatbot (Olson, 2024). The chatbot promised a bereavement fare discount that required submission before the flight. However, it incorrectly advised a passenger that he could apply later. Air Canada argued the chatbot was a "separate legal entity," but the tribunal held the airline responsible for all elements on its webpage. This was a landmark case for all looking to deploy chatbots into production.

AI models process substantial data, posing privacy risks. Banks must carefully evaluate platforms for deploying generative AI. Optimizing organizational data is essential, but complex legacy applications and siloed, poor-quality data often hinder effective training.

Another important consideration is the impact of generative AI on the workforce. As AI systems take over tasks traditionally performed by humans, there is a possibility of job displacement. Roles in customer service, data entry, and certain aspects of financial analysis might be affected. Currently, the emphasis is on enhancing productivity, but this may change as efficiencies are realized.

A number of tech companies are starting to illustrate this, for example, more than a quarter of new code at Google is now generated by AI (Vella, 2024). It is reviewed by engineers before it's added to the code repositories, this massively increases the productivity of developers. This is most certainly a contributory factor in the recent reductions in workforce as companies experience increasing productivity coupled with reduction in costs.

While new roles will be created, it's likely that existing roles will evolve as organizations adopt AI. Support functions like dev ops might take on machine learning operations (MLOps) responsibilities. Although new positions such as prompt engineer will emerge, they probably won't match the number of displaced roles.

## *Conclusion*

The banking sector is in the initial stages of integrating generative AI, with the most impactful applications yet to be realized. This scenario resembles the early days of the Internet, marked by significant enthusiasm but minimal immediate revenue generation. It was only later that companies such as Amazon emerged, expanding into retail and, in 2006, launching the first commercial cloud service. Today cloud services are one the major revenue contributors for Amazon, Microsoft and Google. Gartner predicts a similar hype cycle for generative AI (Jaffri, 2024).

Nonetheless, challenges remain in adopting and implementing generative AI within a highly regulated industry. In a survey conducted by SIMFORM, key business challenges concerning the adoption of Generative AI were identified (Dhaduk, 2024). The survey found that 43.9 percent of organizations cited a lack of expertise and experience in Generative AI, while 39 percent reported difficulties integrating AI into existing workflows. Additionally, 39.6 percent of respondents highlighted concerns about potential bias or inaccuracies in AI output, security vulnerabilities in AI-generated code, and the complexity of model fine-tuning. Other challenges mentioned included uncertainty around return on investment (ROI), costs associated with computing and tools and regulatory uncertainties.

Evaluating the return on investment (ROI) for deploying generative AI requires updated metrics. The increased costs related to adoption, staff training, energy consumption and potential inaccuracies need to be weighed against productivity improvements, the advantages of personalized content and new revenue opportunities or market segments. Additionally, use of AI must align with the environmental, social, and governance (ESG) strategies of the organization.

Despite the risks and concerns there is a realization in many Boardrooms on the benefits of adopting GenAI across the organization and more importantly being out competed by their competitors. Investments by Banks have surged and are expected to continue to grow if they do not.

To fully leverage the benefits of the GenAI revolution, organizations must unlock their internal data repositories, including the 80 percent of unstructured data found in emails, communications, and documents. It is essential that all employees receive appropriate training, understand the associated risks, and that mitigating controls are implemented. Most importantly, employees should be assured that GenAI will enhance their productivity rather than replace them.

## References


*Adopting generative AI in banking—Corporates and Institutions.* (2024, Oct). https://corporates.db.com/publications/White-papers-guides/adopting-generative-ai-in-banking.

*Banking on innovation: How ING uses generative AI to put people first.* (2024, September 5). McKinsey & Company. https://www.mckinsey.com/industries/financial-services/how-we-help-clients/banking-on-innovation-how-ing-uses-generative-ai-to-put-people-first.

*Bloomberg—Are you a robot?* (2023, Oct). https://www.bloomberg.com/company/press/bloomberg-and-riskthinking-ai-launch-physical-risk-indicators-to-reveal-exposure-to-severe-climate-events/.

Buehler, K., Corsi, A., Weintraub, B., Jurisic, M., Siani, A., & Lerner, L. (2024, March 22). *Scaling gen AI in banking: Choosing the best operating model.* McKinsey & Company. https://www.mckinsey.com/industries/financial-services/our-insights/scaling-gen-ai-in-banking-choosing-the-best-operating-model.

Dhaduk, H. (2024, April 17). State of Generative AI in 2024 | Insightful Survey Findings. *Simform—Product Engineering Company.* https://www.simform.com/blog/the-state-of-generative-ai/.

Grudnitski, G.; Osburn, L. Forecasting S&P and Gold Futures Prices: An Application of Neural Networks, *Journal of Futures Markets,* 13, 6, September 1993, pp. 631–43.

Hamann, F. (2023, January 23). *Hyper Personalisation: 5 use case examples in digital banking.* Subaio. https://subaio.com/digital-banking/hyper-personalisation-5-use-case-examples-in-digital-banking.

*Introducing computer use, a new Claude 3.5 Sonnet, and Claude 3.5 Haiku.* (2024, Oct). https://www.anthropic.com/news/3-5-models-and-computer-use.

Jaffri, A. (2024, November 11). *Hype cycle for Artificial Intelligence 2024—AI.* Explore Beyond GenAI on the 2024 Hype Cycle for Artificial Intelligence. https://www.gartner.com/en/articles/hype-cycle-for-artificial-intelligence.

JP Morgan (2024, July). *Quest IndexGPT: Harnessing generative AI for investable indices | J.P. Morgan.* https://www.jpmorgan.com/insights/markets/indices/indexgpt.

Maynard, N. (2024, January 23). *Global Generative AI in Banking Market: 2024–2030.* Juniper Research. https://www.juniperresearch.com/research/fintech-payments/banking/global-generative-ai-in-banking-market/.

Miller, Z. (2024, August 25). *Case Study: How Citi and Numerated are working to transform lending technology through strategic partnership.* Tearsheet. https://tearsheet.co/online-lenders/case-study-how-citi-and-numerated-are-working-to-transform-lending-technology-through-strategic-partnership.

Olson, P. (2024, February 6). Commentary: A US$25 million Hong Kong deepfake scam shows new AI

risks in video calls. *CNA*. https://www.channelnewsasia.com/commentary/deepfake-scam-video-conference-zoom-hong-kong-employee-4103266.
Smigel, L. (2023, October 14). History of AI in Finance—Analyzing Alpha. *Analyzing Alpha*. https://analyzingalpha.com/history-of-ai-in-finance.
Taylor, H. (2016, October 25). *Bank of America launches AI chatbot Erica—here's what it does*. CNBC. https://www.cnbc.com/2016/10/24/bank-of-america-launches-ai-chatbot-erica—heres-what-it-does.htm.
Van Rijmenam Csp, M. (2024, September 18). *Agentic AI: The End of Organizations as We Know It*. Dr Mark Van Rijmenam, CSP | Strategic Futurist Speaker. https://www.thedigitalspeaker.com/agentic-ai-end-organizations-as-we-know/.
Vella, H. (2024, November 8). *Google turns to AI to write new code; Workforce reduced*. Google Turns to AI to Write New Code; Workforce Reduced. https://aibusiness.com/data/google-turns-to-ai-to-write-new-code-workforce-reduced.
Yagoda, M. (2024, February 23). *Airline held liable for its chatbot giving passenger bad advice—what this means for travellers*. https://www.bbc.com/travel/article/20240222-air-canada-chatbot-misinformation-what-travellers-should-know.

# 9. Navigating the Algorithmic Frontier

## *AI in Financial Planning*

Daniel Y.N. Tan

The marriage between finance and technology (FinTech) ushered in a new era of financial planning, with artificial intelligence (AI) becoming more prominent and pervasive at the time of writing this piece. The ability to rapidly process copious amounts of data, identify trends, and make predictive decisions will revolutionize people's financial planning. AI has reshaped the financial services landscape ranging from robo-advisors to financial fraud detection. It goes beyond technological advancements, there is a paradigm shift in decision-making processes and client-advisor interaction (Javaid, 2024).

However, this rapid integration of AI also raises concerns about ethical, social, and regulatory issues. These challenges must be addressed for sustainability and growth, and an understanding AI's potential and limitations of AI is necessary. This contribution delves into AI's evolutionary role in financial planning applications, its benefits and potential risks, and its impact on human financial planners.

## *AI-Powered Financial Planning Tools*

The rise of AI has given birth to a new generation of financial planning tools. These include the personalization of advice, autonomous portfolio management, and risk assessment capabilities. For example, robo-advisors can use algorithms to customize investment portfolios based on an individual's risk tolerance, financial goals, and time horizon (Xie, 2019). Such platforms typically offer lower fees than the equivalent traditional advisory services, making financial advice accessible to a wider population (Fung et al., 2021). To provide greater accuracy in risk assessment and optimize investment strategies, AI-driven tools can be used to analyze market movements and economic data (Jain & Vanzara, 2023).

Besides robo-advisors, other tools that use National Language Processing (NLP) facilitate more intuitive client interactions. For example, chatbots are now widely used to guide users through the financial planning process, answer queries, and provide tailored advice (Rane et al., 2024). To provide deeper insights into market dynamics for clients, sentiment analysis tools analyze trends and social media to assess market sentiment (Huang et al., 2023).

## *Enhancing Efficiency and Accessibility*

A key advantage of AI in financial planning is the automation of time-consuming tasks such as data entry, analysis, and report generation. Financial planners are freed from these dreary tasks to focus on improving client engagement and relationships (Li, 2020). Online platforms and mobile financial planning apps powered by AI can provide personalized guidance at a fraction of the cost of traditional advisory services. This levels the playing field by making such services available to underserved populations, such as people with limited financial resources or in remote areas (Truby et al., 2020).

This is especially critical in emerging economies, where access to financial advisory services and literacy remain low. AI-based applications play a pivotal role in democratizing financial planning. For example, platforms such as Branch and Tala harness AI to assess an individual's creditworthiness via unconventional channels such as mobile use and social media activity (Mondol et al., 2024). Such automation can reduce operational costs and enable financial institutions to extend their services to marginalized communities without compromising profitability.

## *Raising Financial Literacy*

By catering to individual learning paces, styles, and preferences, AI can make financial education more engaging and effective. For example, education technology (EdTech) platforms such as Coursera and Khan Academy apply AI to adjust lesson complexity based on the user's performance to ensure that learning is performed at an optimal level (N. Rane et al., 2023).

The use of interactive platforms to simulate different financial scenarios allows personalized financial education and guidance (Bulathwela et al., 2021). These can help users understand longer-term implications, make better informed decisions about their personal finances, and go a long way to improve financial literacy in a population (Zhai et al., 2021). The gamification of financial planning apps is also an important development. Such an approach can motivate users to engage in financial planning through challenges and rewards to cultivate and sustain an interest in personal financial education.

## *Ethical Considerations and Potential Risks*

The potential benefits of AI are substantial at this stage, and new developments occur almost daily. However, ethical considerations and potential risks associated with its implementation must not be neglected. One primary concern is bias that may be inherent to AI algorithms. An AI algorithm is only as effective as the dataset used to train it. If the data contain societal biases, the resulting financial advice is likely to perpetuate or exacerbate such inequalities (Truby et al., 2020). For example, algorithms trained on historical loan application approval data may inadvertently discriminate against minority groups (Owolabi et al., 2024).

With the greater prevalence of AI usage in financial planning tools, questions about data privacy, security, and transparency are bound to arise (Xie, 2019). With greater autonomy given to AI systems, accountability has become increasingly challenging over time. To address this, appropriate safeguards, such as those proposed by the European Commission, must be in place to protect and prevent the misuse of consumer information (Erdélyi & Goldsmith, 2022). The proper implementation and design of AI systems will improve public confidence and adoption.

## *Overreliance on AI and the Erosion of Human Judgement*

Balancing reliance on AI with critical human oversight is essential. With the increasing sophistication of AI-powered financial planning tools, concerns of potential overreliance on them and the erosion of human judgment need to be addressed. As individuals become more accustomed to obtaining financial advice from AI, they may become less inclined to critically evaluate recommendations (Cao et al., 2024). Contrary to the previous section on AI potentially raising financial literacy levels, this could lead to a decline in financial literacy and higher susceptibility to biased or flawed algorithms (Lauterbach, 2019).

Overreliance on AI may also lead to systemic risks; catastrophic system collapses can occur if widely deployed algorithms fail simultaneously owing to shared coding errors or biases. This phenomenon is known as "algorithmic herding" (Chugh et al., 2024), which underpins the need for human oversight to validate AI recommendations.

As AI has become an increasingly important part of daily life, it has become even more crucial to maintain a balance between leveraging it and retaining the human ability to think critically, discern, and make sound financial decisions. Financial advisors must also be trained to interpret and conceptualize AI outputs to ensure that technology complements human expertise and experience rather than replace it.

## *The Need for Regulation*

With AI evolving at breakneck speeds, new developments are being made daily. This necessitates the development of a set of robust regulations to ensure that development and implementation are conducted responsibly (Brundage et al. 2020). The regulatory framework needs to address issues ranging from algorithmic bias and accountability to transparency and data privacy (Erdélyi & Goldsmith, 2022). Such standards and guidelines are necessary to protect consumers, foster trust in AI-driven financial tools (Brundage et al., 2020) and promote fair competition.

Regulations must also address the ethical implications of predictive analytics. For instance, AI's ability to anticipate a person's financial distress based on spending patterns can be misused by predatory and unscrupulous lenders. As AI challenges cut across borders, cooperation and collaboration among national regulatory bodies to synthesize regulatory standards will be essential to foster a fairer global financial ecosystem (Uzougbo et al., 2024).

## *The Future of AI in Financial Planning*

The trajectory of AI in financial planning indicates greater innovation and integration. AI-powered tools for financial planning will become increasingly sophisticated and will build on their ability to understand an individual's financial needs and tailor advice (Fung et al., 2021). There could also be enhanced security and transparency of financial transactions through the integration of AI with blockchain and distributed ledger technology (Martz, 2018). To realize AI's full potential, ongoing dialogue and collaboration among stakeholders such as policymakers, financial institutions, technology developers, and consumers must take place (Golbin et al., 2020).

Emerging technologies, such as quantum computing, hold the promise of increasing AI capabilities by exponentially increasing their data-processing power. This could enable real-time portfolio optimization even under rapid and complex market conditions (Atadoga et al., 2024). Another area is the integration of AI with augmented reality (AR) and VR (virtual reality). This could create immersive experiences during the financial planning process, helping clients visualize the potential outcomes of their decisions.

## *The Impact of AI on Human Financial Planners*

It is inevitable that the rise of AI in financial planning raises questions about the future of human financial planners and their role. Some fear that human advisors will be replaced by AI, while others predict a collaborative partnership (Salampasis et al., 2017). AI has already automated routine tasks, freeing up the human advisor's precious time to focus on more complex high-touch tasks, such as navigating sensitive or emotional client requirements (Mehrotra, 2019). Advisors can also add value by personalizing and making sense of recommendations churned out by AI algorithms.

This would also call for a shift in skillset, with greater emphasis on emotional intelligence, strategic thinking, and the ability to interpret AI-generated insights into one that is relevant to the client's context (Otieno, 2022). Hybrid advisory models combine AI-driven financial data with human expertise as a sustainable approach to addressing diverse client needs and expectations (Salampasis et al., 2017).

## *Conclusion*

The adoption of AI in financial planning is both an opportunity and challenge. The financial planning landscape has already been transformed by AI and will continue to evolve as it develops. This offers unprecedented ways to enhance operational efficiency, accessibility, and personalization. While the potential for innovation is infinite, responsibility lies with all stakeholders to ensure that AI is used ethically and inclusively.

The associated concerns regarding ethical usage, potential risks, and impact on human financial planners must be addressed and considered. Only by fostering responsible development and implementation can the power of AI be harnessed for a robust, inclusive, and equitable financial future.

### References

Atadoga, N.A., Ike, N.C.U., Asuzu, N.O.F., Ayinla, N.B.S., Ndubuisi, N.N.L., & Adeleye, N.R.A. (2024). The Intersection Of Ai And Quantum Computing In Financial Markets: A Critical Review. *Computer Science & IT Research Journal*, 5(2), 461–472. https://doi.org/10.51594/csitrj.v5i2.816.

Brundage, M., Avin, S., Wang, J., Belfield, H., Krueger, G., Hadfield, G., Khlaaf, H., Yang, J., Toner, H., Fong, R., Maharaj, T., Koh, P.W., Hooker, S., Leung, J., Trask, A., Bluemke, E., Lebensold, J., O'Keefe, C., Koren, M., & Ryffel, T. (2020, April 20). Toward Trustworthy AI Development: Mechanisms for Supporting Verifiable Claims. ArXiv.org. https://doi.org/10.48550/arXiv.2004.07213.

Bulathwela, S., Pérez-Ortiz, M., Holloway, C., & Shawe-Taylor, J. (2021). Could AI Democratise Education? Socio-Technical Imaginaries of an EdTech Revolution. ArXiv.org. https://doi.org/10.48550/arXiv.2112.02034.

Cao, S., Liu, A., & Huang, C.-M. (2024). Designing for Appropriate Reliance: The Roles of AI Uncertainty Presentation, Initial User Decision, and User Demographics in AI-Assisted Decision-Making. ArXiv.org. https://doi.org/10.1145/3637318%7D%2010.1145/3637318%7D.

Chugh, Y., Agrawal, S., Shetty, Y., & Guruprasad, M. (2024, March 21). Algo-Trading and its Impact on Stock Markets. https://journal.ijresm.com/index.php/ijresm/article/view/2965.

Erdélyi, O.J., & Goldsmith, J. (2022). Regulating artificial intelligence: Proposal for a global solution. *Government Information Quarterly*, 39(4), 101748. https://doi.org/10.1016/j.giq.2022.101748.

Fung, G., Polanía, LF., Choi, S T., Wu, V C., & Lawrence, M. (2021, December 1). Editorial: Artificial Intelligence in Insurance and Finance. *Frontiers Media*, 7. https://doi.org/10.3389/fams.2021.795207.

Golbin, I., Rao, A.S., Hadjarian, A., & Krittman, D. (2020). Responsible AI: A Primer for the Legal Community. *2020 IEEE International Conference on Big Data (Big Data)*. https://doi.org/10.1109/bigdata50022.2020.9377738.

Huang, H., Zavareh, A.A., & Mustafa, M.B. (2023). Sentiment Analysis in E-Commerce Platforms: A Review of Current Techniques and Future Directions. *IEEE Journals & Magazine* | IEEE Xplore. https://ieeexplore.ieee.org/abstract/document/10225509/.

Jain, R., & Vanzara, R. (2023). Emerging Trends in AI-Based Stock Market Prediction: A Comprehensive and Systematic Review. *Engineering Proceedings*, 56(1), 254. https://doi.org/10.3390/ASEC2023-15965.

Javaid, H.A. (2024, August 11). The Future of Financial Services: Integrating AI for Smarter, More Efficient Operations. http://mzjournal.com/index.php/MZJAI/article/view/239.

Lauterbach, A. (2019). Artificial intelligence and policy: quo vadis? Digital Policy, Regulation and Governance, 21(3). https://doi.org/10.1108/dprg-09-2018-0054.

Li, Z. (2020, June 1). Analysis on the Influence of Artificial Intelligence Development on Accounting. https://doi.org/10.1109/icbaie49996.2020.00061.

Martz, D. (2018, November 29). How Blockchain Technology and Artificial Intelligence are Revolutionizing FinTech. *CoinAnnouncer*. https://www.coinannouncer.com/how-blockchain-technology-and-artificial-intelligence-are-revolutionizing-fintech/.

Mehrotra, A. (2019, April 1). Artificial Intelligence in Financial Services—Need to Blend Automation with Human Touch. IEEE Xplore. https://doi.org/10.1109/ICACTM.2019.8776741.

Mondol, S., Satpathy, I., Patnaik, B.C.M., Nayak, A., Patnaik, A., & Khang, A. (2024). Impact of AI and Data in Revolutionizing Microfinance in Developing Countries. In *Auerbach Publications eBooks* (pp. 71–87). https://doi.org/10.1201/9781032618845-5.

Otieno, N. (2022, February 1). The Future Role of AI in Finance. Www.worldfinance.com. https://www.worldfinance.com/markets/the-future-role-of-ai-in-finance.

Owolabi, O.S., Uche, P.C., Adeniken, N.T., Ihejirika, C., Islam, R.B., & Chhetri, B.J.T. (2024). Ethical Implication of Artificial Intelligence (AI) Adoption in Financial Decision Making. *Computer and Information Science*, 17(1), 49. https://doi.org/10.5539/cis.v17n1p49.

Rane, N.L., Choudhary, S.P., & Rane, J. (2024). Artificial Intelligence-driven corporate finance: enhancing efficiency and decision-making through machine learning, natural language processing, and robotic process automation in corporate governance and sustainability. *Studies in Economics and Business Relations*, 5(2), 1–22. https://doi.org/10.48185/sebr.v5i2.1050.

Rane, N., Choudhary, S., & Rane, J. (2023). Education 4.0 and 5.0: Integrating Artificial Intelligence (AI) for Personalized and Adaptive Learning. *SSRN Electronic Journal*. https://doi.org/10.2139/ssrn.4638365.

Salampasis, D., Mention, A.-L., & Kaiser, A.O. (2017). Wealth Management in Times of Robo: Towards Hybrid Human-Machine Interactions. *SSRN Electronic Journal*. https://doi.org/10.2139/ssrn.3111996.

Truby, J., Brown, R D., & Dahdal, A. (2020, April 2). Banking on AI: mandating a proactive approach to AI regulation in the financial sector. *Taylor & Francis*, 14(2), 110–120. https://doi.org/10.1080/17521440.2020.1760454.

Uzougbo, N.N.S., Ikegwu, N.C.G., & Adewusi, N. a. O. (2024). Legal accountability and ethical considerations of AI in financial services. *GSC Advanced Research and Reviews*, 19(2), 130–142. https://doi.org/10.30574/gscarr.2024.19.2.0171.

Xie, M. (2019, April 1). Development of Artificial Intelligence and Effects on Financial System. *IOP Publishing*, 1187(3), 032084–032084. https://doi.org/10.1088/1742-6596/1187/3/032084.

Zhai, X., Chu, X., Chai, C.S., Jong, M.S.Y., Istenic, A., Spector, M., Liu, J.-B., Yuan, J., & Li, Y. (2021). A Review of Artificial Intelligence (AI) in Education from 2010 to 2020. *Complexity*, 2021(8812542), 1–18. https://doi.org/10.1155/2021/8812542.

## • *C. AI in Human Resources and Training* •

# 10. Integrating Emotional Intelligence and Artificial Intelligence in HR

MIKE HORNE *and* NICOLE C. JACKSON

The integration of Emotional Intelligence (EI) and Artificial Intelligence (AI) within Human Resources (HR) is becoming an increasingly pivotal strategy for organizations seeking to enhance leadership and enterprise-wide performance. This contribution explores how AI and EI can be skillfully blended to meet many organizational challenges, including refining team dynamics, optimizing decision-making processes, and boosting employee engagement. By examining EI and AI independently and in tandem, HR, other business professionals, and readers can consider, adopt, and practice new approaches that may enable successful performance. Here we anchor our insights into the challenges and opportunities identified by scholars and practitioners working at the intersection of AI, EI, and HR.

Deloitte's (2019) Global Human Capital Trends report highlights the significance of integrating advanced technologies with human-centric values. The report emphasizes that by combining technological advancements with human-centered values, organizations can enhance decision-making processes, boost employee engagement, and cultivate a more adaptable workforce. The authors believe that this shift towards a more human-centric approach was accelerated by the COVID-19 pandemic, highlighting the importance of remote work, employee well-being, and a positive workplace culture.

The concept of EI, as elaborated by Goleman (1998) and extended by Mayer, Salovey, and Caruso (2004), underscores the critical need for professionals to cultivate emotional skills within organizations. These skills may be the needed competency to fuse human-centered values with technology. We will focus on emotionally intelligent approaches, as discussed by Cherniss and Goleman (2001), to outline how organizational leaders can choose blended, innovative AI+EI+ HR strategies to advance an organization's mission and goals.

In this regard, the concept of VUCA, developed by the U.S. Army, presents a relevant temporal framework. VUCA describes living in a variable, uncertain, complex, and ambiguous world (Bennett & Lemoine, 2014). Indeed, today, we are at the frontier of work at the intersection of AI, EI, and HR. Companies that produce long-term value in the age of AI will focus on leading with values of dignity, kindness, and respect—all elements of the human-centered workplace (Horne, 2024). This integration will emphasize operational efficiency and extend our capacity to maintain and sustain organizations that champion and marshal diversity, equity, inclusion, and belonging. As supported by recent research, we know that prioritizing EI within AI systems can culminate in more personalized and emotionally astute processes, stimulating better communication and deeper interpersonal relationships at work and in society (Padios, 2017).

As EI and AI integrative technologies evolve, they will raise new and significant ethical considerations, legal risks, and leadership challenges. How will we develop organizational cultures championing continuous machine learning and AI-powered human innovation while maintaining democratic and participatory environments conducive to innovation? What new skills will HR and other leaders need to support engagement in this new world? Those who lead the fusion of EI and AI will set new standards for the new organizations where people do their best work. To further these ideas, we will explore current HR practices and propose a future roadmap at the intersection of these concepts in the following sections.

## *The Intersectionality of AI, EI and HR*

When AI, EI, and HR are integrated, they have the potential to become drivers of progress. AI offers sophisticated data analytics and decision-making tools; HR measures ensure talent management conforms to strategic objectives. Additionally, EI strengthens interpersonal relationships and thus promotes desirable collaboration behaviors. Can the convergence of these fields lead to enhanced operational efficiency and strategic achievements? How quickly will we arrive at a point when AI, EI and HR demonstrate a combination that profoundly impacts organizational performance? We will explore these questions by defining terms and providing a conceptual integration framework.

## *Connection Between AI and HR*

AI is transforming HR by automating repetitive tasks, providing insights through data analytics, and enhancing the overall efficiency of HR processes. With AI-driven tools, HR departments can simplify recruitment, match candidates with job profiles, and personalize employee experiences based on data trends. However, this shift also necessitates HR professionals to upskill and adapt to work alongside advanced technologies, balancing technological benefits with potential concerns around data privacy and ethical use (Lukaszewski & Stone, 2024; Black & van Esch, 2021; Burnett & Lisk, 2019; Rayhan, 2023).

## *Connection Between AI and EI*

Artificial Intelligence (AI) can be equipped with advanced algorithms to simulate certain aspects of emotional intelligence (EI), such as interpreting facial expressions and analyzing tone of voice. This integration has the potential to significantly improve communication and bolster empathetic decision-making within human resource processes, as highlighted in the previously referenced Deloitte report. However, there is an ongoing debate about whether AI can truly replicate human emotions and the implications of relying on technology for emotional insights. Verma (2023) explores these issues in a comprehensive study on how artificial emotional intelligence can enhance human resource management.

## *Connection Between EI and HR*

Emotional intelligence (EI) is essential for effective leadership and employee engagement, enhancing team dynamics and organizational performance. Leaders with high EI can better understand and connect with their teams, encouraging a supportive environment that drives success (Cherniss & Goleman, 2001). HR departments increasingly prioritize EI in recruitment and talent management, recognizing its role in improving communication and adaptability. Integrating EI into organizational systems and processes supports broader organizational pursuits of adaptability and inclusion. Firm-wide learning programs focusing on EI can cultivate desirable workplace behaviors and yield the benefits of desirable engagement behaviors.

## *The Fusion of AI, HR, and EI*

The fusion of Artificial Intelligence (AI), Human Resources (HR), and Emotional Intelligence (EI) presents an opportunity for organizations seeking to remain competitive and agile. As AI advances, HR departments can leverage these technologies to create more adaptive, employee-centric ecosystems that enhance productivity and promote engagement. Integrating AI in HR functions can lead to more personalized approaches in talent management, improving employee satisfaction and retention. AI's ability to process large datasets allows HR to gain insights into workforce trends, tailoring employee development programs that align with individual EI competencies (Ulrich, 2024).

Moreover, AI-infused EI tools redefine traditional HR processes, from recruitment to performance management. For instance, AI-driven algorithms can assess candidates' emotional intelligence during hiring by analyzing behavioral patterns and linguistic cues in video interviews. These innovations enable HR professionals to make more informed hiring decisions, ensuring that new hires align with organizational values and culture, often underpinned by emotional intelligence. This integration helps bridge the gap between identifying technical capabilities and ensuring cultural fit, developing technically proficient and emotionally intelligent teams (Glodstein, 2014).

The ethical implications, however, of AI and EI integration in HR cannot be overlooked. Concerns regarding data privacy and the potential for bias in AI systems require careful consideration and governance. Current research (e.g., D'Cruz et al., 2022; Telkamp & Anderson, 2022) suggests that organizations should implement robust ethical frameworks and oversight mechanisms to ensure that AI tools in HR are used responsibly and transparently. Furthermore, the role of HR in championing responsible

AI use is becoming increasingly critical, as they must strike a balance between leveraging technology for efficiency and upholding organizational ethics and integrity. In our speaking engagements, we have noticed immense variation in how HR leaders describe and report on their strategies and practices relative to AI; there is significant divergence. This underscores, in part, the urgency to stay ahead of this issue.

As organizations move towards a more automated future, the role of emotional intelligence within AI systems becomes increasingly relevant. Developing AI applications that can understand and respond to human emotions authentically is a complex endeavor. Achieving this requires a multidisciplinary approach, combining insights from psychology, computer science, and HR to develop tools that enhance human interactions rather than replace them. Organizations prioritizing emotional intelligence in AI design can ensure meaningful employee engagement, promoting well-being and fostering a positive workplace culture.

One example of AI resolving low-level employee relations disputes can be seen in the deployment of chatbots programmed to handle grievances or concerns. These AI-driven chatbots provide initial support by offering advice based on company policies, documenting incidents, and suggesting possible resolutions. By automating and streamlining initial interactions, these systems help reduce the workload on HR teams while ensuring that employees' issues are addressed promptly. A study by Malin et al. (2023) highlights the use of AI chatbots in multinational corporations, where they effectively manage certain decision and selection processes.

In conclusion, the successful fusion of AI, HR, and EI hinges on a comprehensive approach that values technological advancement and human-centric values. By employing AI technologies with emotional intelligence insights, HR departments can grow more personalized, emotionally resonant work environments that cater to employee needs and enhance satisfaction. As AI tools handle routine tasks, HR professionals can focus on strategic roles, prioritizing employee engagement and well-being, ultimately contributing to an inclusive and high-performing workplace culture. However, the promise of this integration brings with it a responsibility to navigate ethical concerns, ensuring that AI applications respect privacy and remain free of bias.

## *Ethical Considerations at the Fusion of HR, AI, and EI*

As organizations increasingly integrate AI and EI into HR functions, ethical considerations will continue to emerge. One pressing issue is data privacy, as AI systems often require extensive data collection to function effectively (Fenwick, Molnar, & Frangos, 2024). The use of AI in HR necessitates access to personal data, which, if not appropriately managed, could lead to breaches or misuse. Safeguarding employee data should be a top priority, with stringent policies outlining data handling procedures, consent protocols, and transparent communication about data usage. Based on their professional experiences, the authors are all too familiar with safeguarding and protecting employee data worldwide.

The risk of bias in AI algorithms poses significant ethical challenges. As one study describes, AI models are trained on historical data, which may include biases that reflect past inequities. Without proper oversight, these biases could perpetuate unfair practices in hiring, promotions, or even daily HR operations. Organizations must implement a rigorous auditing process for AI tools to ensure fairness and equity. By doing so, they can identify and correct biases, thus cultivating an inclusive environment that values diversity and equal opportunity (Chowdhury, Joel-Edgar, Dey, Bhattacharya, & Kharlamov, 2022).

The intersection of AI and EI in HR also raises questions about the authenticity of emotional interactions in the workplace. Can AI truly emulate human emotional intelligence, or does it mimic cues without genuine understanding? For example, consider our earlier statements regarding using AI to solve employee relations disputes. This distinction is crucial, as reliance on AI for emotional insights could lead to superficial engagements that lack the depth and empathy intrinsic to human interactions. At this stage of AI-EI fusion, organizations should aim to use AI as a complementary tool rather than a substitute, enabling human HR professionals to focus on relationship-building and nuanced communication.

Furthermore, ethical oversight extends to the decision-making processes influenced by AI applications. Transparency is essential; employees and candidates should understand how AI-driven decisions are made, especially in recruitment or performance evaluations (Rigotti & Fosch-Villaronga, 2024). As one of the author's earliest mentors observed, transparency in organizational life is essential—if it can be found, it will be. Organizational leaders should be prepared to explain the rationale behind AI-generated recommendations and when human intervention is necessary. This transparency promotes trust and ensures AI tools are perceived as allies rather than opaque or authoritarian systems.

Finally, the evolving role of HR professionals in this hybrid environment calls for reevaluating ethical responsibilities. As custodians of AI technology and EI principles, HR must advocate for responsible AI deployment that aligns with organizational values and ethical standards. HR leaders must have the knowledge and skills to navigate these complexities and promote ethical practices that uphold technological and humanistic values. By embedding ethics into the core of AI, HR, and EI integration, organizations can create a future where technology and humanity coexist harmoniously, driving positive outcomes for all stakeholders (Singh & Pandey, 2024; Varma, Dawkins, & Chaudhuri, 2023; Dennis & Aizenberg, 2022).

## *Examples and Illustrations: Practical Implications for Organizational Leaders*

Unilever is a well-documented example of a company that has embraced AI to revolutionize its recruitment and onboarding processes. Their innovative system evaluates candidates during video interviews and games to identify future leaders, offering feedback to every applicant, successful or not. This approach counters traditional job applications' common "ghosting" experience. Meanwhile, "Unabot," a natural language processing bot, assists new employees by providing real-time answers to HR and operational queries, adapting to user-specific data such as location and seniority. Launched in the Philippines and now operational in 36 countries, Unabot represents a successful integration of AI and human interaction in corporate culture, demonstrating the potential of technology to enhance efficiency and engagement in HR practices (Marr, 2019).

Another company that has seamlessly integrated artificial intelligence into its human resources processes is IBM. By implementing Watson, their AI-driven platform, IBM has transformed how it manages talent and employee engagement. Watson assists the HR team by sifting through resumes and identifying candidates with the skills and experience most suited for open positions, significantly reducing the time spent on preliminary interviews. Furthermore, Watson provides employees with personalized learning and career development recommendations by analyzing their career history, performance data, and aspirational goals. This approach aligns with emotional intelligence, creating a more personalized and empathetic working environment where employees feel valued and supported. IBM's commitment to blending AI, emotional intelligence, and HR is a testament to the potential of technology to transform traditional work processes and enhance employee experience (Marr, 2019).

## *Conclusion*

Integrating emotional intelligence (EI) with artificial intelligence (AI) in human resources can change organizational processes and systems, achieve greater efficiency, and develop a rich, supportive culture. As we navigate the balance between technological progress and human-centric values, it is evident that AI's potential can be used to streamline routine tasks, freeing leaders to concentrate on strategic and personal interactions with their teams. This combination creates an environment where technology enhances, rather than supplants, human interaction, establishing it as a crucial strategy for contemporary HR management.

Moreover, the global applicability of EI and AI solutions emphasizes the need for cultural sensitivity and adaptability to deploy these technologies effectively across diverse regions. We hope our contribution highlights the importance of customizing AI applications to meet local cultural norms, facilitating smoother international operations, and respecting individual cultural identities and differences. Organizations can ensure inclusivity and equity across their global operations by adopting universally adaptable and culturally sensitive solutions, setting the stage for sustained innovation and growth.

In the future, ethical considerations will remain at the forefront of successful EI and AI integration. Leaders are tasked with maintaining transparency, fairness, and accountability in AI initiatives, ensuring that technological advancements do not come at the expense of ethical standards or human dignity. Continuous exploration and adherence to ethical frameworks are essential for building trust in AI-driven processes, ensuring they support rather than undermine the values of organizational integrity and human well-being. As the landscape of HR continues to transform, these considerations will be crucial in navigating the challenges and opportunities AI and EI present, ultimately steering organizations toward sustainable and ethically sound success.

## References

Bennett, N., & Lemoine, J.G. (2014). What VUCA means for you. *Harvard Business Review*, 92(1/2), 27–36.
Black, J.S., & van Esch, P. (2021). AI-enabled recruiting in the war for talent. *Business Horizons*, 64(4), 513–524. https://doi.org/10.1016/j.bushor.2021.02.015.

Burnett, J.R., & Lisk, T.C. (2019). The Future of Employee Engagement: Real-Time Monitoring and Digital Tools for Engaging a Workforce. *International Studies of Management & Organization*, 49(1), 108–119.
Cherniss, C., & Goleman, D. (2001). *The emotionally intelligent workplace: How to select for, measure, and improve emotional intelligence in individuals, groups, and organizations.* Jossey-Bass.
Chowdhury, S., Joel-Edgar, S., Dey, P.K., Bhattacharya, S., & Kharlamov, A. (2022). Embedding transparency in artificial intelligence machine learning models: Managerial implications on predicting and explaining employee turnover. *The International Journal of Human Resource Management.*
D'Cruz, P., Du, S., Noronha, E., Parboteeah, K.P., Trittin-Ulbrich, H., & Whelan, G. (2022). Technology, Megatrends, and Work: Thoughts on the Future of Business Ethics. *Journal of Business Ethics*, 180(3), 879–902.
Deloitte. (2019). *Global human capital trends 2019: Leading the social enterprise—Reinvent with a human focus.* Deloitte Insights.
Dennis, M.J., & Aizenberg, E. (2022). The Ethics of AI in Human Resources. *Ethics & Information Technology*, 24(3), 1–3.
Fenwick, Ali & Molnar, Gabor & Frangos, Piper. (2024). Revisiting the role of HR in the age of AI: bringing humans and machines closer together in the workplace. *Frontiers in Artificial Intelligence*, 6.
Glodstein, D. (2014). Recruitment and Retention: Could Emotional Intelligence be the Answer? *Journal of New Business Ideas & Trends*, 12(2), 14–21.
Horne, M. (2024). *The people dividend: Leadership strategies for unlocking employee potential.* Empire Publishers.
Lukaszewski, K.M., & Stone, D.L. (2024). Will the use of AI in human resources create a digital Frankenstein? *Organizational Dynamics*, 53(1), no page identified.
Malin, C., Kupfer, C., Fleiß, J., Kubicek, B., & Thalmann, S. (2023). In the AI of the Beholder—A Qualitative Study of HR Professionals' Beliefs about AI-Based Chatbots and Decision Support in Candidate Pre-Selection. *Administrative Sciences*, 13(11), 231.
Marr, B. (2019). Artificial intelligence in practice: How 50 successful companies used artificial intelligence to solve problems (M. Ward (Ed.)). Wiley.
Mayer, J.D., Salovey, P., & Caruso, D.R. (2004). Emotional intelligence: Theory, findings, and implications. *Psychological Inquiry, 15*(3), 197–215.
Padios, J.M. (2017). Mining the mind: Emotional extraction, productivity, and predictability in the twenty-first century. *Cultural Studies*, 31(2/3), 205–231.
Rayhan, J. (2023). Artificial Intelligence (AI) in Human Resource Management (HRM): A Conceptual Review of Applications, Challenges and Future Prospects. *Globsyn Management Journal, 17*(1/2), 37–52.
Rigotti, C., & Fosch-Villaronga, E. (2024). Fairness, AI & recruitment. *Computer Law & Security Review*, 53, N.PAG.
Singh, A., & Pandey, J. (2024). Artificial intelligence adoption in extended HR ecosystems: enablers and barriers. An abductive case research. *Frontiers in Psychology*, 1–13.
Telkamp, J.B., & Anderson, M.H. (2022). The Implications of Diverse Human Moral Foundations for Assessing the Ethicality of Artificial Intelligence. *Journal of Business Ethics*, 178(4), 961–976.
Ulrich, D. (2024). How are You Doing at AI for HR? A Ten-Item Assessment to Evaluate Your Progress. *Workforce Solutions Review*, 3, 13–16.
Varma, A., Dawkins, C., & Chaudhuri, K. (2023). Artificial intelligence and people management: A critical assessment through the ethical lens. *Human Resource Management Review*, 33(1), 1–11.
Verma, V. (2023). Augmenting Human Resource Management through Artificial Emotional Intelligence: A Comprehensive Exploration of Implementation, Ethical Considerations, and Future Horizons. *Globsyn Management Journal*, 17(1/2), 32–36.

# 11. Efficiency and Ethical Dilemmas in the Hiring and Selection of Employees

Veejay Madhavan

The progressive function of Artificial Intelligence (AI) in business, namely in Human Resources (HR), is a subject of much fascination and discussion. AI can transform HR activities by improving efficiency, precision, and decision-making processes. Nevertheless, this technological progress also gives rise to ethical dilemmas, including concerns about privacy, partiality in decision-making, and the possible replacement of human labor (Boden, 2016; Bostrom & Yudkowsky, 2014). As AI evolves it is essential to maintain a delicate equilibrium between the advantages it offers and the ethical dilemmas it raises.

The integration of AI into HR is not a fad, but rather a significant and disruptive change that is fundamentally altering the way firms handle their personnel. The dichotomous character of AI—its capacity to improve HR operations and the ethical dilemmas it presents—necessitates a sophisticated comprehension and meticulous execution. This contribution seeks to examine the importance of AI in the field of Human Resources (HR), taking into account both its benefits and the challenges it poses through the lens of an actual use case.

## *AI's Place in HR*

The importance of AI in HR is emphasized by the growing dependence on technology to optimize different HR operations. AI is being incorporated into various aspects of business operations, such as recruitment, staff management, and performance evaluations, to enhance efficiency and facilitate informed decision-making. AI-powered recruiting tools may efficiently evaluate large volumes of data to accurately find the most qualified candidates. Similarly, AI-driven performance evaluation systems can offer unbiased assessments of employee performance (Davenport & Ronanki, 2018; Leicht-Deobald et al., 2019). The increasing popularity of this phenomenon underscores the significant impact that AI can have on HR, while also underscoring the importance of addressing the ethical considerations that come with its use.

AI is widely used in HR for recruiting and choosing employees, making it one of the prominent applications of AI in this field. Conventional recruitment techniques frequently require the laborious process of manually reviewing resumes, which is time-consuming and susceptible to human prejudice. AI-driven recruitment solutions can rapidly evaluate extensive amounts of resumes, determining the most appropriate individuals according to predetermined criteria. AI systems can expedite the recruiting process and mitigate bias by prioritizing objective qualifications and experiences over subjective factors (Chamorro-Premuzic et al., 2017).

Members of Gen Z make up the newest recruits to the global, multigenerational workforce. By 2025, the countries that make up the Organization for Economic Co-operation and Development (OECD) predict that Gen Z will account for 27 percent of the workforce (Bloomgarden, 2022). Being raised with technology, they are considered digital natives, and they highly prioritize innovation, adaptability, and purposeful work. The adoption of AI-powered recruitment solutions can meet these desires by offering a tailored and streamlined recruitment experience. AI chatbots can interact with Generation Z candidates in real time, providing responses to their inquiries and assisting them in navigating the application procedure. This not only improves the overall experience for candidates but also enables HR professionals to concentrate on more strategic responsibilities (Sivathanu & Pillai, 2018).

## *Case Study: Role of AI in Recruitment and Selection of ICT Sales Team in Indochina*

This case study examines the utilization of AI in the process of recruiting and selecting an Information and Communication Technology (ICT) sales team in Cambodia and Laos for a prominent provider in the Indo-China region. It explores the particular use of AI, the distinct difficulties encountered in the area, and the insights gained from the utilization of AI in this setting.

AI played a crucial role in automating various components of the recruitment process, such as automated resume screening, candidate matching, and interview scheduling. The HR team of this ICT provider in Indochina utilized AI techniques to efficiently analyze a vast number of resumes, identifying applicants whose qualifications and experiences aligned with the job requirements. The use of this automated screening procedure resulted in a substantial decrease in the duration and exertion needed for the initial selection of candidates.

AI demonstrated its usefulness in the domain of candidate matching. The AI algorithms utilized data analysis to select applicants who possessed the most suitable talents, experiences, and educational backgrounds for the ICT sales roles. In addition, interview scheduling tools powered by artificial intelligence provided flawless coordination between candidates and interviewers, ensuring that the recruitment process advanced seamlessly and quickly.

### Issues Related to the Availability of Skilled Individuals

The recruitment procedure for the ICT sales force in Indo-China encountered various difficulties, specifically related to the availability of skilled individuals. The ICT sector in Cambodia and Laos is characterized by a scarcity of qualified candidates. The limited availability of skilled individuals posed a challenge for AI systems in identifying appropriate candidates for job vacancies.

In addition, the particular criteria for ICT sales positions added complexity to the recruitment process. The positions required a rare blend of technical expertise and sales savvy, which was not readily available in the talent pool. Consequently, the AI algorithms encountered difficulties in aligning resumes with the job requirements, resulting in a success rate that was lower than anticipated in discovering appropriate individuals.

### Job Description Bias

Another notable obstacle that emerged from the recruitment process was the existence of biases in the job descriptions produced by the hiring managers. The AI algorithms' performance was influenced by prejudices, specifically gender bias. For example, specific job descriptions had wording and requirements that unintentionally gave preference to male candidates rather than female prospects. The AI's candidate matching procedure exhibited this bias, leading to a distorted selection of applicants.

### Lessons Learnt and Follow Up Measures Implemented

The utilization of AI in the recruitment process for the ICT sales team in Indochina revealed numerous significant insights. First and foremost, it emphasized the importance of meticulously evaluating the wording and criteria employed in job descriptions. To alleviate biases, it was crucial to employ inclusive terminology and guarantee that the criteria utilized are both pertinent and unbiased.

Furthermore, the difficulties associated with finding suitable personnel highlight the significance of consistently improving AI algorithms to enhance comprehension and adjustment to the particular demands of various positions. This entailed integrating input from hiring managers and candidates to enhance the precision and efficiency of AI technologies.

To rectify the biases that were detected in the recruitment process, several measures were implemented. The job descriptions underwent a thorough examination and modification process to remove any language or criteria that displayed bias towards a certain gender. This required the cooperation of HR specialists and line managers to guarantee that the job descriptions were equitable and encompassing.

In addition, the AI algorithms were enhanced to identify and address biases. This involved training the algorithms using a variety of datasets and incorporating tools to identify and rectify biases. By employing this approach, the AI systems become more proficient in delivering impartial candidate recommendations.

## *Future of AI in a Multicultural and Multigenerational Workplace*

The potential of AI in HR recruiting and selection for a workforce that is diverse in terms of culture and age is highly promising and can bring about significant changes. AI technologies are becoming more widely used to increase efficiency, mitigate bias, and enhance decision-making in HR operations. AI-driven solutions can examine extensive quantities of data to pinpoint the most suitable applicants, simplify the recruitment procedure, and deliver impartial performance assessments.

AI can promote diversity and inclusion in a multicultural workforce by eliminating human prejudices during the recruitment process. AI algorithms can be programmed to prioritize objective criteria, like as talents and experience, instead of subjective elements that could result in prejudice. This can lead to a broader range of skilled individuals and a more equitable recruitment procedure. AI can accommodate the distinct requirements and inclinations of various age cohorts in a multigenerational workforce. AI-powered chatbots can offer immediate help to tech-savvy younger Gen Z candidates, while also providing tailored advice to older candidates who may need additional coaching. The ability of AI to adapt makes it a valuable tool in enhancing the inclusivity and supportiveness of the recruitment process.

Nevertheless, it is imperative to acknowledge and tackle ethical problems and potential biases inherent in AI systems. Regular monitoring and enhancement of AI systems are essential to guarantee their fairness and effectiveness. Through the responsible utilization of AI, HR professionals may establish a recruitment process that is fair and effective, resulting in advantages for both employers and employees.

## *Conclusion*

The case study of the ICT sales force in Indochina underscores the revolutionary capacity of AI in recruitment, as well as the problems and ethical considerations it presents. AI has the potential to improve efficiency and precision in the process of selecting candidates. However, it is essential to overcome biases and fine-tune algorithms to ensure fair recruitment processes. Through assimilating knowledge and executing proactive strategies, firms may harness the power of AI to construct teams that are both diverse and high-performing. Although AI has notable benefits in the field of HR, it is crucial to take into account ethical concerns, including privacy and the potential biases inherent in algorithms. It is crucial to prioritize the security and transparency of data management and consistently enhance AI systems to uphold trust and justice. By carefully weighing the advantages of artificial intelligence (AI) against ethical concerns, firms can improve their human resources (HR) processes and establish a workplace that is both more efficient and fair.

### References

Bloomgarden, K. (2022). Gen Z and the end of work as we know it. *World Economic Forum annual Meeting.* Davos: World Economic Forum. https://www.weforum.org/agenda/2022/05/gen-z-don-t-want-to-work-for-you-here-s-how-to-change-their-mind/.

Boden, M.A. (2016). AI: Its nature and future. Oxford University Press.

Bostrom, N., & Yudkowsky, E. (2011). The ethics of artificial intelligence In: Frankish K, Ramsey WM, eds. *The Cambridge Handbook of Artificial Intelligence.* Cambridge University Press; 2014:316–334. http://www.nickbostrom.com/ethics/artificial-intelligence.pdf.

Chamorro-Premuzic, T., Akhtar, R., Winsborough, D., & Sherman, R.A. (2017)." The datafication of talent: How technology is advancing the science of human potential. *Current Opinion in Behavioral Sciences, 18, 13–16.* DOI:10.1016/j.cobeha.2017.04.007.

Davenport, T.H., & Ronanki, R. (2018). Artificial intelligence for the real world. *Harvard Business Review, 96(1), 108–116.* DOI: 10.4236/ojbm.2022.101026.

Hunkenschroer, A.L., & Luetge, C. (2022). Ethics of AI-enabled recruiting and selection: A review and research agenda. *Journal of Business Ethics, 178, 977–1007.* DOI:10.1007/s10551-022-05049-6.

Leicht-Deobald, U., Busch, T., Schank, C., Weibel, A., Scherer, A., & Wildhaber, I. (2019). The challenges of algorithm-based HR decision-making for personal integrity. *Journal of Business Ethics, 160(2), 377–392.* DOI:10.1007/s10551-019-04204-w.

Raghavan, M., Barocas, S., Kleinberg, J., & Levy, K. (2020). Mitigating bias in algorithmic hiring: Evaluating claims and practices. *Proceedings of the 2020 Conference on Fairness, Accountability, and Transparency, 469–481.* DOI:10.1145/3351095.3372828.

Sivathanu, B., & Pillai, R. (2018). Smart HR 4.0—how industry 4.0 is disrupting HR. *Human Resource Management International Digest, 26(4), 7–11* DOI:10.1108/HRMID-04-2018-0059.

Tambe, P., Cappelli, P., & Yakubovich, V. (2019). Artificial intelligence in human resources management: Challenges and a path forward. *California Management Review, 61(4), 15–42.* DOI: 10.2139/ssrn.3263878.

Zhai, Y., Zhang, L., & Yu, M. (2024). AI in human resource management: Literature review and research implications. *Journal of the Knowledge Economy.* https://doi.org/10.1007/s13132-023-01631-z.

# 12. Artificial Intelligence for Recruiters

SAMEEKSHA SAHNI

Recruitment is a crucial function of human resources, typically a labor-intensive process characterized by large amounts of data and the requirement for prompt and precise decision-making. The advent of artificial intelligence and machine learning technologies marks the beginning of a new age in which these difficulties may be tackled with greater precision and efficiency. Large language models (LLMs) and generative artificial intelligence aim to instill confidence in the accuracy of the recruiting process by converting a manual, error-prone procedure into a data-driven, streamlined approach.

## *The Hiring Process Without AI*

Conventional hiring techniques are sequential and primarily dependent on manual procedures, which take time and expose human error. Posting job positions, screening prospects, interviewing, debriefing, and drafting job offers may all take much time and frequently lead to a protracted time-to-hire and a less-than-ideal candidate experience. This inefficiency calls for a change to more advanced, automated methods to manage the rising complexity and volume of recruiting responsibilities (Jesuthasan, 2022).

## *Breakdown of Traditional Recruitment Tasks*

- Posting Job Openings: Requires around 20–30 minutes for each job.
- Candidate Screening: This process involves reviewing resumes and doing background checks, often taking 20–30 minutes for each candidate.
- Conducting candidate interviews typically requires a time commitment of 20–60 minutes per applicant.
- Debriefing Interviews: This can take 20–40 minutes per applicant.
- Generating employment offers: Allocating an additional 15–30 minutes for each candidate.

These essential administrative chores significantly deplete recruiters' time, reducing their availability for cultivating relationships and engaging in strategic planning. AI recruiting technologies may free HR personnel from monotonous activities, enabling them to concentrate on the strategic parts of their roles. This frequently leads to a less than ideal applicant experience and a longer time-to-hire, which ultimately causes candidates to lose interest and withdraw from the hiring process.

## *Transforming Unstructured Information into Structured Data*

One of the main difficulties in conventional hiring is turning unstructured interview material into organized data fit for analysis. Particularly those based on Large Language Models (LLMs), generative artificial intelligence engines automate the transcription and analysis of interview talks, collecting important information and transforming it into a structured format much valued for recruiting systems. Driven by artificial intelligence, this change not only removes the need for hand data input but also greatly lowers administrative overhead, hence improving the accuracy and efficiency of the hiring process (Brynjolfsson et al., 2021).

## *Main Use Cases of AI in Recruitment*

AI technologies offer many applications in the recruitment process, transforming how organizations identify, attract, and engage potential candidates.

**Identifying and Attracting Talent**. AI recruiting technologies effectively search large databases and sites such LinkedIn, GitHub, and Stack Overflow to find people that fit certain job requirements, including passive applicants who may be open to appropriate offers but are not actively searching for new possibilities. Beyond conventional approaches, advanced data analytics and machine learning algorithms offer rapid and effective candidate identification, hence extending the scope (Sivaraman & Sivasankaran, 2023). To match individuals with pertinent job opportunities, for example, artificial intelligence algorithms can examine candidates' talents, experiences, and professional activity. This technique may provide a whole picture of candidates' qualifications by include assessing their social media activity, attendance in professional forums, and contributions to open-source projects.

**AI for Headhunting**. Using algorithms and natural language processing to examine large amounts of data from social media profiles, professional networks, and other sources, AI greatly improves headhunting by pointing out high-impact opportunities fit for passive applicants. For finding co-founders, IT experts, and C-suite executives—all of which are especially important—this approach is especially successful because using artificial intelligence, companies may find candidates with the right credentials, access a larger talent pool, and interact with them with customized messaging. Based on their career path, recent professional performance, and other behavioral traits, AI driven headhunting technologies may evaluate applicants' probability of being open to new possibilities.

**Automated Outreach and Engagement**. AI-driven systems customize and automate outreach messaging, hence improving applicant involvement and increasing the pool of talent. Generative artificial intelligence personalizes interactions to ensure relevance and effect, therefore strengthening ties and enhancing the recruiting experience (Kaplan & Haenlein, 2020). Depending on candidates' characteristics and prior contacts, automated systems may send customized emails and messages, therefore improving the efficiency and appeal of communication. This degree of customizing may greatly raise response rates and keep possible applicants engaged all through the hiring process.

**Job Description Generation and Interview Preparation**. AI helps create thorough job descriptions and pertinent interview questions, therefore guaranteeing that job ads and interviews are customized to fairly evaluate applicants' fit. This simplified method guarantees high-quality candidate assessment and saves time (Lee et al., 2021). By means of analysis of prior successful job announcements, artificial intelligence may propose enhancements, therefore assuring accurate and enticing job descriptions for appropriate individuals. AI may also create interview questions unique to the position and in line with the corporate principles and culture, therefore enabling interviewers to produce more consistent and perceptive responses.

**AI-Based Interviews and Sentiment Analytics**. By means of sentiment analytics and preparatory screenings, AI systems measure emotional tones and engagement levels during interviews, therefore offering deeper information on applicants' fit beyond conventional evaluation criteria. More complex assessments made possible by this technology also assist to quickly identify the most qualified applicants (Bogen & Rieke, 2021). To evaluate candidates' confidence, honesty, and general fit for the position, AI driven interview systems may examine several facets of their responses—including verbal and nonverbal signals. By helping to identify any differences between candidates' stated opinions and their actual feelings, sentiment analysis offers even another level of assessment.

**Efficient Screening and Candidate Matching**. Among the most time-consuming parts of hiring is screening applicants. By offering a bestmatch score for prospects, artificial intelligence recruiters drive effective screening. AI recruiters make sure that only the most qualified applicants are shortlisted for additional review by considering several elements including talents, experience, and cultural fit. This skill improves the caliber of employees in addition to saving time. Quickly sorting through hundreds of resumes and applications, artificial intelligence algorithms may find the best applicants depending on specified criteria. This automatic screening system guarantees that top-notch applicants are found fast and helps recruiters to have less work.

**Predictive Analytics and Workforce Planning.** Predictive analytics helps artificial intelligence recruiters to project recruiting demand, find skill gaps, and create proactive recruitment plans. Through industry trends and historical data analysis, artificial intelligence recruiters enable companies to keep ahead of their personnel needs. This capacity enables businesses to guarantee they have the correct personnel in place to fulfill their corporate objectives and more properly allocate their staff. Predictive analytics may reveal future recruiting requirements, thereby enabling companies to develop and carry out plans fit for their long-term goals. This forward-looking strategy guarantees businesses' competitiveness in a fast-changing employment environment.

## *Advantages of AI in Recruitment*

**Enhanced Efficiency**. AI significantly accelerates many aspects of the recruitment process, from initial candidate screening to final hiring stages. Automation of routine tasks like scheduling interviews and data entry reduces the time spent on administrative duties, allowing recruiters to focus on more strategic activities.

**Improved Candidate Matching**. By analyzing vast data, AI can more accurately match candidates with job vacancies based on skills, experiences, and cultural fit. This precision helps in minimizing the chances of a bad hire and optimizes the recruitment process by identifying the most promising candidates early on.

**Support for Diversity and Inclusion**. AI tools can be designed to ignore demographic factors such as age, gender, and ethnicity, focusing solely on skills and qualifications. This helps in reducing unconscious biases that might occur during manual resume screening, promoting a more diverse and inclusive workplace.

**Predictive Analytics**. AI can forecast future hiring needs based on company growth and turnover rates, allowing HR departments to proactively recruit and fill talent gaps before they impact the business. This strategic approach aids in better workforce planning and management.

**Cost Reduction**. Automating the recruitment process with AI can lead to significant cost savings by reducing the need for extensive HR personnel to conduct initial screenings and by shortening the time to hire, which lowers the cost per hire and operational costs associated with lengthy recruitment processes.

## *Potential Pitfalls of AI in Recruitment*

**Algorithmic Bias**. If not carefully managed, AI systems can perpetuate existing biases found in training data. For instance, if the historical data used to train AI systems has gender or racial biases, the AI's decision-making might also become biased, leading to discriminatory hiring practices.

**Loss of Personal Touch.** Over-reliance on AI can make the recruitment process feel impersonal to candidates. Human interaction is crucial in understanding nuanced aspects of a candidate's profile and building a relationship that encourages them to join an organization.

**Privacy Concerns.** AI handles vast amounts of personal data, posing significant data privacy and security risks. Candidates' personal information could be exposed to cybersecurity risks or misused if not effectively managed.

**Dependence on Data Quality.** AI's effectiveness is heavily dependent on the quality and volume of the data on which it is trained. Poor quality data can lead to inaccurate conclusions and poor hiring decisions.

**Resistance to Change.** HR departments and recruitment teams may resist adopting AI technologies. Concerns about job displacement or distrust in AI's capabilities can hinder the integration of these technologies into existing recruitment processes.

**Transparency and Accountability.** AI systems can sometimes be "black boxes," where the decision-making process is not transparent. This can create accountability issues, especially if a candidate challenges a decision made by the AI system.

**Regulatory and Ethical Issues.** As AI in recruitment evolves, it faces a growing number of regulatory challenges. Ensuring compliance with labor laws and regulations regarding automated decision-making processes remains a critical concern.

## *Responsible Use of AI in Recruitment*

It is essential to adopt responsible AI practices to maximize the benefits and mitigate the pitfalls of AI in recruitment. This involves ensuring transparency, fairness, and accountability in AI systems. Companies must implement strategies to audit AI algorithms for biases and make necessary adjustments regularly. Additionally, training HR professionals on AI tools and fostering a culture of continuous improvement can help better integrate AI into recruitment processes.

**Transparency and Explainability**. Transparency in AI decision-making processes is crucial. AI systems should be designed to explain their decisions clearly, allowing HR professionals to understand how conclusions were reached. This helps build trust and ensures that AI decisions can be scrutinized and validated.

**Regular Audits and Bias Mitigation**. Regular audits of AI systems are necessary to detect and mitigate biases. Organizations can continuously monitor AI outputs and compare them against predefined fairness criteria to ensure their AI tools promote equitable hiring practices.

**Data Privacy and Security**. To protect candidates' personal information, strict data privacy and security measures must be implemented. This includes adhering to regulations such as GDPR and ensuring that data handling practices are transparent and secure.

**Human-AI Collaboration.** AI should be viewed as a tool to augment human decision-making, not replace it. Ensuring that human recruiters remain actively involved in the recruitment process helps maintain the personal touch and nuanced understanding that AI might lack. This collaborative approach leverages the strengths of both AI and human intelligence.

**Ethical AI Development.** Developing ethical AI systems involves incorporating ethical considerations into every stage of AI development. This includes setting ethical guidelines for data usage, ensuring diversity in AI development teams, and fostering an inclusive approach to AI design and implementation.

## *Conclusion*

AI in recruitment presents both significant opportunities and notable challenges. As these tools become more integrated into HR practices, professionals must ensure they are used responsibly, enhancing decision-making while maintaining fairness and transparency. Future research should focus on developing AI systems that support these goals, driving the evolution of recruitment in the age of artificial intelligence.

In conclusion, AI has the potential to revolutionize recruitment by improving efficiency, enhancing candidate matching, and supporting diversity and inclusion. However, careful consideration must be given to AI adoption's ethical and practical challenges. By adopting responsible AI practices, organizations can harness the power of AI to build a more effective, fair, and inclusive recruitment process.

## References

Bogen, M., & Rieke, A. (2021). Help wanted: An examination of hiring algorithms, equity, and bias. *Upturn*.

Brynjolfsson, E., Rock, D., & Syverson, C. (2021). The productivity paradox of artificial intelligence: Research evidence from the U.S. manufacturing sector. *Journal of Economic Perspectives*, 33(3), 3–30.
Jesuthasan, R. (2022). Reinventing jobs: A 4-step approach for applying automation to work. *Harvard Business Review Press*.
Kaplan, A., & Haenlein, M. (2020). Rulers of the world, unite! The challenges and opportunities of artificial intelligence. *Business Horizons*, 63(1), 37–50.
Kelly, J. (2023). AI recruiting will be a game changer. *Forbes*.
Lee, I., Lee, K.S., & Lee, H.J. (2021). Exploring the role of AI in recruitment: A systematic review. *Journal of Business Research*, 123, 81–91.
Sivaraman, V., & Sivasankaran, N. (2023). AI in recruitment: A game changer in talent acquisition. *Journal of Human Resource Management*, 6(2), 45–60.

# 13. Crisis-Ready Leadership

## *How AI Elevates the Power of Mindfulness*

Emmanuel G. Zara, Jr.

Can having a mindfulness practice make a better leader? How can AI help in enhancing a leader's practice of mindfulness? Over the years, the study of mindfulness and leadership has gained significant attention. Researchers have explored how practicing mindfulness can enhance leadership traits, helping leaders manage organizations more effectively. This piece delves into mindfulness and its role in helping leaders navigate the complexities of managing teams during challenging times, particularly during the COVID-19 pandemic. As it reveals how a mindfulness practice can help leaders exercise better leadership during crises, it will also discuss how AI can support leaders to initiate or enhance this practice for better leadership in general.

## *Leadership and Mindfulness*

Mindfulness has emerged as a powerful tool for leader development. Mindfulness is defined as state-mindfulness or trait-mindfulness. State-mindfulness involves "paying attention in a particular way; on purpose, in the present moment, and non-judgmentally" (Rupprecht et al., 2019). Trait-mindfulness, on the other hand, is the type of mindfulness that can be developed to achieve state-mindfulness. It is achieved through practices like daily mindfulness meditations (Sisk, 2017), loving-kindness meditations (Auten, 2017), and yoga (Gordon, 2013). Structured programs like Mindfulness-Based Stress Reduction (MBSR) and Mindfulness-Based Cognitive Therapy (MBCT) also enhance this trait (Rupprecht et al., 2019).

Studies have shown that mindfulness benefits leaders by enhancing self-care, self-reflection, self-regulation, and self-awareness (Frizzel et al., 2016; Rupprecht et al., 2019). It increases awareness of physical sensations, emotions, thoughts, and non-productive habits (Frizzel et al., 2016) while also reducing stress, which followers perceive as effective leadership behavior (Wasylkiw, et al., 2015). Because mindfulness enhances awareness, sensitivity to the environment, openness to new information, adaptability to change, and organizational transformation (Aviles & Dent, 2015; King & Haar, 2017; Rupprecht et al., 2019), it helps leaders adapt to varying situations, leading to positive performance (Arendt et al., 2019; Baron, et al., 2018), better decision-making (Smollan, 2013), and improved strategic planning (Ahlvik, 2019).

## *Leader Mindfulness and Its Effects on Employees*

Effective human resource management focuses on employee development to enhance individual, team, and organizational performance (Amina & Hadi, 2020). Leader mindfulness can be a factor in improving relationships with others and promoting well-being among subordinates (Aviles & Dent, 2015; Rupprecht et al., 2019). It enhances a leader's capability to deeply listen, be empathic, express empathic concern, take multiple perspectives, and exhibit the ability to respond flexibly towards employees (Auten, 2017; Frizzel et al., 2016). These capabilities then contribute to employee job satisfaction, psychological need satisfaction, and reduced emotional exhaustion (Schuh et al., 2019), leading to better work-life balance (Pinck & Sonnentag, 2018). Mindfulness fosters higher-quality relationships between leaders and followers through emotional attention and self-regulation (Amina & Hadi, 2020). It increases work engagement and performance through genuineness, authenticity, justice, and lesser feelings of emotional drain (Pinck & Sonnentag, 2018).

## *Mindfulness in Times of Crisis*

Crises like the COVID-19 pandemic pose significant challenges for leaders in steering their organizations and protecting their employees. Economic downturns, like those seen during COVID-19, resulted in numerous layoffs, downsizing, and involuntary unemployment, severely affecting industries like hospitality, food, manufacturing, retail, travel, and trade (International Labor Organization, 2020). Companies faced reduced sales income and had to cut jobs, wages, and hours, impacting micro, small, and medium-sized enterprises.

When organizations face these challenges, employees are affected. Crises can disrupt career progressions, impede job search strategies, and delay competency development for career advancements (International Labor Organization, 2020). Fear and panic, leading to mental health issues like anxiety and depression, are common during crises (Tusl et al., 2021). Indian and Chinese employees reportedly experienced work-life strain due to distractions at home and increased depression rates (Restubog et al., 2020).

Mindfulness has been shown to be an effective practice during crises, particularly during the COVID-19 pandemic. For example, mindfulness mediated the relationship between the fear of COVID-19 and mental health issues (Belen, 2021). Individuals who practiced mindfulness were less affected by the fear and its negative consequences (Belen, 2021). In fact, meditation and mindfulness were encouraged as supportive practices for healthcare professionals, patients, caregivers, and the public during the pandemic as practicing mindfulness was considered a way to strengthen mental resilience against fear and other mental health problems (Gotink et al., 2016). Long-term meditation practitioners showed increased activity in brain areas associated with emotion regulation, aiding in better emotional control (Majeed et al., 2020).

## *New Findings on Leader Mindfulness in Times of Crisis*

This piece highlights the results of a study among four leaders who actively practice trait mindfulness and use it as an effective leadership tool amidst uncertainty and ambiguity. It presents how mindfulness enhanced their traits, characteristics, and behaviors, enabling them to relate effectively to their followers during crises.

Four leaders were chosen to participate in this study given their similarities in profile. The participants were in middle management positions, supervising at least two subordinates, and had a trait mindfulness practice of at least one year. They held Human Resources and Learning and Development roles in a Business Process Outsourcing (BPO) company, Management Consulting companies, and a leading financial company. These leaders were chosen for the study based on community, peer, or self-nomination and had a mean score of at least 5 in the 15-item Mindfulness Attention Awareness Scale (MAAS). They were individually interviewed, and their lived experiences were analyzed using Interpretative Phenomenological Analysis (IPA).

The interviews resulted in three main themes illustrating how a leader's mindfulness practice shapes their relationship with employees during a crisis, particularly the pandemic.

The first theme describes the leader-participants' focus on doing better work to secure jobs threatened by irrelevance during the lockdown in March 2020. This approach acknowledges job insecurity and the need for innovation and process improvement. All four participants reported how they made sure they provided relevance in the work of their employees during this challenging event. Re-establishing clarity of team purpose provided direction amidst the chaos.

The second theme highlights how leaders built personal relationships with employees through frequent interactions, even while working remotely. They regularly reached out to discuss personal struggles, alleviating feelings of isolation and building trust. Empathy, optimism, honesty, and transparent communication were crucial in these relationships (Boyatzis and McKee, 2005). Through this frequent communication, the leaders were able to develop deeper trust with their subordinates enabling courage to show their vulnerability without fear of being judged as weak or incompetent.

The third theme involves the leaders' regular self-examination, demonstrating humility and self-awareness which the participants all reported as a result of their mindfulness practice. Enhanced self-awareness helped the leader-participants be sensitive to their environment, correct inappropriate goals, and promote employee well-being. This introspection allowed leaders to manage emotions and behaviors more effectively (Goleman, 2014; Stedham and Skaar, 2019).

## *The Role of AI in Initiating and Enhancing a Leader's Mindfulness Practice*

Given the new findings on how a mindfulness practice of a leader can help in leading employees during crisis, the use of AI, through its innovative tools and techniques can support a leader's journey into mindfulness whether one is just starting the practice or may already have one but wishes to enhance it.

For one, AI aids in stress level analysis, sleep and rest patterns, varied human emotional states through machine learning algorithms able to customize based on individual requires. Through AI apps, leaders can personalize, for example, their preferences on time, situations, and contexts on when they need to calm down, meditate, or regulate their emotions. These apps could program situational patterns one experiences that enable the need to practice, for example, meditation. These apps can be voice activated whenever necessary.

Through smart watches and other gadgets where AI is embedded as a means of biofeedback analysis, physiological activities that pertain to emotional highs and lows can be detected which can then signal an individual to regulate emotional states in real-time. These on-demand feedback can then individualize a practice based on an analysis of these data.

AI can also detect prompts of emotional distress through one's tone of voice through speech or even through text messages sent while in conversations. Data collected from these experiences allow for better development of recommended mindfulness practices as meditation, body scanning, breathing exercises, or even silence breaks to regulate intense moods.

AI can also act as a virtual coach for the leader during times that conversations with another person is not readily available or if the topic is sensitive enough to be talked about with even a friend or family. Through AI virtual coaching tools, conversations about what is going on in a leader's mind can be utilized without fear of judgement.

For leaders with a long-time practice of mindfulness, AI can study the effectiveness of one's journey through time. The analysis can be based on what data can be gathered through machine learning as discussed above. This can then be given as feedback and provide areas of opportunities on how to enhance further the use of mindfulness in particular situations where challenges are still common.

AI can also be a source of resources on how to start a practice, how to do one, and how to keep one. Through a library of AI curated mindfulness resources on tools, a leader can access various journals, studies, and evidences on why one needs to start, continue, and keep with the practice.

Lastly, leaders can also share their own success stories on mindfulness by creating content on tools and methodologies that aided them in improving their leadership capacities. This can then be shared with their teams to build a culture where mindfulness is an integral part of business operations.

## *Conclusion and Call to Action*

The new finding on mindful leadership during crisis aims to contribute to the research on mindfulness practice among leaders, particularly in the Philippines and during a crisis like the COVID-19 pandemic. IPA was used to show how a mindfulness practice among leaders enhances the relationship of the leader with employees in a time of uncertainty. Based on the results, leaders with a mindfulness practice exhibited focus on doing better and meaningful work, built deep and personal relationships with their direct reports, and enhanced their self-awareness as leaders to relate to employees during crisis. Leader vulnerability in a time of crisis is also an emerging theme where it is seen as a manner of building relationships. Awareness and attention are kept at the center of mindfulness which leads to non-judgement of the self and others.

As a way of applying the lessons of the study in the context of leadership both in the public and private sectors of the Philippines and other countries, it is recommended that leaders are trained to initiate and sustain a mindfulness practice in their daily leader activities, incorporating skill development in leadership programs. AI can provide individual tools to make a leader's mindfulness practice accessible, collecting data to track and enhance the experience. With AI, platforms can be developed for immediate feedback on emotional levels as leaders manage the team collectively as well as the individuals within the group. AI can provide leaders with virtual coaching and knowledge-based resources on mindfulness to help them lead effectively while being compassionate to themselves particularly in times when crisis challenges their problem solving and decision-making capabilities.

It is time to recognize that mindfulness can aid in developing resilience during times of crisis. As leaders become central in navigating through a volatile, uncertain, complex, and ambiguous environment, AI can be incorporated in developing skills for crisis leadership. Through AI, mindfulness practice can be initiated and enhanced by providing real-time, individualized, and flexible tools to improve leadership capacities in times of crisis. This enables leaders to build teams that are resilient, mentally strong, and holistically healthy. This also ensures that leaders prioritize the well-being of everyone they manage. It is important to invest and extend support in the development of more effective leaders who will role model compassion, adaptability, and strength in trying times.

## References


Ahlvik, C. (2019). The Power of Awareness Unlocking the Potential of Mindfulness in Organizations.

Amina, Hadi, W. and F. (2020). Th Effect of Leader Mindfulness on Employee Job Performance: Investigating the Mediating and Moderating role of Leader-Member Exchange and organization Culture. *Journal of Behavioral Science*, 31(2), 12–26.

Arendt, J.F.W., Verdorfer, A.P., & Kugler, K.G. (2019). Mindfulness and leadership: Communication as a behavioral correlate of leader mindfulness and its effect on follower satisfaction. *Frontiers in Psychology*, 10 (MAR). https://doi.org/10.3389/fpsyg.2019.00667.

Auten, D.A. (2017). Supervisor mindfulness and its association with leader-member exchange (Master's thesis). Portland State University—PDX Scholar. https://doi.org/10.15760/etd.5546.

Aviles, P.R., & Dent, E.B. (2015). The role of mindfulness in leading organizational transformation: A systematic review. *The Journal of Applied Management and Entrepreneurship*, 20(3), 31–55. https://doi.org/10.9774/gleaf.3709.2015.ju.00005.

Belen, H. (2021). Fear of COVID-19 and mental health: The role of mindfulness during times of crisis. *International Journal of Mental Health and Addiction*. https://doi.org/10.1007/s11469-020-00470-2.

Boyatzis, R.E., & McKee, A. (2005). *Resonant leadership: Renewing yourself and connecting with others through mindfulness, hope, and compassion*. Harvard Business Review Press.

Frizzell, D., Hoon, S., & Banner, D. (2016). A phenomenological investigation of leader development and mindfulness meditation. *Journal of Social Change*, 8 (1), 14–25. https://doi.org/10.5590/JOSC.2016.08.1.02.

Goleman, D. (2008). *Working with emotional intelligence*. Bantam Doubleday Dell Publishing Group.

Gordon, T. (2013). Theorizing yoga as a mindfulness skill. *Procedia—Social and Behavioral Sciences*, 84 (2013), 1224–1227. https://doi.org/10.1016/j.sbspro.2013.06.733.

International Labour Organization. (2020). *COVID-19 and the world of work: Impact and policy responses*.

Majeed, M., Irshad, M., Fatima, T., Khan, J., & Hassan, M.M. (2020). Relationship between problematic social media usage and employee depression: A moderated mediation model of mindfulness and fear of COVID-19. *Frontiers in Psychology*, 11(December), 1–13. https://doi.org/10.3389/fpsyg.2020.557987.

Pinck, A.S., & Sonnentag, S. (2018). Leader mindfulness and employee well-being: The mediating role of transformational leadership. *Mindfulness*, 9, 884–896. https://doi.org/10.1007/s12671-017-0828-5.

Restubog, S.L.D., Ocampo, A.C.G., & Wang, L. (2020, June 1). Taking control amidst the chaos: Emotion regulation during the COVID-19 pandemic. *Journal of Vocational Behavior*. Academic Press Inc. https://doi.org/10.1016/j.jvb.2020.103440.

Rupprecht, S., Falke, P., Kohls, N., Tamdjidi, C., Wittmann, M., & Kersemaekers, W. (2019). Mindful leader development: How leaders experience the effects of mindfulness training on leader capabilities. *Frontiers in Psychology*, 10, 1081. https://doi.org/10.3389/fpsyg.2019.01081.

Schuh, S.C., Zheng, M.X., Xin, K.R., & Fernandez, J.A. (2019). The interpersonal benefits of leader

mindfulness: A serial mediation model linking leader mindfulness, leader procedural justice enactment, and employee exhaustion and performance. *Journal of Business Ethics*, 156(4). https://doi.org/10.1007/s10551-017-3610-7.
Sisk, D.A. (2017). Mindfulness and Its Role in Psychological Well-Being. *Journal of Psychology Research*, *7*(10). https://doi.org/10.17265/2159-5542/2017.10.002.
Stedham, Y., & Skaar, T.B. (2019). Mindfulness, trust, and leader effectiveness: A conceptual framework. *Frontiers in Psychology*, 10 (JULY). https://doi.org/10.3389/fpsyg.2019.01588.
Tusl, M., Kerksieck, P., Brauchli, R., & Bauer, G.F. (2021). Perceived impact of the COVID-19 crisis on work and private life and its association with mental well-being and self-rated health in German and Swiss employees: A cross-sectional study. *To Be Submitted*, 1–21.
Wasylkiw, L., Holton, J., Azar, R., & Cook, W. (2015). The impact of mindfulness on leadership effectiveness in a health care setting: A pilot study. *Journal of Health Organization and Management*, 29(7), 893–911. https://doi.org/10.1108/JHOM-06-2014-0099.

## • *D. AI in Services* •

# 14. Real-Time Insights from AI Training in Indian Hospitality

Smrite Goudhaman

The restaurant industry is one of the most dynamic sectors, where shifting consumer preferences and fast-paced operational environments demand high employee adaptability and skill (Mehta & Awasthi, 2019). Toscano, an Italian restaurant chain in India, embodies these challenges. With a goal to expand from 15 to 30 locations in 2 years (2023–25), Toscano needed a robust strategy to maintain operational standards while adapting to an increasingly competitive landscape. This contribution explores Toscano's unique experiment with AI-based training, examining both the promises of AI in hospitality and the practical challenges faced along the way.

## *Introduction: AI's Role in Employee Development and Retention*

The rise of AI in the workplace has brought new dimensions to employee training, especially in high-turnover industries like hospitality. Toscano's leadership identified two main challenges: enhancing employee performance and increasing retention. These goals were essential to handle the expansion without compromising quality. With AI's potential to deliver customized, adaptive training, Toscano aimed to bridge skill gaps and engage employees through personalized learning paths (Ibrahim, 2012).

## *Critical Success Variables in Hospitality and AI's Role in Enhancing Them*

Success in the hospitality industry hinges on three key factors: employee engagement, operational efficiency, and retention. These elements directly impact customer service quality, brand loyalty, and operational resilience, which are essential for Toscano's sustained growth and competitive edge. Research supports that employee engagement is pivotal in customer-facing industries like hospitality, where motivated and satisfied staff can elevate customer experiences significantly (Cope et al., 2021). Toscano's decision to adopt an AI-based approach to employee engagement aimed to tackle these core industry success factors by delivering interactive, adaptive training.

## *Employee Engagement and Operational Efficiency*

In high-touch environments, engaged employees contribute substantially to overall operational efficiency. Toscano's AI-powered training modules like "Greeting & Seating the Guest" and "Telephone Handling Etiquettes" recorded high engagement levels, with 99 percent completion in the initial rollout. Studies by Mehta and Awasthi (2019) indicate that adaptive, real-time learning improves engagement by allowing employees to interact with relevant content, thus sustaining their interest. AI tools create interactive learning environments that appeal to diverse learning styles, as supported by Buck and Morrow (2018), and help maintain a high engagement rate essential for service consistency in hospitality.

## *Performance Improvement Through Personalized Learning*

Performance is a critical measure in hospitality, where skill levels directly impact service quality. Toscano's AI modules included real-time feedback mechanisms, allowing employees to adjust and enhance their learning methods. Performance metrics revealed a 10–15 percent improvement in quiz scores among employees who engaged in simulations, highlighting AI's ability to reinforce training in practical settings. This aligns with findings from Tymon Jr., Stumpf, and Smith (2011), who argue that immediate feedback and practice-driven learning significantly improve employee performance in dynamic industries.

## *Retention Through Continuous Learning and Recognition*

Addressing high turnover rates remains crucial in hospitality, as the sector often faces retention challenges that impact service continuity (Abdullah, 2020). Toscano's AI-based training model promotes continuous learning and recognition, directly linking skill development to job satisfaction. Employee surveys confirmed that the personalized and adaptable nature of the AI training made them feel more valued, which correlates with higher retention rates. Studies on retention suggest that employees are more likely to remain with organizations that invest in their growth and acknowledge their contributions (Pan & Froese, 2023; Rousseau & ten Have, 2022).

## *The Pilot Test and Results*

Toscano initiated a pilot AI-based training program to enhance the essential skills of frontline staff, covering 17 locations and involving 100 employees. This group comprised 1 Deputy General Manager, 10 Managers, 2 Senior Managers, 10 Assistant

Managers, and 77 Senior Executives. The program featured modules like "Greeting & Seating the Guest," "Telephone Handling Etiquettes," and "Table Service Etiquettes," each incorporating AI-driven assessments to adapt content based on individual learning needs. Metrics such as adoption, engagement, performance scores, and satisfaction ratings were tracked throughout the program to measure the effectiveness of this approach.

## *Insights from the First Month: Adoption and Engagement Metrics*

The first month, September 2024, focused on measuring adoption and initial engagement. Toscano achieved an impressive 99 percent training completion rate within the first two weeks, which highlights the effectiveness of the interactive and adaptive design of the AI modules. This high adoption aligns with research indicating that AI-based adaptive learning platforms can increase employee engagement, especially in fast-paced industries like hospitality (Deloitte Insights, 2020).

Completion times also improved over the course of the month, starting at 10 minutes in Week 1 and dropping to 6 minutes by Week 3, reflecting growing familiarity and efficiency in using the AI-based system. Such consistency suggests that the user-friendly, adaptive nature of AI training contributed to sustained engagement, a factor identified as crucial for long-term success in the hospitality industry (Mehta & Awasthi, 2019).

## *Engagement Through Quizzes, Surveys, and Collaborative Learning*

Engagement was tracked through quiz participation, survey responses, and contributions to the discussion board. Quiz participation rates rose from 74 percent in Week 1 to 100 percent by Week 4, reflecting increasing interest and interaction with the training material. This trend suggests that employees found the content relevant and accessible, which is essential in industries that rely heavily on customer-facing staff (Cope et al., 2021). Surveys also saw high participation, starting at 87 percent in Week 1 and reaching 100 percent by Weeks 3 and 4, indicating strong interest in feedback mechanisms and further reinforcing the program's relevance (Buck & Morrow, 2018).

The discussion board proved to be an effective tool for collaborative learning, with employees actively sharing insights and discussing scenarios. Over the second month, employees contributed over 1,300 comments to the discussion boards, showing an active exchange of ideas and support among team members. Research highlights that collaborative digital learning environments can significantly enhance engagement and knowledge retention, particularly in team-oriented industries like hospitality (Rousseau & ten Have, 2022).

## *Insights from the Second Month: Enhancing Employee Performance*

In October, Toscano shifted the focus from adoption to performance improvement. The AI modules included interactive quizzes and simulated exercises with real-time feedback, which enabled employees to correct and adapt their learning strategies immediately.

Quiz scores initially averaged 84 percent in Week 1 but saw a gradual decline, reaching 63 percent by Week 4. This fluctuation suggests that while employees were initially engaged and retained information well, ongoing reinforcement may be necessary to ensure consistent comprehension of complex concepts (Pan & Froese, 2023).

Despite the drop in quiz scores, satisfaction ratings from surveys remained high, with employees rating their experience between 4.8 and 4.9 out of 5 weekly. This consistently positive feedback underscores the effectiveness of personalized, interactive training in enhancing employee morale and confidence, especially when adapting to new technologies (Zhang et al., 2023). Toscano's experience aligns with research that suggests sustained engagement and reinforcement are key to maximizing the impact of AI-driven learning tools (Mehta & Awasthi, 2019).

## *Addressing Retention Challenges with AI Training*

A broader objective of Toscano's AI-based training was to reduce high turnover rates by fostering a continuous learning culture. The hospitality industry is known for high turnover, impacting service consistency and employee morale. Toscano's pilot program demonstrated that AI-enabled training could positively influence retention by creating a work environment that prioritizes development and feedback (Tymon Jr., Stumpf, & Smith, 2011). Survey feedback indicated that employees felt more valued and supported due to the training, which aligns with industry findings that link professional development to improved retention (Abdullah, 2020).

The personalized learning approach helped employees see the relevance of their training to their job roles, increasing job satisfaction—a critical factor for retention in high-turnover sectors. According to research, ongoing development opportunities help employees feel invested in their roles, reducing the likelihood of turnover (Berezina et al., 2019). While AI training alone may not completely solve turnover issues, it plays a crucial role in supporting Toscano's broader retention strategies.

## *Limitations and Challenges*

Despite positive outcomes, Toscano encountered challenges during the pilot. Technological resistance emerged as an initial barrier, especially among employees unfamiliar with AI-driven interfaces. Toscano addressed this issue by simplifying the training schedule and offering technical support, helping employees adapt to the platform (Pan & Froese, 2023). Resistance to new technology is a well-documented challenge, and studies emphasize the importance of providing accessible support to ease transitions (Riemer et al., 2019).

Time constraints also posed a challenge, particularly during peak operational hours. To minimize disruptions, Toscano optimized the training modules for shorter completion times. Data privacy concerns arose as well, with some employees expressing reservations about the platform's access to performance metrics. In response, Toscano ensured transparency in data handling practices and strictly adhered to privacy protocols, a necessary step to build trust and align with ethical standards in AI use (Rousseau & ten Have, 2022).

## *Conclusion and Recommendations from the Pilot Test*

The pilot AI-based training program provided Toscano with key insights on how AI can enhance employee engagement, skill development, and retention across its front-line staff. Throughout this pilot, Toscano aimed to address critical hospitality competencies through modules that focused on greeting guests, telephone etiquette, and table service, utilizing AI-driven personalization to make learning more relevant and impactful.

With strong engagement metrics and performance improvements in specific areas, the pilot has demonstrated the potential of AI to drive learning in a dynamic restaurant environment. However, the data also highlighted areas where adjustments were needed to further enhance training efficacy and alignment with operational demands. The following recommendations and adjustments are based on lessons learned, ensuring Toscano's AI training approach remains adaptive, accessible, and effective for future expansion.

## *Adjustments Based on Pilot Lessons*

1. **Enhanced Support for Technology Adoption**: Early in the pilot, some employees, especially those less familiar with AI tools, expressed discomfort with the training platform. This initial resistance was mitigated by simplifying training modules, providing real-time support, and offering hands-on technical assistance, particularly during peak training weeks. These actions aligned with findings from similar AI training environments, where support is crucial for technology adoption and sustained engagement (Rousseau & ten Have, 2022).

2. **Customized Training Schedules for Peak Times**: One challenge was balancing training demands with busy restaurant hours. Initially, the training averaged 10–12 minutes per module, but feedback indicated that employees found this challenging during high-traffic periods. As a result, Toscano optimized training modules to be completed in shorter durations, with average completion times reducing to about 6–7 minutes by the second month. This adjustment facilitated smoother integration of training within daily responsibilities and ensured uninterrupted service quality (Abdullah, 2020).

3. **Increased Emphasis on Core Content Reinforcement**: The performance data revealed a decline in quiz scores, from an average of 84 percent in early weeks to 63 percent by the fourth week. This trend suggested that employees would benefit from periodic refresher modules on core concepts. Toscano introduced additional support materials and suggested weekly "knowledge recap" sessions, a strategy aimed at reinforcing knowledge retention without overwhelming employees during their work shifts (Berezina et al., 2019).

4. **Boosting Engagement through Collaborative Learning**: One of the most successful components was the discussion board, where employees contributed over 1,300 comments across four weeks. Toscano encouraged this engagement by highlighting top contributors and fostering a collaborative environment, a strategy known to improve team learning and morale. The recommendation for future training is to continue fostering collaborative spaces, as team-based learning

enhances retention and builds a culture of mutual support and problem-solving (Mehta & Awasthi, 2019).

## *Recommendations for Expansion*

1. **Scaled Technical Assistance**: For future rollouts across additional locations, Toscano should continue providing technical support to ease the adoption of AI training. Assigning tech ambassadors at each location or providing digital guides can ensure employees have immediate assistance, further decreasing resistance to AI platforms (Cope et al., 2021).
2. **Shortened and Reinforced Training Sessions**: Given the time constraints faced in hospitality environments, future AI training modules should maintain short, focused sessions (5–7 minutes) with periodic reinforcement. This structure aligns with Toscano's findings and industry best practices, where short training segments reduce cognitive load and allow employees to immediately apply learned skills in their daily roles (Pan & Froese, 2023).
3. **Continuous Feedback Mechanisms**: Toscano observed high survey satisfaction, with an average score of 4.8 to 4.9 out of 5. To maintain engagement, Toscano should continue soliciting feedback regularly, particularly around module content and usability. Ensuring a loop of continuous feedback allows for rapid improvements and keeps the training aligned with employee expectations and business needs.
4. **Targeted Support for Key Areas of Performance Decline**: Quiz scores indicated areas where knowledge reinforcement could be beneficial. Implementing targeted follow-ups or mini-tests for lower-performing modules (such as "Taking Food & Beverage Orders" or "Service of Wines") can help stabilize and improve performance consistency across all training areas, addressing critical competencies for front-line staff (Tymon Jr., Stumpf, & Smith, 2011).

Toscano's pilot of AI-based training highlights a transformative step in Indian hospitality, demonstrating how adaptive, personalized learning can enhance employee engagement, skill proficiency, and retention in a competitive environment. The pilot insights reveal a promising path forward, where continual improvements in technology adoption, targeted support, and customized training schedules align Toscano's workforce with operational demands—positioning the organization to scale its AI training successfully across future locations.

### References

Abdullah, A. (2020). Employee development and organizational retention. *Journal of Human Resource Development.*

Berezina, K., Ciftci, O., & Cobanoglu, C. (2019). Robots, Artificial Intelligence, and Service Automation in Restaurants. In S. Ivanov & C. Webster (Eds.), *Robots, Artificial Intelligence, and Service Automation in Travel, Tourism and Hospitality* (pp. 185–219). Emerald Publishing Limited. Retrieved from https://www.emerald.com/insight/content/doi/10.1108/978-1-78756-687-320191010/full/html.

Buck, S., & Morrow, S. (2018). AI-driven adaptive training for improved employee engagement. *Journal of Technology and Training.*

Cope, J., et al. (2021). Transformative potential of AI-based training in hospitality. *Training Industry Insights.*

Deloitte Insights. (2020). AI adoption trends in global hospitality sectors. Retrieved from https://www2.deloitte.com/global/en/insights/industry/2020-hospitality-ai-trends.html.
Ibrahim, A. (2012). AI-driven solutions in HRM. *Advances in Artificial Intelligence for Business.*
Mehta, V., & Awasthi, A. (2019). Operational efficiency in the restaurant industry. *Journal of Business and Restaurant Management.*
Pan, Y., & Froese, F. (2023). AI and employee engagement in dynamic industries. *Management and Organizational Review.*
Riemer, F., et al. (2019). Data privacy in AI applications for human resources. *Data and Privacy Journal.*
Rousseau, J., & ten Have, M. (2022). Overcoming resistance to AI in training. *Journal of Organizational Change Management.*
Tymon Jr., W.G., Stumpf, S.A., & Smith, R.R. (2011). Employee retention strategies in high-turnover sectors. *Journal of Applied Business Strategy.*
Zhang, W., et al. (2023). Real-time feedback in AI-driven training environments. *International Journal of Interactive Learning.*

# 15. Adapting AI-Based Feng Shui Application for Real Estate Enterprises in Vietnam

DANG THI THANH TAM *and* MICKEY P. MCGEE

The following provides an examination of three key areas: technological integration (development of the conceptual idea for an advanced application that integrates AI with traditional Feng Shui principles); socio-cultural integration (real estate buyers' analysis on perception and acceptance, including buyers' preference factors, cultural resonance of AI-based Feng Shui); and business impact (assess how AI-based Feng Shui contributes to the competitive advantage and business performance). Specifically, this piece explores (1) if the integration of AI-based with traditional Feng Shui principles will lead to more accurate and efficient recommendations of real estate property, and (2) if the AI-based application with Feng Shui approach might provide a unique perspective that addresses buyers' emotional and psychological needs, leading to higher engagement, preference and loyalty.

Feng Shui, which literally means wind and water, is an ancient Eastern doctrine involving the arrangement of buildings, objects, and space in an environment to achieve living harmony and balance. In real estate, it is believed that a property or land with good Feng Shui would promote good health, wealth, prosperity, and luck to its owners (Cho, 2020b). Originally, Feng Shui has root in early Taoist philosophy and remains popularly adopted today, influencing spaces across China, Asia, and even in Western cultures along with the current globalization trends of socio-cultural integration. In the West, Wu (Wu, 2019) emphasizes that Feng Shui practice has been used widely for years by architects, interior designers, and landscape architects as a design guideline for pursuing harmony in living environment. Overall, Feng Shui is popularly used as it is highlighted as one of the main determinants of making decisions for real estate purchases, as stated by Gibler and Nelson (Gibler & Nelson, 1998).

Vietnam is not an exception among the use of Feng Shui in real estate. In Vietnamese context, people's habits in lifestyles and real estate purchases are highly impacted by Chinese culture. Taking a step back to historical background, the periods of Chinese conquests and Nguyen dynasty reveal shifts on cultural influences. Lately, despite for efforts of the French colonialists disrupting the Vietnam's cultural links with China, Vietnamese mindsets continued to reflect the foundational Chinese philosophical and religious doctrines of Confucianism (appeared in structured layouts with central axes and symmetry), Taoism (Feng Shui), and Buddhism (sense of continuity between life and the afterlife) (Vladimirovna & Chu, 2019). Real estate, a part of living space and environment, is considered spiritual goods market in Vietnam (Nghiem-Phu, 2022), which is influenced by these mindsets.

*Cultural and historical context:* Feng Shui, a traditional Chinese practice, has been deeply embedded in Vietnamese culture for centuries that there is high rate of Vietnamese people continue to adhere to Feng Shui principles when making decisions about property design, construction, and purchase. According to a survey by the Vietnamese Royal Institute of Feng Shui, with 2,516 out of 2,568 real estate buyers in 2018 showing concern about Feng Shui in their purchasing decisions, this suggests that Feng Shui holds significant value in the real estate market, with approximately 98 percent of buyers influenced by it (Truong Thinh, 2020).

*Technological context:* Emerging AI technologies, such as machine learning, data analytics, and predictive modelling, have the potential to transform traditional practices by providing more precise, efficient, and scalable solutions. Integrating AI with Feng Shui might potentially modernize the practice, making it more accessible and appealing to contemporary buyers, while preserving its cultural significance.

*Business context:* Real estate market in Vietnam is characterized by rapid urban development and a growing middle class with increasing disposable income. This economic growth has spurred demand for residential, commercial, and industrial properties. However, the market is also highly competitive, with numerous local and international players vying for market share. In this competitive landscape, offering technologically unique and culturally resonant services, such as AI-based Feng Shui, might possibly provide an advantage for enhancing buyers' preferences.

The Feng Shui Institute survey highlights a clear demand for reliable Feng Shui guidance in real estate, with almost all (98%) buyers showing interest in its influence on their property or land. In another survey result, only 30 percent of these buyers were consulted in this field, while few admitted to receiving satisfactory advice (1%) (T.R.I., 2024). The traditional practice of visiting Feng Shui masters combining with searching multiple online channels are complicated and time consuming for buyers. This gap suggests several critical insights and opportunities for the research on AI-based Feng Shui solutions in real estate to gain high personal impact of Feng Shui for shifting in buyers' preferences and satisfaction.

The high level of concern for Feng Shui underlines the potential for considering an AI-based Feng Shui application to appeal to a broad audience, especially in Vietnam, where cultural and spiritual practices play a substantial role in purchasing behaviors. For this reason, the study is to explore the potential of integrating AI-based Feng Shui into the practices of real estate enterprises in Vietnam, to enhance buyers' preferences.

The significance of the AI-based Feng Shui involves several areas: cultural relevance and appeal, technological innovation, market differentiation, improvement of buyers' decision-making, and business impact.

1. As Feng Shui holds significant cultural importance in Vietnam, by incorporating AI-based Feng Shui, real estate enterprises could appeal to the cultural beliefs and preferences of Vietnamese buyers, thereby increasing buyer satisfaction and trust.
2. Leveraging AI to modernize Feng Shui practices represents a fusion of tradition and technology. Such innovation could streamline and enhance the

accuracy of Feng Shui assessments, making them more accessible and reliable for both real estate enterprises and buyers.

3. Real estate companies that adopt AI-based Feng Shui could differentiate themselves from competitors by offering unique, culturally resonant services, that could attract a niche market segment that values traditional Feng Shui principles in property selection and design.

4. AI-based Feng Shui tools could provide data-driven insights, predictive analytics, and personalized property recommendations, enabling more informed decision-making in property purchasing. This could lead to better alignment of properties with market demand and buyers' preferences, which lead to greater loyalty and word-of-mouth referrals.

5. By attracting more buyers who value Feng Shui, real estate enterprises could potentially increase sales and profitability. The business impact extends to the broader real estate market, contributing to its growth and dynamism.

This exploratory study of the AI-based application in real estate using Feng Shui principles underscores the potential of AI-based Feng Shui to transform real estate practices in Vietnam. By marrying cultural traditions with cutting-edge technology, real estate enterprises could gain buyers' preference and satisfaction, foster sales and profitability for enterprises, and contribute to the sustainable growth of the real estate market.

## *Feng Shui, Feng Shui Principles, and Its Impacts on Real Estate*

Previous researchers have well mentioned about the concept of Feng Shui since early times. The idea of Feng Shui is to create a more comfortable living experience for people by examining how their home is positioned in an environment and how objects are arranged within it (Cho, 2020b). In addition, in the western world, Feng Shui is often translated as Geomancy (Lee, 1986). The key evidence, that makes above concept of Feng Shui well persuadable with a large population of users is the consistence of "six scientific theories and principles that have endured more than two thousand years of tests and experiments" (Sang, 2013). These principles are (1) The Yin/Yang Principles, (2) The Five Elements, (3) The Eight Trigrams, (4) The Directions, (5) The Solar System, and (6) The Environment.

Feng Shui has significant impacts on the real estate area in terms of pricing and interior design. The pricing impact of Feng Shui practice involves the buyers' willing to pay higher price for a good Feng Shui living environment (Lin & Chen, 2012). Furthermore, a study of Ahmadnia et al. (Ahmadnia et al., 2012) showed that the proper architectural or interior design of the property is essential to the feeling of comfort and safety in the home or workplace.

## *AI Technologies in Real Estate*

Recently, researchers have uncovered useful findings in AI-based approach in real estate. Conway & Architecture (Conway & Architecture, 2018) attempted to study how AI is applied, and the actionable opportunities for the AI-based application in real estate. Lorenz (Felix Lorenz, 2021) also studied rental pricing of residential market using AI-based methods in their interpretability in real estate. The most recent publication of Baur et al. (Baur et al., 2023) has discussed the concept of AI-based models for automatic evaluation based on property descriptions. From the above citations, the adaption of AI has positive impacts on real estate industry including identifying investment opportunities and valuation.

## *Theorical Frameworks: Sociotechnical Systems Theory (STS) and Cultural Capital Theory*

The combination of STS and Cultural Capital Theory attempts to provide a framework for the effective integration of technology within a cultural context. It posits that the successful implementation of technology, such as AI, requires a harmonious balance between the technical (application) and social (buyers, culture) components while also recognizing the value of cultural assets (Feng Shui). The research is grounded in these two theories that are well-described by theory owners Trist (Trist, 1981) & Bourdieu (Bourdieu, 1986) in their publications.

Feng Shui impacts the real estate industry in terms of price valuation and property design. However, the AI-based application in real estate based on Feng Shui principles has not been investigated. This new application has the potential to provide a tool that is more convenient, quick, and user friendly that will be able to add value to the business area of real estate. The application, if feasible, will potentially help not only real estate investors to identify the proper land or property quickly and efficiently, but also for sales force and brokers to recommend the suitable products for their customers, and hence, to increase sales influence.

The idea of AI-based application in real estate using Feng Shui principles is a new idea in the real estate sector. Moreover, due to the challenges in pursuing big dataset that needs deeper investigation on technical aspects, this study area shall be limited at the behavior research only, that means exploring consumer insights to understand how preferable the application will potentially be. Further development will be conducted to refine and complete the application for commercialization purposes.

## *Conclusion*

The AI-based Feng Shui application is designed and implemented with a careful balance of technical capabilities and cultural sensitivity. By modernizing Feng Shui practices through AI, the application not only leverages advanced technology but also preserves and enhances its cultural significance in real estate transactions. Such integration helps real estate enterprises in Vietnam appeal to buyers who value cultural practices, thereby increasing their market position and competitive advantage. In addition, the sociotechnical and cultural dimensions work together to create a solution that is technologically advanced, socially relevant, and culturally respectful. Overall, this approach would provide a solid foundation for analyzing how AI-based Feng Shui could enhance buyers' preferences in the real estate sector in Vietnam while keeping the research focused and manageable.

## References

Ahmadnia, H., Gholizadeh, M., Bavafa, M., & Rahbarianyazd, R. (2012). *Art of Feng Shui and its Relationship With Modern Interior Design*. academia.edu.

Baur, K., Rosenfelder, M., & Lutz, B. (2023). Automated real estate valuation with machine learning models using property descriptions. *Expert Systems with Applications*, *213*, 119147. https://doi.org/10.1016/J.ESWA.2022.119147.

Bourdieu, P. (1986). The forms of capital. In J. Richardson (Ed.), *Handbook of Theory and Research for the Sociology of Education* (pp. 241–258). Greenwood.

Cho, A. (2020b). How to Use Feng Shui for Wealth and Prosperity in Your Home. http://www.Thespruce.Com. https://www.thespruce.com/feng-shui-money-tips-and-wealth-1275336.

Conway, J., & Architecture, B.A. (2018). *Artificial Intelligence and Machine Learning: Current Applications in Real Estate*.

Felix Lorenz. (2021). Dissertation_Felix_Lorenz_Pflichtexemplar. *Library of University of Regensburg*.

Gibler, K., & Nelson, S. (1998). Consumer Behavior Applications to Real Estate. *Academia.Edu*.

Lee, S. (1986). FS definitions and concepts. *Cornell University ProQuest Dissertations Publishing*.

Lin, C.-C., & Chen, C.-L. (2012). An Estimation of the Impact of Feng-Shui on Housing Prices in Taiwan: A Quantile Regression Application (Vol. 15, Issue 3).

Nghiem-Phu, B. (2022). Consumer behaviour towards purchasing Feng Shui goods: An empirical study from Vietnam. *Journal of Asian Finance*, *9*(1), 83–0092. https://doi.org/10.13106/jafeb.2022.vol9.no1.0083.

Sang, L. (2013). The Principles of Feng Shui—Larry Sang—Google Books. Books.Google.Com.

T.R.I. (2024). *Hiểu biết phong thuỷ—Lợi thế lớn của môi giới bất động sản*. Viện Nghiên Cứu Bất Động Sản (Real Estate Research and Training Institute). https://tri.edu.vn/hieu-biet-phong-thuy/.

Trist, E. (1981). The evolution of sociotechnical systems: A conceptual framework and an action research program. In A.H. Van De Ven & W.F. Joyce (Eds.), *Perspectives on Organization Design and Behavior* (pp. 19–75).

Truong Thinh. (2020). Phú Mỹ Gold City mang vượng khí đón Phú quý | Báo Dân trí. *Dan Tri Newspaper*. https://dantri.com.vn/kinh-doanh/phu-my-gold-city-mang-vuong-khi-don-phu-quy-20201118084003622.htm.

Vladimirovna, G., & Chu, H. (2019). The gardens and architecture of the imperial burial complexes of Vietnam as a result of the influence of Chinese culture on the countries of Southeast Asia. *Amazonia Investiga*, *8*(21), 491–499.

Wu, S.-J. (2019). *Feng Shui: A Comparison of the Original Concept and Its Current Westernized Version*. proquest.com

# 16. Artificial Stupidity

## *Limitations of AI in Enhancing Customer Experience*

Sandeep Arora

In today's digital era, businesses are increasingly leveraging Artificial Intelligence (AI) to revolutionize customer service in contact centers. AI technologies promise enhanced efficiency, personalized interactions, and seamless customer experiences. However, while AI excels at handling routine tasks and providing quick responses, it often lacks the human touch that customers value. This gap has led to the exploration of "Artificial Stupidity"—the intentional design of AI systems to exhibit human-like imperfections. By making AI interactions feel more natural and engaging in a "carefully careless" manner, artificial stupidity aims to improve customer satisfaction in modern digital contact centers.

This piece explores the evolution of AI in customer service, delves into the concept of artificial stupidity, examines its applicability in digitally enabled contact centers, and discusses recent developments in the field. Through a humanized yet technical lens, we will uncover how businesses can harness artificial stupidity optimally to deliver exceptional customer experiences.

### *Evolution of AI in Customer Service*

The landscape of customer service has transformed dramatically over the past few decades. Traditional call centers have evolved into sophisticated digital contact centers, capable of handling interactions across multiple channels such as voice calls, emails, live chats, social media, and mobile apps. Customers now expect quick, personalized responses and seamless experiences across all touchpoints (Salesforce, 2020).

AI has been a driving force behind this evolution. By automating routine tasks and providing instant responses, AI has enabled businesses to handle higher volumes of customer inquiries efficiently. AI-powered chatbots and virtual assistants have become commonplace, offering 24/7 support and reducing wait times (IBM, 2020). However, as AI systems become more advanced, there is a growing recognition of the need to make them more relatable and human-like to fully meet customer expectations.

## *Historical and Theoretical Foundations*

The term "artificial stupidity" might sound counterintuitive, but it captures a crucial aspect of AI development: intentionally incorporating human-like imperfections into AI systems to enhance user experience. This concept emerged from the realization that perfectly rational AI systems can feel unnatural and unengaging to users.

In the gaming industry of the 1980s and 1990s, developers noticed that AI opponents who played flawlessly were predictable and less fun. By programming these AI opponents to make mistakes—missing shots, making poor strategic decisions—they created more challenging and enjoyable experiences for players (Laird & van Lent, 2001). This practice underscored the value of embedding human-like flaws into AI behavior, laying the groundwork for artificial stupidity.

While "artificial stupidity" is not a formally established term in academic literature, it encapsulates a recognized strategy in AI design aimed at enhancing user experience. The concept is discussed under various terminologies, such as humanized AI, imperfect AI, or anthropomorphic design (Duffy, 2003; Fink, 2012).

Artificial stupidity has evolving and emerging influences from:

*Psychology.* Suggest humans are more comfortable interacting with entities that exhibit familiar, human-like behaviors (Epley et al., 2007). This includes not only physical appearance but also behavioral traits such as making mistakes or expressing emotions.

*Technology.* Advancements in AI and machine learning have enabled developers to manipulate AI behavior more precisely, making it possible to introduce intentional imperfections (Goodfellow et al., 2016).

*Culture.* Media and popular culture have also influenced the concept. Depictions of AI in films and literature often highlight the importance of human-like qualities in machines to facilitate better human-machine relationships (Kang, 2011).

Theoretical foundations of artificial stupidity come from the following:

*The Uncanny Valley and User Comfort.* The Uncanny Valley theory posits that as a robot or virtual agent becomes more human-like, there is a point where it appears, but not entirely, human, leading to discomfort (Mori, 1970). Introducing imperfections helps avoid this valley by ensuring AI remains comfortably within the realm of the familiar (Kätsyri et al., 2015).

*Anthropomorphism in AI.* Anthropomorphism involves attributing human characteristics to non-human entities. This psychological tendency can enhance user engagement with AI systems (Epley et al., 2007). By incorporating human-like flaws, developers tap into users' innate tendencies to relate to machines on a human level (Duffy, 2003).

*Cognitive Load and User Engagement.* Introducing imperfections in AI can reduce cognitive load on users by making interactions more predictable and less intimidating (Sweller, 2011). When AI exhibits human-like behavior, users may find it easier to understand and predict its actions, leading to smoother interactions (Kim et al., 2019).

## *Current Frameworks and Applications*

Evolving rapidly in terms of theory and practice, current frameworks and applications of artificial stupidity include:

*Humanizing Chatbots and Virtual Assistants.* Modern chatbots and virtual assistants are designed to mimic human conversational patterns, including the use of colloquialisms, humor, and occasional misunderstandings.

- *Conversational AI Design.* Developers create chatbots that intentionally misunderstand queries or ask for clarification to simulate a human conversation (Shum et al., 2018). This approach can make interactions feel more natural and engaging (Ciechanowski et al., 2019).
- *Emotional Intelligence in AI.* Incorporating emotional responses, such as expressing empathy or apologizing for errors, enhances the user experience (Peters et al., 2019). For example, an AI assistant might say, "I'm sorry, I didn't get that. Could you please rephrase?"

*Adaptive Learning Systems.* In educational technology, AI-driven tutoring systems sometimes intentionally make mistakes to encourage student participation and critical thinking.

- *Promoting Active Learning.* By simulating errors, AI tutors prompt students to correct them, reinforcing learning outcomes (VanLehn, 2011). This method leverages the "teachable agent" concept, where students learn by teaching or correcting the AI (Chase et al., 2009).
- *Personalized Learning Paths.* AI systems might intentionally limit their capabilities to match the student's proficiency level, gradually increasing complexity as the student progresses (Koedinger et al., 2013).

*Gaming AI.* In video games, AI opponents are programmed with limitations to provide a balanced and enjoyable experience.

- *Difficulty Adjustment.* AI opponents may exhibit strategic flaws, miss shots, or make tactical errors to prevent player frustration and maintain engagement (Yannakakis & Togelius, 2018).
- *Behavioral Diversity.* Introducing randomness and imperfections in AI behavior creates more dynamic and less predictable gameplay (Holmgård et al., 2018).

*Social Robotics.* Robots designed for social interaction often exhibit human-like imperfections to improve approachability and acceptance.

- *Expressive Behaviors.* Robots may display emotions, such as confusion or delight, to connect with users on an emotional level (Breazeal, 2003).
- *Nonverbal Communication.* Incorporating human-like gestures and body language, including imperfections, enhances the naturalness of interactions (Fong et al., 2003).

Applying artificial stupidity frameworks and applications to digital contact centers results in the following:

*Creating Natural and Engaging Interactions.* In digital contact centers, applying *artificial stupidity* involves designing AI systems that communicate in ways that feel natural to customers. This can be achieved through:

- *Conversational Language.* Using everyday language, including slang or colloquial expressions, to make interactions more relatable (Ciechanowski et al., 2019).

- *Expressing Uncertainty.* Allowing AI to admit when it doesn't understand something and asking for clarification, which mirrors human conversation patterns (Liu & Sundar, 2018).
- *Incorporating Small Talk.* Engaging in light, context-appropriate small talk can enhance the customer experience by making interactions feel more personal (Zhou et al., 2020).

*Building Trust and Rapport.* When AI systems display that they are not infallible, it can build trust with customers. Admitting limitations and offering to connect the customer with a human agent when necessary demonstrates transparency and respect for the customer's needs (Lee & See, 2004). Customers appreciate honesty and are more likely to feel valued when the AI acknowledges its boundaries.

*Improving Customer Satisfaction.* By mirroring human conversation patterns and behaviors, AI can reduce customer frustration that often arises from rigid, scripted responses. Humanized AI interactions can lead to higher customer satisfaction by making customers feel heard and understood (Kim et al., 2019). This approach can transform customer service from a transactional experience into a relational one.

## *Case Studies*

There are a growing number of industry examples including:

*Retail industry.* A major retail company introduced an AI chatbot designed to use a friendly, conversational tone and acknowledge when it didn't have all the answers. For example, if the chatbot didn't understand a query, it might respond, "Oops, I didn't catch that. Could you please rephrase?" This approach led to increased customer engagement and positive feedback, as customers found the interactions more personable and less mechanical (Retail Customer Experience, 2020).

*Telecommunications industry.* A telecommunications firm deployed an AI assistant capable of handling complex queries but programmed it to express uncertainty when appropriate. By admitting when it couldn't solve an issue and seamlessly transferring customers to human agents, the company reduced customer frustration and improved overall service ratings (Telecom Insights, 2021).

## *Good Practices*

For companies considering the implementation of artificial stupidity in contact centers, here are five good practices to consider:

*Integrate Human Oversight.* Ensure that human agents are available to intervene when AI reaches its limitations. This hybrid model combines the efficiency of AI with the empathy and problem-solving skills of human agents, providing a comprehensive customer service experience (Accenture, 2020).

*Prioritize Data Privacy and Ethics.* Adhere to data protection regulations and implement robust security measures. Transparency with customers about how their data is used and protected builds trust and demonstrates a commitment to ethical practices (European Union Agency for Cybersecurity, 2020).

*Continuous Monitoring and Improvement.* Regularly monitor AI interactions and gather customer feedback to refine and improve the system. Updating machine learning models to reflect evolving customer preferences ensures that the AI remains effective and aligned with customer needs (IDC, 2021).

*Balance Efficiency with Relatability.* Finding the right balance between making AI efficient and making it relatable is crucial. Introducing too many imperfections can hinder performance, while too few may make interactions feel robotic (Seeger et al., 2021).

*Personalize Interactions.* Use customer data to personalize interactions but do so ethically and transparently. Personalization fosters a stronger connection with customers, enhancing satisfaction and loyalty (Smith & Anderson, 2020).

- *Using Everyday Language.* AI systems should communicate using language that customers naturally use. Avoiding technical jargon and using simple sentences make interactions more accessible and enjoyable.
- *Showing Empathy.* Acknowledging customer emotions and expressing a willingness to help can make a significant difference. Phrases like "I'm sorry to hear that you're experiencing this issue" can make interactions more comforting (Liu & Sundar, 2018).
- *Adaptive Communication.* Adjusting communication style based on the customer's behavior and preferences shows attentiveness. If a customer seems confused, the AI can provide simpler explanations or offer additional assistance (Kim et al., 2019).

## Concerns

Despite the inherent logic of applying artificial stupidity, companies should consider these concerns:

*Risk of Misinterpretation.* Some critics argue that intentional imperfections may confuse users or reduce trust in AI systems (Marcus & Davis, 2019).

*Balancing Efficiency and Relatability.* There is an ongoing debate on how to balance the benefits of AI efficiency with the need for human-like interactions (Seeger et al., 2021).

*Ethical Considerations.* There are ethical concerns related to deliberately designing AI to deceive or manipulate customers, even if the intent is to make interactions more natural. Misrepresenting AI capabilities can be seen as dishonest, raising questions about transparency and integrity (Floridi et al., 2018).

*Regulatory Compliance.* Deliberate imperfections might lead to non-compliance with industry regulations, especially in sectors where accuracy and reliability are critical, such as finance or healthcare. Regulatory bodies may have strict guidelines on the use of AI, and intentional errors could violate these standards (European Commission, 2020).

## Conclusion

Incorporating human-like imperfections into AI systems—embracing artificial stupidity—offers a promising approach to enhancing customer experiences in digital contact centers. By making AI interactions more natural and personable, businesses can build stronger relationships with their customers, leading to increased satisfaction and loyalty. As AI technology continues to advance, focusing on the human element will be crucial in delivering exceptional customer service. By applying best practices and embracing recent developments in humanized AI, organizations can fully harness the power of AI while providing seamless and satisfying customer experiences.

## References


Accenture. (2020). *Reinventing Customer Service with AI.* Retrieved from https://www.accenture.com.

Breazeal, C. (2003). Toward sociable robots. *Robotics and Autonomous Systems*, 42(3–4), 167–175. https://doi.org/10.1016/S0921-8890(02)00373-1.

Chase, C.C., Chin, D.B., Oppezzo, M.A., & Schwartz, D.L. (2009). Teachable agents and the protégé effect: Increasing the effort towards learning. *Journal of Science Education and Technology*, 18(4), 334–352. https://doi.org/10.1007/s10956-009-9180-4.

Ciechanowski, L., Przegalinska, A., Magnuski, M., & Gloor, P. (2019). In the shades of the uncanny valley: An experimental study of human–chatbot interaction. *Future Generation Computer Systems*, 92, 539–548. https://doi.org/10.1016/j.future.2018.01.055.

Duffy, B.R. (2003). Anthropomorphism and the social robot. *Robotics and Autonomous Systems*, 42(3–4), 177–190. https://doi.org/10.1016/S0921-8890(02)00374-3.

Epley, N., Waytz, A., & Cacioppo, J.T. (2007). On seeing human: A three-factor theory of anthropomorphism. *Psychological Review*, 114(4), 864–886. https://doi.org/10.1037/0033-295X.114.4.864.

European Union Agency for Cybersecurity. (2020). *Guidelines on Data Protection and Privacy in AI Systems*. Retrieved from https://www.enisa.europa.eu.

Fong, T., Nourbakhsh, I., & Dautenhahn, K. (2003). A survey of socially interactive robots. *Robotics and Autonomous Systems*, 42(3–4), 143–166. https://doi.org/10.1016/S0921-8890(02)00372-X.

Fink, J. (2012). Anthropomorphism and human likeness in the design of robots and human-robot interaction. In *Social Robotics* (pp. 199–208). Springer. https://doi.org/10.1007/978-3-642-34103-8_20.

Floridi, L., Cowls, J., Beltrametti, M., Chatila, R., Chazerand, P., Dignum, V., ... & Schafer, B. (2018). AI4People—An ethical framework for a good AI society: Opportunities, risks, principles, and recommendations. *Minds and Machines*, 28(4), 689–707. https://doi.org/10.1007/s11023-018-9482-5.

Goodfellow, I., Bengio, Y., & Courville, A. (2016). *Deep Learning*. MIT Press.

Holmgård, C., Liapis, A., Togelius, J., & Yannakakis, G.N. (2018). Automated playtesting with procedural personas through MCTS with evolved heuristics. *IEEE Transactions on Games*, 11(4), 352–362. https://doi.org/10.1109/TG.2018.2801639.

IBM. (2020). *AI-Powered Customer Service Solutions*. Retrieved from https://www.ibm.com.

IDC. (2021). *AI in Customer Service: Continuous Improvement Strategies*. Retrieved from https://www.idc.com.

Kang, J. (2011). The immersive whodunit: A study of film noir, detective fiction, and artificial intelligence in *Blade Runner*. *Journal of Popular Film and Television*, 39(2), 79–85. https://doi.org/10.1080/01956051.2011.555253.

Kätsyri, J., Förger, K., Mäkäräinen, M., & Takala, T. (2015). A review of empirical evidence on different uncanny valley hypotheses. *Frontiers in Psychology*, 6, 390. https://doi.org/10.3389/fpsyg.2015.00390.

Kim, Y., Glassman, M., Bartholomew, M., & Hur, E. (2019). Creating an educational maker space for the promotion of collaborative learning. *International Journal of Social Media and Interactive Learning Environments*, 6(1), 44–62. https://doi.org/10.1504/IJSMILE.2018.090274.

Koedinger, K.R., Brunskill, E., Baker, R.S., McLaughlin, E.A., & Stamper, J. (2013). New potentials for data-driven intelligent tutoring system development and optimization. *AI Magazine*, 34(3), 27–41. https://doi.org/10.1609/aimag.v34i3.2484.

Laird, J.E., & van Lent, M. (2001). Human-level AI's killer application: Interactive computer games. *AI Magazine*, 22(2), 15–25. https://doi.org/10.1609/aimag.v22i2.1554.

Lee, J.D., & See, K.A. (2004). Trust in automation: Designing for appropriate reliance. *Human Factors*, 46(1), 50–80. https://doi.org/10.1518/hfes.46.1.50_30392.

Liu, S., & Sundar, S.S. (2018). Should machines express sympathy and empathy? *Computers in Human Behavior*, 97, 130–139. https://doi.org/10.1016/j.chb.2019.02.023.

Marcus, G., & Davis, E. (2019). *Rebooting AI: Building Artificial Intelligence We Can Trust*. Pantheon Books.

Mori, M. (1970). The uncanny valley. *Energy*, 7(4), 33–35. (Translated into English in 2012). https://doi.org/10.1109/MRA.2012.2192811.

Peters, C., Castellano, G., & de Freitas, S. (2019). An exploration of user engagement in HCI. Journal on Multimodal User Interfaces, 13(1), 1–5. https://doi.org/10.1007/s12193-019-00293-8.

Retail Customer Experience. (2020). *Humanizing Chatbots to Boost Engagement.* Retrieved from https://www.retailcustomerexperience.com.
Salesforce. (2020). *State of Service Report.* Retrieved from https://www.salesforce.com.
Seeger, A.M., Pfeiffer, J., & Heinzl, A. (2021). Texting with human-like conversational agents: Designing for anthropomorphism. *Journal of the Association for Information Systems,* 22(2), 291–324. https://doi.org/10.17705/1jais.00658.
Shneiderman, B. (2020). Human-centered artificial intelligence: Reliable, safe & trustworthy. *International Journal of Human–Computer Interaction,* 36(6), 495–504. https://doi.org/10.1080/10447318.2020.1741118.
Shum, H.Y., He, X., & Li, D. (2018). From Eliza to XiaoIce: Challenges and opportunities with social chatbots. *Frontiers of Information Technology & Electronic Engineering,* 19(1), 10–26. https://doi.org/10.1631/FITEE.1700826.
Smith, A., & Anderson, M. (2020). Personalization in AI-Powered Customer Service. *International Journal of Information Management,* 52, 102069. https://doi.org/10.1016/j.ijinfomgt.2019.10.002.
Sweller, J. (2011). Cognitive load theory. *Psychology of Learning and Motivation,* 55, 37–76. https://doi.org/10.1016/B978-0-12-387691-1.00002-8.
Telecom Insights. (2021). *AI in Telecom Customer Service: Balancing Automation and Human Touch.* Retrieved from https://www.telecominsights.com.
VanLehn, K. (2011). The relative effectiveness of human tutoring, intelligent tutoring systems, and other tutoring systems. *Educational Psychologist,* 46(4), 197–221. https://doi.org/10.1080/00461520.2011.611369.
Yannakakis, G.N., & Togelius, J. (2018). *Artificial Intelligence and Games.* Springer. https://doi.org/10.1007/978-3-319-63519-4.
Zhou, L., Gao, J., Li, D., & Shum, H.-Y. (2020). T.he Design and Implementation of XiaoIce, an Empathetic Social Chatbot. *Computational Linguistics,* 46(1), 53–93. https://doi.org/10.1162/coli_a_00368.

# • *E. AI in Healthcare* •

## 17. Mammography AI Can Cost Patients Extra. Is It Worth It?*

MICHELLE ANDREWS

As I checked in at a Manhattan radiology clinic for my annual mammogram in November, the front desk staffer reviewing my paperwork asked an unexpected question: Would I like to spend $40 for an artificial intelligence analysis of my mammogram? It's not covered by insurance, she added.

I had no idea how to evaluate that offer. Feeling upsold, I said no. But it got me thinking: Is this something I should add to my regular screening routine? Is my regular mammogram not accurate enough? If this AI analysis is so great, why doesn't insurance cover it?

I'm not the only person posing such questions. The mother of a colleague had a similar experience when she went for a mammogram recently at a suburban Baltimore clinic. She was given a pink pamphlet that said: "You Deserve More. More Accuracy. More Confidence. More power with artificial intelligence behind your mammogram." The price tag was the same: $40. She also declined.

In recent years, AI software that helps radiologists detect problems or diagnose cancer using mammography has been moving into clinical use. The software can store and evaluate large datasets of images and identify patterns and abnormalities that human radiologists might miss. It typically highlights potential problem areas in an image and assesses any likely malignancies. This extra review has enormous potential to improve the detection of suspicious breast masses and lead to earlier diagnoses of breast cancer.

While studies showing better detection rates are extremely encouraging, some radiologists say, more research and evaluation are needed before drawing conclusions about the value of the routine use of these tools in regular clinical practice.

"I see the promise and I hope it will help us," said Etta Pisano, a radiologist who

---

*Originally published as Michelle Andrews, "Mammography AI Can Cost Patients Extra. Is It Worth It?" *Kaiser Health News*, https://kffhealthnews.org/news/article/artificial-intelligence-mammography-extra-cost/view/republish/ (January 10, 2024). Kaiser Health News is a nonprofit news service covering health issues. It is an editorially independent program of the Kaiser Family Foundation that is not affiliated with Kaiser Permanente.

is chief research officer at the American College of Radiology, a professional group for radiologists. However, "it really is ambiguous at this point whether it will benefit an individual woman," she said. "We do need more information."

The radiology clinics that my colleague's mother and I visited are both part of RadNet, a company with a network of more than 350 imaging centers around the country. RadNet introduced its AI product for mammography in New York and New Jersey last February and has since rolled it out in several other states, according to Gregory Sorensen, the company's chief science officer.

Sorensen pointed to research the company conducted with 18 radiologists, some of whom were specialists in breast mammography and some of whom were generalists who spent less than 75 percent of their time reading mammograms. The doctors were asked to find the cancers in 240 images, with and without AI. Every doctor's performance improved using AI, Sorensen said.

Among all radiologists, "not every doctor is equally good," Sorensen said. With RadNet's AI tool, "it's as if all patients get the benefit of our very top performer."

But is the tech analysis worth the extra cost to patients? There's no easy answer.

"Some people are always going to be more anxious about their mammograms, and using AI may give them more reassurance," said Laura Heacock, a breast imaging specialist at NYU Langone Health's Perlmutter Cancer Center in New York. The health system has developed AI models and is testing the technology with mammograms but doesn't yet offer it to patients, she said.

Still, Heacock said, women shouldn't worry that they need to get an additional AI analysis if it's offered.

"At the end of the day, you still have an expert breast imager interpreting your mammogram, and that is the standard of care," she said.

About 1 in 8 women will be diagnosed with breast cancer during their lifetime, and regular screening mammograms are recommended to help identify cancerous tumors early. But mammograms are hardly foolproof: They miss about 20 percent of breast cancers, according to the National Cancer Institute.

The FDA has authorized roughly two dozen AI products to help detect and diagnose cancer from mammograms. However, there are currently no billing codes radiologists can use to charge health plans for the use of AI to interpret mammograms. Typically, the federal Centers for Medicare & Medicaid Services would introduce new billing codes and private health plans would follow their lead for payment. But that hasn't happened in this field yet and it's unclear when or if it will.

CMS didn't respond to requests for comment.

Thirty-five percent of women who visit a RadNet facility for mammograms pay for the additional AI review, Sorensen said.

Radiology practices don't handle payment for AI mammography all in the same way.

The practices affiliated with Boston-based Massachusetts General Hospital don't charge patients for the AI analysis, said Constance Lehman, a professor of radiology at Harvard Medical School who is co-director of the Breast Imaging Research Center at Mass General.

Asking patients to pay "isn't a model that will support equity," Lehman said, since only patients who can afford the extra charge will get the enhanced analysis. She said she believes many radiologists would never agree to post a sign listing a charge for AI analysis because it would be off-putting to low-income patients.

Sorensen said RadNet's goal is to stop charging patients once health plans realize the value of the screening and start paying for it.

Some large trials are underway in the United States, though much of the published research on AI and mammography to date has been done in Europe. There, the standard practice is for two radiologists to read a mammogram, whereas in the States only one radiologist typically evaluates a screening test.

Interim results from the highly regarded MASAI randomized controlled trial of 80,000 women in Sweden found that cancer detection rates were 20 percent higher in women whose mammograms were read by a radiologist using AI compared with women whose mammograms were read by two radiologists without any AI intervention, which is the standard of care there.

"The MASAI trial was great, but will that generalize to the U.S.? We can't say," Lehman said.

In addition, there is a need for "more diverse training and testing sets for AI algorithm development and refinement" across different races and ethnicities, said Christoph Lee, director of the Northwest Screening and Cancer Outcomes Research Enterprise at the University of Washington School of Medicine.

The long shadow of an earlier and largely unsuccessful type of computer-assisted mammography hangs over the adoption of newer AI tools. In the late 1980s and early 1990s, "computer-assisted detection" software promised to improve breast cancer detection. Then the studies started coming in, and the results were often far from encouraging. Using CAD at best provided no benefit, and at worst reduced the accuracy of radiologists' interpretations, resulting in higher rates of recalls and biopsies.

"CAD was not that sophisticated," said Robert Smith, senior vice president of early cancer detection science at the American Cancer Society. Artificial intelligence tools today are a whole different ballgame, he said. "You can train the algorithm to pick up things, or it learns on its own."

Smith said he found it "troubling" that radiologists would charge for the AI analysis.

"There are too many women who can't afford any out-of-pocket cost" for a mammogram, Smith said. "If we're not going to increase the number of radiologists we use for mammograms, then these new AI tools are going to be very useful, and I don't think we can defend charging women extra for them."

*KFF Health News is a national newsroom that produces in-depth journalism about health issues and is one of the core operating programs at KFF—an independent source of health policy research, polling, and journalism.*

# 18. The AI Paradox

## *Promises of Precision and Pitfalls in Healthcare*

Upal Roy

## *The Promises of AI in "Healthcare 5.0"*

The healthcare industry is at the forefront of a radical and transformative shift towards "Healthcare 5.0." The paradigm shift would expand the operational boundaries of "Healthcare 4.0" to leverage and adopt more patient-centric digital wellness. Healthcare 5.0 focuses on real-time patient monitoring, ambient control and wellness, and privacy compliance through data-driven and assisted technologies. "Healthcare 5.0" represents a new horizon of patient-centric delivery systems with emphasis on Artificial intelligence (AI) powered diagnostic support, personalized medicine, expedited bench side to bedside innovations, and the Internet of Things (IoT) enabled wearable device for vital tracking, smart homes and connected medical devices for remote management. Although AI systems have existed for more than fifty years, over the past decade, these black-box AI systems have become pervasive and utilized to make decisions within crucial healthcare systems, which necessitates trustable, explainable, responsible AI as a critical component.

While emerging technologies are paving the path towards the "Healthcare 5.0" journey, it is critical to reflect that the ultimate goal of a healthcare system is to improve overall patient outcomes. This process involves active participation of patients in their health management and emerging "Preventive Care" model. This "Preventive Care" model would lead to a fundamental shift of healthcare delivery focus from "tertiary prevention" (interventions or treatment post disease development) to "primary prevention" (prevention through early detection and lifestyle interventions).

"Healthcare 4.0" induced a paradigm shift in diagnostics through the integration of technology laying a stronger foundation. However, "Healthcare 5.0" is poised to revolutionize this domain further, emphasizing precision, personalization, and prevention. With digitalization of records, Electronic Health Records (EHRs) and advanced imaging became the norm, enabling data-driven insights and ushering in the dawn of using computer vision to detect abnormalities in medical images with higher accuracy comparable with human experts. Integrating data on behavioral factors and environmental factors for comprehensive assessments now AI can predict disease outbreaks and identify high-risk populations in a way which would become more precise and personalized day by day.

Precision medicine otherwise known as personalized medicine, has seen significant advancements from "Healthcare 4.0" to "5.0." "Healthcare 4.0" popularized the concept of personalized medicine, but its implementation was often limited due to data silos, technological constraints, and a fragmented healthcare system. With a deeper integration of technology, data, and clinical expertise to deliver truly personalized care, precision medicine takes center stage in "Healthcare 5.0."

The convergence of real-world data (RWD) and digital twin technology is a hallmark of Healthcare 5.0. This powerful combination is transforming how we understand, predict, and manage health. "Healthcare 4.0" laid the groundwork; however, "Healthcare 5.0" is characterized by a more comprehensive and integrated approach towards data-driven healthcare. RWD encompasses a broader spectrum of data, including data from wearable devices (IoT), patient-generated health data (Claims, Longitudinal Access and Adjudication Data and Electronic Medical Reports), disease registry data and real-world evidence from clinical trials. Leveraging that enriched diverse source of data could feed to a "'Digital Twins" based systems which could generate virtual representations of patients to simulate disease progression and treatment responses enabling personalized treatment plans and predicting intervention outcomes. A healthcare digital twin is a sophisticated computer model that simulates the health and well-being of an individual, disease pathway or an entire healthcare ecosystem that incorporates all its components and their dynamic interactions.

We have observed uptake and evolution of Robotic Surgery under "Healthcare 4.0" focused primarily on achieving state of the art surgical precision and minimal invasiveness and offering enhanced dexterity and visualization, as experts (specialized surgeons) controlled systems. Under "Healthcare 5.0" the area is expected to have more evolved Human—computer interaction (HCI) maturity with surgical precision to include autonomy, intelligence, and patient-centric care increasing capability. This would expand healthcare's horizon to a wider range of procedures, including various minimally invasive surgeries, complex reconstructions, and even remote surgeries. The overall ecosystem is poised to evolve with autonomous robotic systems and AI-driven optimal surgical plans, and haptic feedback improving overall surgical outcomes.

The hallmark of "Healthcare 5.0" is a blurring distinction between clinical and non-clinical pathways, as technology increasingly enables collaboration and integration between the two. Clinicians and administrators would work together to optimize workflows and resource utilization. While the convergence of clinical and non-clinical pathways offers significant opportunities for innovation and improved in-patient care, it also poses challenges, as the primary requirement would be significant change-management to overcome traditional silos, along with digital upskilling and fostering a collaborative culture.

## *The Perils of AI in "Healthcare 5.0"*

The transition from "Healthcare 4.0 to 5.0," while promising, is fraught with challenges and potential perils. Over the past decade, artificial intelligence (AI) models have gained significant acceptance to automate decision-making processes; however, the opaqueness of these models has posed a question mark on their applications in safety-critical systems. Most AI systems are still considered "black box" models that lack explainability. There is an increasing trend to develop Responsible AI systems to augment adoption across domains. Responsible AI systems leverage methods such as attention mechanisms and surrogate models to drive transparency and explainability, to augment bias detection, fairness in decision-making processes facilitating interpretability and adversarial robustness.

The "new normal" post-pandemic healthcare business paradigm is currently being propelled by advances in Machine learning (ML), Deep learning (DL), Federated Learning (FL) and more recently Generative AI (GenAI) based healthcare AI systems. Applications of those "black box" AI algorithms are powerful in performing accurate predictions for healthcare systems and delivery. However, explainability remains one of the most arguable issues surrounding the implementation. Even though AI-powered systems have been found to outperform humans in certain analytical tasks, their lack of explainability has sparked criticism. To guarantee that medical AI delivers on its promises, developers, healthcare professionals, and regulatory bodies must be made aware of the constraints and limits of opaque algorithms in medical AI, as well as stimulate interdisciplinary collaboration in the future to develop explainable discrimination-free AI systems.

In the journey of "Healthcare 4.0 to 5.0," there would be increasing reliance on digital health records and AI systems which makes patient data more vulnerable to breaches, increasing demand of stringent requirements for data privacy and security. Balancing data sharing for research and innovation with stringent privacy regulations is complex and may sometimes prove be counterproductive.

The digital divide, the widen gap between those with access to technology and those without, may also present significant challenges. As Healthcare 5.0 progresses with increased reliance on AI, IoT, and other advanced technologies, the digital divide is likely to widen. Individuals, especially in rural and underserved communities, who lack access to high-speed internet and digital devices lead to disparities in health outcomes, social isolations for those without access to digital healthcare tools being at a disadvantage. The digital divide may also impose a significant burden on healthcare systems increasing risk of operational and infrastructural cost and inefficient care delivery. If not managed strategically, these would lead to exacerbated health disparities, inequitable healthcare access, and incomplete data availability and biased research for policy formulation.

For a critical domain like healthcare, there are multiple stakeholder perspectives and a varying degree of need for Responsible AI implementation. Black-box AI systems could lead to unacceptable healthcare decisions. Regulatory oversight and user distrust of opaque AI systems represent challenges to the industry in the adoption of complex AI systems. In the absence of proper change management and upskilling plans, "Healthcare 5.0" automation and AI could lead to job losses in the healthcare sector, impacting workforce morale and patient care. The demands of "Healthcare 5.0" require substantial investments in technology infrastructure, which may be challenging for small scale healthcare organizations.

# 19. As AI Eye Exams Prove Their Worth, Lessons for Future Tech Emerge*

HANNAH NORMAN

Christian Espinoza, director of a Southern California drug-treatment provider, recently began employing a powerful new assistant: an artificial intelligence algorithm that can perform eye exams with pictures taken by a retinal camera. It makes quick diagnoses, without a doctor present.

His clinics, Tarzana Treatment Centers, are among the early adopters of an AI-based system that promises to dramatically expand screening for diabetic retinopathy, the leading cause of blindness among working-age adults and a threat to many of the estimated 38 million Americans with diabetes.

"It's been a godsend for us," said Espinoza, the organization's director of clinic operations, citing the benefits of a quick and easy screening that can be administered with little training and delivers immediate results.

His patients like it, too. Joseph Smith, who has Type 2 diabetes, recalled the cumbersome task of taking the bus to an eye specialist, getting his eyes dilated, and then waiting a week for results. "It was horrible," he said. "Now, it takes minutes."

Amid all the buzz around artificial intelligence in health care, the eye-exam technology is emerging as one of the first proven use cases of AI-based diagnostics in a clinical setting. While the FDA has approved hundreds of AI medical devices, adoption has been slow as vendors navigate the regulatory process, insurance coverage, technical obstacles, equity concerns, and challenges of integrating them into provider systems.

The eye exams show that the AI's ability to provide immediate results, as well as the cost savings and convenience of not needing to make an extra appointment, can have big benefits for both patients and providers. Of about 700 eye exams conducted during the past year at Espinoza's clinics, nearly one-quarter detected retinopathy, and patients were referred to a specialist for further care.

Diabetic retinopathy results when high blood sugar harms blood vessels in the retina. While managing a patient's diabetes can often prevent the disease—and there are treatments for more advanced stages—doctors say regular screenings are crucial for catching symptoms early. An estimated 9.6 million people in the U.S. have the disease.

---

*Originally published as Hannah Norman, "As AI Eye Exams Prove Their Worth, Lessons for Future Tech Emerge." *Kaiser Health News*, https://kffhealthnews.org/news/article/artificial-intelligence-ai-eye-exams-diabetic-retinopathy-innovation/ (March 27, 2024). Kaiser Health News is a nonprofit news service covering health issues. It is an editorially independent program of the Kaiser Family Foundation that is not affiliated with Kaiser Permanente.

The three companies with FDA-approved AI eye exams for diabetic retinopathy—Digital Diagnostics, based in Coralville, Iowa; Eyenuk of Woodland Hills, California; and Israeli software company AEYE Health—have sold systems to hundreds of practices nationwide. A few dozen companies have conducted research in the narrow field, and some have regulatory clearance in other countries, including tech giants like Google.

Digital Diagnostics, formerly Idx, received FDA approval for its system in 2018, following decades of research and a clinical trial involving 900 patients diagnosed with diabetes. It was the first fully autonomous AI system in any field of medicine, making its approval "a landmark moment in medical history," said Aaron Lee, a retina specialist and an associate professor at the University of Washington.

The system, used by Tarzana Treatment Centers, can be operated by someone with a high school degree and a few hours of training, and it takes just a few minutes to produce a diagnosis, without any eye dilation most of the time, said John Bertrand, CEO of Digital Diagnostics.

The setup can be placed in any dimly lit room, and patients place their face on the chin and forehead rests and stare into the camera while a technician takes images of each eye.

The American Diabetes Association recommends that people with Type 2 diabetes get screened every one to two years, yet only about 60 percent of people living with diabetes get yearly eye exams, said Robert Gabbay, the ADA's chief scientific and medical officer. The rates can be as low as 35 percent for people with diabetes age 21 or younger.

In swaths of the U.S., a shortage of optometrists and ophthalmologists can make appointments hard to schedule, sometimes booking for months out. Plus, the barriers of traveling to an additional appointment to get their eyes dilated—which means time off work or school and securing transportation—can be particularly tricky for low-income patients, who also have a higher risk of Type 2 diabetes.

"Ninety percent of our patients are blue-collar," said Espinoza of his Southern California clinics, which largely serve minority populations. "They don't eat if they don't work."

One potential downside of not having a doctor do the screening is that the algorithm solely looks for diabetic retinopathy, so it could miss other concerning diseases, like choroidal melanoma, Lee said. The algorithms also generally "err on the side of caution" and over-refer patients.

But the technology has shown another big benefit: Follow-up after a positive result is three times as likely with the AI system, according to a recent study by Stanford University.

That's because of the "proximity of the message," said David Myung, an associate professor of ophthalmology at the Byers Eye Institute at Stanford. When it's delivered immediately, rather than weeks or even months later, it's much more likely to be heard by the patient and acted upon.

Myung launched Stanford's automated teleophthalmology program in 2020, originally focusing on telemedicine and then shifting to AI in its Bay Area clinics. That same year, the National Committee for Quality Assurance expanded its screening standard for diabetic retinopathy to include the AI systems.

Myung said it took about a year to sift through the Stanford health system's cybersecurity and IT systems to integrate the new technology. There was also a learning curve, especially for taking quality photos that the AI can decipher, Myung said.

"Even with hitting our stride, there's always something to improve," he added.

The AI test has been bolstered by a reimbursement code from the Centers for Medicare & Medicaid Services, which can be difficult and time-consuming to obtain for breakthrough devices. But health care providers need that government approval to get reimbursement.

In 2021, CMS set the national payment rate for AI diabetic retinopathy screenings at $45.36—quite a bit below the median privately negotiated rate of $127.81, according to a recent New England Journal of Medicine AI study. Each company has a slightly different business model, but they generally charge providers subscription or licensing fees for their software.

The companies declined to share what they charge for their software. The cameras can cost up to $20,000 and are either purchased separately or wrapped into the software subscription as a rental.

The greater compliance with screening recommendations that the machines make possible, along with a corresponding increase in referrals to specialists, makes it worthwhile, said Lindsie Buchholz, clinical informatics lead at Nebraska Medicine, which in mid-December began using Eyenuk's system.

"It kind of helps the camera pay for itself," she said.

Today, Digital Diagnostics' system is in roughly 600 sites nationwide, according to the company. AEYE Health said its eye exam is used by "low hundreds" of U.S. providers. Eyenuk declined to share specifics about its reach.

The technology continues to advance, with clinical studies for additional cameras—including a handheld imager that can screen patients in the field—and looking at other eye diseases, like glaucoma. The innovations put ophthalmology alongside radiology, cardiology, and dermatology as specialties in which AI innovation is happening fast.

"They are going to come out in the near future—cameras that you can use in street medicine—and it's going to help a lot of people," said Espinoza.

*KFF Health News is a national newsroom that produces in-depth journalism about health issues and is one of the core operating programs at KFF—an independent source of health policy research, polling, and journalism.*

# 20. Forget Ringing the Button for the Nurse. Patients Now Stay Connected by Wearing One*

PHIL GALEWITZ

HOUSTON—Patients admitted to Houston Methodist Hospital get a monitoring device about the size of a half-dollar affixed to their chest—and an unwitting role in the expanding use of artificial intelligence in health care.

The slender, battery-powered gadget, called a BioButton, records vital signs including heart and breathing rates, then wirelessly sends the readings to nurses sitting in a 24-hour control room elsewhere in the hospital or in their homes. The device's software uses AI to analyze the voluminous data and detect signs a patient's condition is deteriorating.

Hospital officials say the BioButton has improved care and reduced the workload of bedside nurses since its rollout last year.

"Because we catch things earlier, patients are doing better, as we don't have to wait for the bedside team to notice if something is going wrong," said Sarah Pletcher, system vice president at Houston Methodist.

But some nurses fear the technology could wind up replacing them rather than supporting them—and harming patients. Houston Methodist, one of dozens of U.S. hospitals to employ the device, is the first to use the BioButton to monitor all patients except those in intensive care, Pletcher said.

"The hype around a lot of these devices is they provide care at scale for less labor costs," said Michelle Mahon, a registered nurse and an assistant director of National Nurses United, the profession's largest U.S. union. "This is a trend that we find disturbing," she said.

The rollout of BioButton is among the latest examples of hospitals deploying technology to improve efficiency and address a decades-old nursing shortage. But that transition has raised its own concerns, including about the device's use of AI; polls show the public is wary of health providers relying on it for patient care.

In December 2022 the FDA cleared the BioButton for use in adult patients who are

---

*Originally published as Phil Galewitz, "Forget Ringing the Button for the Nurse. Patients Now Stay Connected by Wearing One," *Kaiser Health News*, https://kffhealthnews.org/news/article/hospital-artificial-intelligence-patient-monitoring-biobutton-houston/view/republish/ (May 8, 2024). Kaiser Health News is a nonprofit news service covering health issues. It is an editorially independent program of the Kaiser Family Foundation that is not affiliated with Kaiser Permanente.

not in critical care. It is one of many AI tools now used by hospitals for tasks like reading diagnostic imaging results.

In 2023, President Joe Biden directed the Department of Health and Human Services to develop a plan to regulate AI in hospitals, including by collecting reports of patients harmed by its use.

The leader of BioIntelliSense, which developed the BioButton, said its device is a huge advance compared with nurses walking into a room every few hours to measure vital signs. "With AI, you now move from 'I wonder why this patient crashed' to 'I can see this crash coming before it happens and intervene appropriately,'" said James Mault, CEO of the Golden, Colorado-based company.

The BioButton stays on the skin with an adhesive, is waterproof, and has up to a 30-day battery life. The company says the device—which allows providers to quickly notice deteriorating health by recording more than 1,000 measurements a day per patient—has been used on more than 80,000 hospital patients nationwide in the past year.

Hospitals pay BioIntelliSense an annual subscription fee for the devices and software.

Houston Methodist officials would not reveal how much the hospital pays for the technology, though Pletcher said it equates to less than a cup of coffee a day per patient.

For a hospital system that treats thousands of patients at a time—Houston Methodist has 2,653 non–ICU beds at its eight Houston-area hospitals—such an investment could still translate to millions of dollars a year.

Hospital officials say they have not made any changes in nurse staffing and have no plans to because of implementing the BioButton.

Inside the hospital's control center for virtual monitoring on a recent morning, about 15 nurses and technicians dressed in scrubs sat in front of large monitors showing the health status of hundreds of patients they were assigned to monitor.

A red checkmark next to a patient's name signaled the AI software had found readings trending outside normal. Staff members could click into a patient's medical record, showing patients' vital signs over time and other medical history. These virtual nurses, if you will, could contact nurses on the floor by phone or email, or even dial directly into the patient's room via video call.

Nutanben Gandhi, a technician who was watching 446 patients on her monitor that morning, said that when she gets an alert, she looks at the patient's health record to see if the anomaly can be easily explained by something in the patient's condition or if she needs to contact nurses on the patient's floor.

Oftentimes an alert can be easily dismissed. But identifying signs of deteriorating health can be tough, said Steve Klahn, Houston Methodist's clinical director of virtual medicine.

"We are looking for a needle in a haystack," he said.

Donald Eustes, 65, was admitted to Houston Methodist in March for prostate cancer treatment and has since been treated for a stroke. He is happy to wear the BioButton.

"You never know what can happen here, and having an extra set of eyes looking at you is a good thing," he said from his hospital bed. After being told the device uses AI, the Montgomery, Texas, man said he has no problem with its helping his clinical team. "This sounds like a good use of artificial intelligence."

Patients and nurses alike benefit from remote monitoring like the BioButton, said Pletcher of Houston Methodist.

The hospital has placed small cameras and microphones inside all patient rooms enabling nurses outside to communicate with patients and perform tasks such as helping with patient admissions and discharge instructions. Patients can include family members on the remote calls with nurses or a doctor, she said.

Virtual technology frees up on-duty nurses to provide more hands-on help, such as starting an intravenous line, Pletcher said. With the BioButton, nurses can wait to take routine vital signs every eight hours instead of every four, she said.

Pletcher said the device reduces nurses' stress in monitoring patients and allows some to work more flexible hours because virtual care can be done from home rather than coming to the hospital. Ultimately it helps retain nurses, not drive them away, she said.

Sheeba Roy, a nurse manager at Houston Methodist, said some members of the nursing staff were nervous about relying on the device and not checking patients' vital signs as often themselves. But testing has shown the device provides accurate information.

"After we implemented it, the staff loves it," Roy said.

Serena Bumpus, chief executive officer of the Texas Nurses Association, said her concern with any technology is that it can be more burdensome on nurses and take away time with patients.

"We have to be hypervigilant in ensuring that we are not leaning on this to replace the ability of nurses to critically think and assess patients and validate what this device is telling us is true," Bumpus said.

Houston Methodist this year plans to send the BioButton home with patients so the hospital can better track their progress in the weeks after discharge, measuring the quality of their sleep and checking their gait.

"We are not going to need less nurses in health care, but we have limited resources and we have to use those as thoughtfully as we can," Pletcher said. "Looking at projected demand and seeing the supply we have coming, we will not have enough to meet demand, so anything we can do to give time back to nurses is a good thing."

*KFF Health News is a national newsroom that produces in-depth journalism about health issues and is one of the core operating programs at KFF—an independent source of health policy research, polling, and journalism.*

# 21. ¿Cómo Se Dice?*

## *California Loops In AI to Translate Health Care Information*

PAULA ANDALO

Tener gripe, tener gripa, engriparse, agriparse, estar agripado, estar griposo, agarrar la gripe, coger la influenza. In Spanish, there are at least a dozen ways to say someone has the flu—depending on the country.

Translating "cardiac arrest" into Spanish is also tricky because "arresto" means getting detained by the police. Likewise, "intoxicado" means you have food poisoning, not that you're drunk.

The examples of how translation could go awry in any language are endless: Words take on new meanings, idioms come and go, and communities adopt slang and dialects for everyday life.

Human translators work hard to keep up with the changes, but California plans to soon entrust that responsibility to technology.

State health policy officials want to harness emerging artificial intelligence technology to translate a broad swath of documents and websites related to "health and social services information, programs, benefits and services," according to state records. Sami Gallegos, a spokesperson for California's Health and Human Services Agency, declined to elaborate on which documents and languages would be involved, saying that information is "confidential."

The agency is seeking bids from IT firms for the ambitious initiative, though its timing and cost is not yet clear. Human editors supervising the project will oversee and edit the translations, Gallegos said.

Agency officials said they hope to save money and make critical health care forms, applications, websites, and other information available to more people in what they call the nation's most linguistically diverse state.

The project will start by translating written material. Agency Secretary Mark Ghaly said the technology, if successful, may be applied more broadly.

"How can we potentially not just transform all of our documents, but our websites, our ability to interact, even some of our call center inputs, around AI?" Ghaly asked during an April briefing on AI in health care in Sacramento.

---

*Originally published as Paula Andalo, "¿Cómo Se Dice? California Loops in AI to Translate Health Care Information," *Kaiser Health News*, https://kffhealthnews.org/news/article/california-artificial-intelligence-translate-health-information-language/ (June 18, 2024). Kaiser Health News is a nonprofit news service covering health issues. It is an editorially independent program of the Kaiser Family Foundation that is not affiliated with Kaiser Permanente.

But some translators and scholars fear the technology lacks the nuance of human interaction and isn't ready for the challenge. Turning this sensitive work over to machines could create errors in wording and understanding, they say—ultimately making information less accurate and less accessible to patients.

"AI cannot replace human compassion, empathy, and transparency, meaningful gestures and tones," said Rithy Lim, a Fresno-based medical and legal interpreter for 30 years who specializes in Khmer, the main language of Cambodia.

Artificial intelligence is the science of designing computers that emulate human thinking by reasoning, problem-solving, and understanding language. A type of artificial intelligence known as generative AI, or GenAI, in which computers are trained using massive amounts of data to "learn" the meaning of things and respond to prompts, is driving a wave of investment, led by such companies as Open AI and Google.

AI is quickly being integrated into health care, including programs that diagnose diabetic retinopathy, analyze mammograms, and connect patients with nurses remotely. Promotors of the technology often make the grandiose claim that soon everyone will have their own "AI doctor."

AI also has been a game changer in translation. ChatGPT, Google's Neural Machine Translation, and Open Source are not only faster than older technologies such as Google Translate, but they can process huge volumes of content and draw upon a vast database of words to nearly mimic human translation.

Whereas a professional human translator might need three hours to translate a 1,600-word document, AI can do it in a minute.

Arjun "Raj" Manrai, an assistant professor of biomedical informatics at Harvard Medical School and the deputy editor of New England Journal of Medicine AI, said the use of AI technology represents a natural progression in medical translation, given that patients already use Google Translate and AI platforms to translate for themselves and their loved ones.

"Patients are not waiting," he said.

He said GenAI could be particularly useful in this context.

These translations "can deliver real value to patients by simplifying complex medical information and making it more accessible," he said.

In its bidding documents, the state says the goal of the project is to increase "speed, efficiency, and consistency of translations, and generate improvements in language access" in a state where 1 in 3 people speak a language other than English, and more than 200 languages are spoken.

In May 2023, the state Health and Human Services Agency adopted a "language access policy" that requires its departments to translate all "vital" documents into at least the top five languages spoken by Californians with limited English proficiency. At the time, those languages were Spanish, Chinese, Tagalog, Vietnamese, and Korean.

Examples of vital documents include application forms for state programs, notices about eligibility for benefits, and public website content.

Currently, human translators produce these translations. With AI, more documents could be translated into more languages.

A survey conducted by the California Health Care Foundation late last year found that 30 percent of Spanish speakers have difficulty explaining their health issues and concerns to a doctor, compared with 16 percent of English speakers.

Health equity advocates say AI will help close that gap.

"This technology is a very powerful tool in the area of language access," said Sandra R. Hernández, president and CEO of the foundation. "In good hands, it has many opportunities to expand the translation capability to address inequities."

But Hernández cautioned that AI translations must have human oversight to truly capture meaning.

"The human interface is very important to make sure you get the accuracy and the cultural nuances reflected," she said.

Lim recalled an instance in which a patient's daughter translated preoperative instructions to her mother the night before surgery. Instead of translating the instructions as "you cannot eat" after a certain hour, she told her mom, "You should not eat."

The mother ate breakfast, and the surgery had to be rescheduled.

"Even a few words that change meaning could have a drastic impact on the way people consume the information," said Sejin Paik, a doctoral candidate in digital journalism, human-computer interaction, and emerging media at Boston University.

Paik, who grew up speaking Korean, also pointed out that AI models are often trained from a Western point of view. The data that drives the translations filters languages through an English perspective, "which could result in misinterpretations of the other language," she said. Amid this fast-changing landscape, "we need more diverse voices involved, more people thinking about the ethical concepts, how we best forecast the impact of this technology."

Manrai pointed to other flaws in this nascent technology that must be addressed. For instance, AI sometimes invents sentences or phrases that are not in the original text, potentially creating false information—a phenomenon AI scientists call "hallucination" or "confabulation."

Ching Wong, executive director of the Vietnamese Community Health Promotion Project at the University of California–San Francisco, has been translating health content from English into Vietnamese and Chinese for 30 years.

He provided examples of nuances in language that might confuse AI translation programs. Breast cancer, for instance, is called "chest cancer" in Chinese, he said.

And "you" has different meanings in Vietnamese, depending on a person's ranking in the family and community. If a doctor uses "you" incorrectly with a patient, it could be offensive, Wong said.

But Ghaly emphasized that the opportunities outweigh the drawbacks. He said the state should "cultivate innovation" to help vulnerable populations gain greater access to care and resources.

And he was clear: "We will not replace humans."

*KFF Health News is a national newsroom that produces in-depth journalism about health issues and is one of the core operating programs at KFF—an independent source of health policy research, polling, and journalism.*

# 22. "Dr. Google" Meets Its Match*

## *Dr. ChatGPT*

Andrew Leonard

As a fourth-year ophthalmology resident at Emory University School of Medicine, Riley Lyons' biggest responsibilities include triage: When a patient comes in with an eye-related complaint, Lyons must make an immediate assessment of its urgency.

He often finds patients have already turned to "Dr. Google." Online, Lyons said, they are likely to find that "any number of terrible things could be going on based on the symptoms that they're experiencing."

So, when two of Lyons' fellow ophthalmologists at Emory came to him and suggested evaluating the accuracy of the AI chatbot ChatGPT in diagnosing eye-related complaints, he jumped at the chance.

In June, Lyons and his colleagues reported in medRxiv, an online publisher of health science preprints, that ChatGPT compared quite well to human doctors who reviewed the same symptoms—and performed vastly better than the symptom checker on the popular health website WebMD. And despite the much-publicized "hallucination" problem known to afflict ChatGPT—its habit of occasionally making outright false statements—the Emory study reported that the most recent version of ChatGPT made zero "grossly inaccurate" statements when presented with a standard set of eye complaints.

The relative proficiency of ChatGPT, which debuted in November 2022, was a surprise to Lyons and his co-authors. The artificial intelligence engine "is definitely an improvement over just putting something into a Google search bar and seeing what you find," said co-author Nieraj Jain, an assistant professor at the Emory Eye Center who specializes in vitreoretinal surgery and disease.

But the findings underscore a challenge facing the health care industry as it assesses the promise and pitfalls of generative AI, the type of artificial intelligence used by ChatGPT: The accuracy of chatbot-delivered medical information may represent an improvement over Dr. Google, but there are still many questions about how to integrate this new technology into health care systems with the same safeguards historically applied to the introduction of new drugs or medical devices.

---

*Originally published as Andrew Leonard, "'Dr. Google' Meets Its Match: Dr. ChatGPT," *Kaiser Health News*, https://kffhealthnews.org/news/article/chatgpt-chatbot-google-webmd-symptom-checker/ (September 12, 2023). Kaiser Health News is a nonprofit news service covering health issues. It is an editorially independent program of the Kaiser Family Foundation that is not affiliated with Kaiser Permanente.

The smooth syntax, authoritative tone, and dexterity of generative AI have drawn extraordinary attention from all sectors of society, with some comparing its future impact to that of the internet itself. In health care, companies are working feverishly to implement generative AI in areas such as radiology and medical records.

When it comes to consumer chatbots, though, there is still caution, even though the technology is already widely available—and better than many alternatives. Many doctors believe AI-based medical tools should undergo an approval process similar to the FDA's regime for drugs, but that would be years away. It's unclear how such a regime might apply to general-purpose AIs like ChatGPT.

"There's no question we have issues with access to care, and whether or not it is a good idea to deploy ChatGPT to cover the holes or fill the gaps in access, it's going to happen and it's happening already," said Jain. "People have already discovered its utility. So, we need to understand the potential advantages and the pitfalls."

The Emory study is not alone in ratifying the relative accuracy of the new generation of AI chatbots. A report published in Nature in early July by a group led by Google computer scientists said answers generated by Med-PaLM, an AI chatbot the company built specifically for medical use, "compare favorably with answers given by clinicians."

AI may also have better bedside manner. Another study, published in April by researchers from the University of California–San Diego and other institutions, even noted that health care professionals rated ChatGPT answers as more empathetic than responses from human doctors.

Indeed, a number of companies are exploring how chatbots could be used for mental health therapy, and some investors in the companies are betting that healthy people might also enjoy chatting and even bonding with an AI "friend." The company behind Replika, one of the most advanced of that genre, markets its chatbot as, "The AI companion who cares. Always here to listen and talk. Always on your side."

"We need physicians to start realizing that these new tools are here to stay and they're offering new capabilities both to physicians and patients," said James Benoit, an AI consultant. While a postdoctoral fellow in nursing at the University of Alberta in Canada, he published a study in February reporting that ChatGPT significantly outperformed online symptom checkers in evaluating a set of medical scenarios. "They are accurate enough at this point to start meriting some consideration," he said.

Still, even the researchers who have demonstrated ChatGPT's relative reliability are cautious about recommending that patients put their full trust in the current state of AI. For many medical professionals, AI chatbots are an invitation to trouble: They cite a host of issues relating to privacy, safety, bias, liability, transparency, and the current absence of regulatory oversight.

The proposition that AI should be embraced because it represents a marginal improvement over Dr. Google is unconvincing, these critics say.

"That's a little bit of a disappointing bar to set, isn't it?" said Mason Marks, a professor and MD who specializes in health law at Florida State University. He recently wrote an opinion piece on AI chatbots and privacy in the Journal of the American Medical Association. "I don't know how helpful it is to say, 'Well, let's just throw this conversational AI on as a band-aid to make up for these deeper systemic issues,'" he said to KFF Health News.

The biggest danger, in his view, is the likelihood that market incentives will result in AI interfaces designed to steer patients to particular drugs or medical services.

"Companies might want to push a particular product over another," said Marks. "The potential for exploitation of people and the commercialization of data is unprecedented."

OpenAI, the company that developed ChatGPT, also urged caution.

"OpenAI's models are not fine-tuned to provide medical information," a company spokesperson said. "You should never use our models to provide diagnostic or treatment services for serious medical conditions."

John Ayers, a computational epidemiologist who was the lead author of the UCSD study, said that as with other medical interventions, the focus should be on patient outcomes.

"If regulators came out and said that if you want to provide patient services using a chatbot, you have to demonstrate that chatbots improve patient outcomes, then randomized controlled trials would be registered tomorrow for a host of outcomes," Ayers said.

He would like to see a more urgent stance from regulators.

"One hundred million people have ChatGPT on their phone," said Ayers, "and are asking questions right now. People are going to use chatbots with or without us."

At present, though, there are few signs that rigorous testing of AIs for safety and effectiveness is imminent. In May, Robert Califf, the commissioner of the FDA, described "the regulation of large language models as critical to our future," but aside from recommending that regulators be "nimble" in their approach, he offered few details.

In the meantime, the race is on. In July, *The Wall Street Journal* reported that the Mayo Clinic was partnering with Google to integrate the Med-PaLM 2 chatbot into its system. In June, WebMD announced it was partnering with a Pasadena, California-based startup, HIA Technologies Inc., to provide interactive "digital health assistants." And the ongoing integration of AI into both Microsoft's Bing and Google Search suggests that Dr. Google is already well on its way to being replaced by Dr. Chatbot.

*This article was produced by KFF Health News, which publishes California Healthline, an editorially independent service of the California Health Care Foundation.*

*KFF Health News is a national newsroom that produces in-depth journalism about health issues and is one of the core operating programs at KFF—an independent source of health policy research, polling, and journalism. Learn more about KFF.*

# • *F. AI in Industrial Ecosystems* •

# 23. The Strategic Integration of AI in Testing, Inspection and Certification

Courtlin Holt-Nguyen

The Testing, Inspection and Certification (TIC) industry, valued at $250 billion in 2023 (Wadhwani & Ambekar, 2024), remains largely reliant on manual processes despite technological advances in other sectors (Yuen et al., 2022). This piece argues that the industry's continued dependence on traditional methods is unsustainable given growing global trade volumes, rising quality expectations, and complex supply chains. However, successful AI implementation requires both careful consideration of potential risks and proper preparation.

## *The Necessity of AI Adoption in the TIC Industry*

### Limitations of Traditional TIC Processes

Traditional TIC processes rely heavily on human inspectors whose performance can degrade over time due to fatigue and variable attention. This is particularly problematic in tasks requiring constant vigilance, such as visual inspections on production lines.

*Fatigue and Attention Variability:* Research shows that maintaining attention during monotonous tasks leads to mental fatigue, reducing accuracy and increasing error rates over time. This phenomenon, known as vigilant attention decline, is well-documented in industrial settings where human inspectors are required to perform repetitive tasks for extended periods. Fatigue results in a depletion of attentional resources, leading to diminished performance and increased risk of missed defects (Langner & Eickhoff, 2013).

*Human Error in Visual Inspection:* Human inspectors are prone to errors caused by fatigue, distraction, and subjective judgment. These factors can lead to inconsistent inspection results, missed defects, and compromised product quality. For example, manual inspection often results in false negatives (missed defects), which can have serious implications for manufacturers due to product returns and damage to brand reputation (Knop et al., 2017).

*Physical Limitations:* Human inspectors can only work effectively for a limited number of hours before their performance deteriorates (Kujawinska & Vogt, 2015). This is especially true when the work environment requires the human inspector to work in abnormally low temperatures, as is the case for fresh produce inspections.

## Economic Pressure for Enhanced Efficiency

As products become more complex, particularly in industries like automotive, electronics, and manufacturing, the cost of maintaining quality control rises significantly. Traditional TIC processes struggle to keep pace with this complexity due to their reliance on manual inspection and outdated methods. The increasing intricacy of components and systems necessitates more detailed inspections, which in turn drives up costs.

*Product Complexity and Cost:* According to Budiono et al. (2021), the growing complexity of modern manufacturing systems requires more sophisticated quality control measures, which are often labor-intensive and costly. This economic pressure forces companies to seek scalable solutions that can handle higher volumes of inspections without proportional increases in cost.

High warranty and recall costs necessitate proactive defect detection: Warranty claims and product recalls represent significant financial risks for companies, especially when defects are identified after products have been released into the market. Traditional TIC methods often rely on reactive quality control approaches, which only catch defects at later stages of production or after products have been distributed. Real-time data analytics can significantly reduce warranty and recall costs by enhancing the accuracy and efficiency of fault detection, prediction, and management. By employing machine learning and statistical pattern recognition, manufacturers can better analyze warranty claims and identify recurring issues that might not be apparent through traditional methods. This can significantly reduce the incidence of "trouble not found" (TNF) cases, which can account for up to 38 percent of warranty costs (Hand et al., 2024).

## The Growing Capabilities of AI Technologies

AI-driven computer vision systems achieve superior defect detection rates compared to human inspection: AI-based computer vision systems outperform human inspectors in defect detection, particularly in industrial settings where product complexity is high, and component sizes are minuscule, such as the semiconductor industry (Muniandi, 2024). Through a combination of machine learning and computer vision, AI systems offer improved detection, yield enhancement, and production efficiency (Peker et al., 2023).

AI's adaptive detection parameters allow dynamic quality control for evolving standards: By leveraging machine learning algorithms and real-time data analysis, AI systems can dynamically adjust to evolving conditions, leading to improved fault detection and process optimization (Kyriakopoulos et al., 2023). This adaptability not only minimizes disruptions but also ensures consistent product quality, ultimately boosting manufacturing performance. In industries where product specifications frequently change, adaptability allows for continuous quality improvement without the need for manual recalibration.

Additionally, Smart Active Sampling (SAS), based on AI/ML models, is a technique that enhances quality assurance by selecting specific samples for inspection, improving prediction accuracy while reducing the number of samples needed, up to five times fewer than traditional random sampling in some situations. By detecting quality violations earlier, SAS minimizes scrap production and lowers inspection costs (Clemens et al., 2023).

## *Advantages of AI in the TIC Industry*

### Enhanced Accuracy and Consistency

The adoption of AI technologies in the TIC industry has significantly enhanced accuracy and consistency in quality control processes. AI-driven systems, particularly those using computer vision and machine learning algorithms, can detect defects with far greater precision than human inspectors. For example, McKinsey research found that AI-driven quality testing can increase defect detection rates by up to 90 percent, compared to traditional human inspection methods (McKinsey & Company 2017). Similarly, Arora (2023), found that AI-powered defect detection using advanced image processing techniques significantly reduced the time and errors associated with traditional manual inspections. This automation led to more consistent and accurate assessments of product quality, ensuring that defects were identified and addressed promptly. Furthermore, the ability of AI systems to continuously learn and adapt through feedback aids in ensuring compliance with evolving regulatory requirements in industries such as medical device manufacturing (Khinvasara et al., 2024).

### AI Inspection Systems Identify Defects Missed by Humans

The use of deep learning for surface defect inspection has significantly improved detection accuracy compared to traditional methods that rely on human inspectors. One of the challenges of training an AI system to identify defects can be a lack of high-quality training data. However, generative models, such as Variational Auto-Encoders (VAEs) and Generative Adversarial Networks (GANs), can generate realistic training samples, addressing the limitations of small defect datasets (He et al., 2024). Additionally, Kim (2023), found that employing transfer learning and data augmentation techniques allows an AI system to effectively overcome the challenge of insufficient defect image data. Consequently, AI systems can achieve higher accuracy and better model generalization, making them potentially more effective than human inspectors in identifying defects in manufacturing processes.

### Consistent Application of Quality Standards

The consistent application of quality standards across multiple facilities is a key advantage of AI in the TIC industry, ensuring uniformity in product quality and compliance. In traditional TIC processes, human inspectors may vary in their interpretation of standards, leading to inconsistencies across different locations or production lines. However, AI-driven systems eliminate this variability by applying the same algorithms and detection criteria uniformly across all facilities. This ensures that quality standards are consistently met, regardless of geographic location or local workforce expertise. According to the TIC Council (2024), AI technologies enhance the ability to standardize compliance procedures globally by providing a harmonized framework for conformity assessments, which is essential for ensuring that products meet regulatory requirements across borders. This uniformity is particularly critical in industries with global supply chains, where maintaining consistent product quality across multiple sites is a challenge. Moreover, AI's ability to process vast amounts of data in real-time allows for continuous monitoring and immediate adjustments to ensure compliance with evolving standards. AI-based systems can analyze thousands of data points simultaneously and apply the same rigorous inspection protocols across different facilities without human error or fatigue. This not only improves the accuracy of inspections but also ensures that every facility adheres to the same high standards, reducing the risk of non-compliance and product recalls. By automating these processes, companies can ensure that their products consistently meet both internal quality benchmarks and external regulatory requirements.

## *Disadvantages of AI in the TIC Industry*

### Risks of Automation Complacency

The over-reliance on AI in the TIC industry presents a significant risk by reducing cognitive engagement and oversight by human inspectors. When AI systems are perceived as highly reliable, there is a tendency for human operators to become complacent. This phenomenon, known as automation complacency, can lead to inadequate monitoring of the system's performance, increasing the likelihood of critical errors being overlooked. Research by Harbarth et al. (2024) highlights that automation complacency occurs when individuals over-trust automated systems and fail to critically evaluate their outputs, leading to suboptimal decision-making and performance failures. In the context of TIC, this could mean that inspectors may become overly dependent on AI-driven detection systems, neglecting their role in verifying the system's accuracy or identifying potential anomalies that the AI might miss, as observed in other industries including healthcare and aviation (Liu, 2023).

Furthermore, studies have shown that automation bias, the tendency to favor automated decisions over human judgment, can exacerbate this issue. According to Horowitz & Kahn (2024), automation bias occurs when users accept AI recommendations without seeking confirmatory evidence, effectively transferring responsibility for decision-making onto the machine. This over-reliance can erode human skills and judgment over time, a phenomenon known as deskilling, where inspectors may lose their ability to detect defects independently due to prolonged reliance on AI systems (Ross & Spates, 2020). To reduce these risks, it is important to maintain a balanced approach where human inspectors remain actively engaged in the quality control process. Implementing human-in-the-loop systems, where humans monitor and adjust AI recommendations, when necessary, can help ensure that inspectors retain their critical oversight role while benefiting from the efficiency of AI technologies (Sele & Chugunova, 2024). By fostering a collaborative relationship between humans and AI, TIC companies can mitigate the risks of automation complacency while enhancing overall inspection accuracy.

### Process Optimization Prerequisites

The implementation of AI in the TIC industry can be significantly hindered by poorly optimized workflows, which reduce the potential benefits of AI technologies. AI systems are designed to enhance efficiency, accuracy, and scalability, but these advantages are only fully realized when integrated into well-structured and optimized processes. Without optimizing processes, automation may simply accelerate existing inefficiencies, resulting in wasted resources and potential financial losses (Harbour, 2008). Furthermore, automating processes without prior optimization can result in a loss of process knowledge, as employees may become overly reliant on automated systems. This overreliance can hinder the ability to make informed decisions and adapt to changes in the business environment (Eulerich et al., 2023). Therefore, a clear understanding and management of processes are essential for maximizing the effectiveness of automation and achieving substantial quality improvements and cost reductions.

## *Conclusion*

In summary, the adoption of AI in the TIC industry represents a critical strategic imperative that organizations cannot afford to ignore. While the benefits of enhanced accuracy, consistency, and efficiency are compelling, successful implementation requires a thoughtful approach that addresses both prerequisites and potential risks. Organizations must first optimize their existing processes before automation and maintain robust human oversight to prevent automation complacency. When properly implemented with these considerations in mind, AI technologies can transform traditional TIC processes, enabling organizations to meet the growing demands of increasingly complex global supply chains while maintaining high quality standards and regulatory compliance. The future of TIC lies not in choosing between human expertise and AI, but in their strategic integration

### References

Arora, M. (2023). AI-Driven Industry 4.0: Advancing Quality Control through Cutting-Edge Image Processing for Automated Defect Detection. *International Journal of Computer Science and Mobile Computing*, 12(8), 16–32. https://doi.org/10.47760/ijcsmc.2023.v12i08.003.

Budiono, H.D.S., Nurcahyo, R., & Habiburrahman, M. (2021). Relationship between manufacturing complexity, strategy, and performance of manufacturing industries in Indonesia. *Heliyon*, 7(6), e07225. https://doi.org/10.1016/j.heliyon.2021.e07225.

Eulerich, M., Waddoups, N., Wagener, M., & Wood, D.A. (2023). The Dark Side of Robotic Process Automation (RPA): Understanding Risks and Challenges with RPA. *Accounting Horizons*, 38(2), 143–152. https://doi.org/10.2308/horizons-2022-019.

Hand, J., Hall, S., Carr, M., & Worm, J. (2024). Warranty Repairs Reimagined through Machine Learning and Statistical Pattern Recognition (Part 1). *SAE Technical Papers on CD-ROM/SAE Technical Paper Series*. https://doi.org/10.4271/2024-01-5071.

Harbarth, L., Gößwein, E., Bodemer, D., & Schnaubert, L. (2024). (Over)Trusting AI Recommendations: How System and Person Variables Affect Dimensions of Complacency. *International Journal of Human-Computer Interaction*, 1–20. https://doi.org/10.1080/10447318.2023.2301250.

Harbour, J.L. (2008). Increasing Efficiency: A Process-Oriented Approach. *Performance Improvement Quarterly*, 6(4), 92–114. https://doi.org/10.1111/j.1937-8327.1993.tb00607.x.

He, Y., Li, S., Wen, X., & Xu, J. (2024). A Survey on Surface Defect Inspection Based on Generative Models in Manufacturing. *Applied Sciences*, 14(15), 6774. https://doi.org/10.3390/app14156774.

Heistracher, C., Casas, P., Stricker, S., Weißenfeld, A., Schall, D., & Kemnitz, J. (2023). *Should I Sample it or Not? Improving Quality Assurance Efficiency Through Smart Active Sampling.* Vienna University of Technology. https://doi.org/10.1109/iecon51785.2023.10311778.

Hopf, K., Müller, O., Shollo, A., & Thiess, T. (2023). Organizational Implementation of AI: Craft and Mechanical Work. *California Management Review*, 66(1), 23–47. https://doi.org/10.1177/00081256231197445.

Horowitz, M.C., & Kahn, L. (2024). Bending the Automation Bias Curve: A Study of Human and AI-Based Decision Making in National Security Contexts. *International Studies Quarterly*, 68(2). https://doi.org/10.1093/isq/sqae020.

Khinvasara, T., Ness, S., & Shankar, A. (2024). Leveraging AI for Enhanced Quality Assurance in Medical Device Manufacturing. *Asian Journal of Research in Computer Science*, 17(6), 13–35. https://doi.org/10.9734/ajrcos/2024/v17i6454.

Kim, D. (2023). Implementation of an AI Deep Learning-based Inspection System for Detecting Defective Vehicle Action Loaders. *The Transactions of the Korean Institute of Electrical Engineers*, 72(12), 1714–1721. https://doi.org/10.5370/kiee.2023.72.12.1714.

Knop, K., Ingaldi, M., & Smilek-Starczynowska, M. (2017). Reduction of Errors of the Conformity Assessment During the Visual Inspection of Electrical Devices. In *Lecture notes in mechanical engineering* (pp. 857–867). https://doi.org/10.1007/978-3-319-68619-6_83.

Kujawińska, A., & Vogt, K. (2015). Human Factors in Visual Quality Control. *Management and Production Engineering Review*, 6(2), 25–31. https://doi.org/10.1515/mper-2015-0013.

Kyriakopoulos, C.A., Gialampoukidis, I., Vrochidis, S., & Kompatsiaris, I. (2023). Adaptive Quality Diagnosis Framework for Production Lines in a Smart Manufacturing Environment. *Machines*, 11(4), 499. https://doi.org/10.3390/machines11040499.

Langner, R., & Eickhoff, S.B. (2013). Sustaining attention to simple tasks: a meta-analytic review of the neural mechanisms of vigilant attention. *Psychological bulletin*, 139(4), 870–900. https://doi.org/10.1037/a0030694.

Liu, P. (2023). Reflections on Automation Complacency. *International Journal of Human-Computer Interaction*, 1–17. https://doi.org/10.1080/10447318.2023.2265240.

Muniandi, N.B. (2024). Integration of AI to Increase Efficiency in Smart Semi-Conductor Wafer Inspection Systems. *Journal of Electrical Systems*, 20(9s), 3009–3016. https://doi.org/10.52783/jes.5855.

Peker, K., Yıldız, M., Turgay, S., İleri, G., & Teknolojileri, D. (2023). Improving Quality Inspections with Image Analysis and Artificial Intelligence. *Accounting and Corporate Management*, 5(11). https://doi.org/10.23977/acccm.2023.051120.

Ross, P., & Spates, K. (2020). Considering the Safety and Quality of Artificial Intelligence in Health Care. *The Joint Commission Journal on Quality and Patient Safety*, 46(10), 596–599. https://doi.org/10.1016/j.jcjq.2020.08.002.

Sele, D., & Chugunova, M. (2024). Putting a human in the loop: Increasing uptake, but decreasing accuracy of automated decision-making. *PLoS ONE*, 19(2), e0298037. https://doi.org/10.1371/journal.pone.0298037.

Smartening up with artificial intelligence (AI)—*What's in it for Germany and its industrial sector? (2017). McKinsey & Company.* https://www.mckinsey.com/~/media/mckinsey/industries/semiconductors/our%20insights/smartening%20up%20with%20artificial%20intelligence/smartening-up-with-artificial-intelligence.ashx.

TIC Council. (2024). *Trustworthiness of Artificial Intelligence Recommendations of the TIC sector.* https://www.tic-council.org/application/files/9117/1326/9285/White_Paper_on_Trustworthiness_of_Artificial_Intelligence.pdf.

Wadhwani, P., & Ambekar, A. (2024). *Testing, Inspection and Certification (TIC) services Market size—by service (Testing, inspection, certification), by type (In-house, outsource), by application, Forecast 2024-2032.* In Global Market Insights Inc. https://www.gminsights.com/industry-analysis/testing-inspection-certification-tic-market.

Yuen, N.H.Y., Tang, F., & Li, C.H. (2022). Adoption of Industry 4.0 in the TIC Industry: Systematic Review. *2022 IEEE International Conference on Industrial Engineering and Engineering Management (IEEM)*, 8, 1169–1176. https://doi.org/10.1109/ieem55944.2022.9989827.

# 24. Brushstrokes of Innovation

## *AI's Palette in the Paints and Coatings Industry*

Pratik N. Shah

In a world breaking into digital innovation, there is a force of change across all industries with artificial intelligence. While most people think of AI in high-tech fields, application in paints and coatings is finding its way into change-making and influencing how firms develop products, interact with customers, and operate (Holmes, 2020; Mungoli, 2023). Historically positioned in manufacturing and chemistry, this industry now sees AI as a tool that can augment creativity, precision, and sustainability. Applying AI enables a company to create completely customized, interactive experiences that virtually animate customers' color choices in real time, eventually increasing confidence and satisfaction with final selections (Ramachandran, 2024).

The capabilities of artificial intelligence that extend beyond the customer journey enhancement processes include facilitating production, enhancing supply chain management, and sustainable practices by optimizing resource utilization and reducing the environmental impact of the firm's activities (Lakshmi V, 2023; Park et al., 2023). Advanced articulations of machine learning techniques, augmented reality, and data-driven materials science create a new level of precision and efficiency in paint formulation and quality control in artificial intelligence through which companies can innovate to levels unprecedented in 2024 (Shukla, 2024).

Here we will analyze artificial intelligence's promise, challenges, and practical applications in paints and coatings to spot how such a transformation may alter customer experience and industry standards. We, therefore, consider, in our search for the potential AI-driven tools that might offer to personalize and optimize every stage of the paint lifecycle, the challenges that companies face while embracing AI, from concerns with data privacy to the high cost of implementing such practice (Ngozwana, 2018; Piduru, 2023). Armed with such technology capabilities through artificial intelligence, it is no wonder that the exciting trend in paints and coatings can now be well appreciated as they edge forward.

With industries now beginning to realize AI's full potential, the paints and coatings sector is slowly uncovering specific applications that will revolutionize processes and customer engagement to reposition product innovation.

## *Promises of AI in the Paints and Coatings Industry*

AI offers transformative potential across the paints and coatings industry by improving customer engagement, speeding up research and development, increasing operational efficiency, enhancing sustainability, and optimizing supply chain management. Below are some of the key areas where AI is making significant strides.

1. **Enhanced Customer Experience**
   - **Personalized Color Recommendations**

AI-based color recommendation tools use algorithms that enable them to scan and analyze the trends and preferences of customers. They would, therefore, leverage large language models, meaning that for any client's style and emotional needs, they would have suitable paint color recommendations accordingly. AI considers a comprehensive database of colors and customer histories to deliver personalized recommendations; therefore, customers are satisfied and successful in making purchases (Shukla, 2024). Personalization also increases brand loyalty and drives repeat sales as customers feel their tastes are acknowledged.

- **Visual Simulation and Augmented Reality Integration**

With AR, customers can view colors in their space before making a final decision. By utilizing AI-enhanced AR, consumers can select a hue and virtually apply it to walls to see what it would look like at different times of day under different lighting conditions. With this technology, the fantasy of desire meets reality to provide greater customer confidence and minimize post-purchase dissatisfaction (Ramachandran, 2024). The choice of paint can now be an experience, as the customer can test shades without creating a mess or paying for samples.

2. **Streamlined Product Development**
   - **Accelerated R&D with Machine Learning**

It is revolutionizing how R&D in paints and coatings is done by simulating chemical reactions and predicting how the new formulation will behave. Properties such as durability, color stability, and adhesion are estimated without exhaustive testing through machine learning. Fast-track formulation development accelerates time-to-market for new products, thereby hastening innovation that aligns with customer requirements and market trends. This further enhances the research stage as companies can now target the most essential features of the products, paints, and coatings to be soundly produced and environmentally compliant (Mungoli, 2023).

- **Predictive Quality Control**

AI quality control systems monitor temperature, humidity, and viscosity during production to maintain batch consistency. With real-time tracking, issues are discovered quickly enough for companies to adjust the processes before flaws are committed. Predictive quality control reduces waste and boosts efficiency, resulting in better product quality—making each batch meet the brand's standards. Significant benefits of mass production include catching and rectifying issues at an early production stage (Lakshmi V, 2023).

3. **Operational Efficiency & Predictive Maintenance**
   - **Automated Quality Control**

Artificially intelligent vision systems have computer algorithms that can detect product inaccuracies at the highest accuracy. Bubbles, texture misalignment, or even labeling mistakes become discoverable by AI vision systems that human inspectors may have overlooked. With fewer defects, brand consistency, and reliability, automated quality control must be considered. Early catching of defects cuts costly rework and waste in materials and labor (Shukla, 2024).

   - **Predictive Machinery Maintenance**

These models monitor key equipment metrics, such as temperature, pressure, and vibration, to enable predictive maintenance and minimize downtime. For example, AI can analyze historical data on machinery to predict their maintenance requirements and avoid halting production lines. This proactive maintenance reduces repair costs, extends equipment life, and minimizes production interruptions, which is crucial in large-scale manufacturing because even a tiny delay in the manufacturing process is very cost-intensive (Lakshmi, 2023).

4. **Sustainability & Waste Reduction**
   - **Optimize Resource Utilization**

AI optimizes raw material consumption and use and finds better ways to be efficient. This results in minimal waste and improvements in formulations to reduce manufacturing environmental impacts. AI tools also motivate energy-efficient manufacturing by monitoring electricity usage and suggesting adjustments based on savings (Bouchrika, 2022). Utilizing AI in resource optimization to meet regulations and consumer demands is essential in a sustainable industry.

   - **Eco-Friendly Product Design**

AI also promotes the development of legal eco-friendly paints that reduce the impact of environmental degradation. It provides low–VOC ingredients that ensure safe air quality inside the space and meet the requirements for green products. In this manner, by promoting the development of green products, AI helps companies compete since environmental concerns are constantly growing (Lakshmi V, 2023).

5. **Improved Supply Chain and Inventory Management**
   - **Demand Forecasting**

AI-demand-based forecasting is the process wherein previous sales, economic trends, and seasonal patterns are considered to forecast future product demands. Better inventory management helps companies avoid overproduction, potential stockouts, and excess stock. The production schedule will align with market demand to adapt quickly to customer preferences and seasonal changes (Park et al., 2023).

   - **Logistics Optimization**

AI improves logistics by developing optimal routes for transporting commodities, minimizing costs and environmental impacts. AI tools optimize routes and schedules with analytical data regarding fuel usage, traffic, and delivery periods. Efficient logistics improve organizational efficiency by improving timely deliveries and removing customer irritations (Yang & Shen, 2022).

These applications show the vast potential of AI in the paints and coatings industry. AI can transform everything from customer engagement through product innovation to operational efficiency and environmental sustainability. Integrating AI will help companies enhance decision-making and cost-cutting and improve customer experience, bringing loyalty and growth (Shukla, 2024).

Together, these advancements form a new frontier of customer satisfaction, process efficiency, and innovation for the paints and coatings industry. While AI presents many advantages and efficiencies, realizing such potential comes with challenges. Companies planning to exploit AI's value must contend with various critical problems. The following section examines the hurdles—data privacy and resource investments—that paint and coatings companies must overcome to make AI a reality.

## *Perils and Challenges of AI in the Paints and Coatings Industry*

1. **Data Privacy and Security Concerns**

- Customer Data Sensitivity: AI personalization tools rely on a large amount of customer data and raise questions as to the privacy of such data. Poor handling can easily jeopardize customer trust and attract regulatory obstacles (Ngozwana, 2018).
- Cybersecurity Risks: AI systems integrated into manufacturing are susceptible to cyberattacks. Such unauthorized access may break up production and harm the company's reputation (Ngozwana, 2018)

2. **Quality of Data and Model Accuracy**

- Dependence on High-Quality Data: The reliability of AI is strongly dependent on high-quality, consistent data, which is difficult to obtain from legacy systems. Inconsistent data would plummet the accuracy and effectiveness of AI insights (Park et al., 2023).
- Bias in Data: AI models are likely to reflect the biases that exist in data they are trained on, which would skew up recommendations and forecasts that hamper customer experience (Park et al., 2023).

3. **High Initial Investment and ROI Uncertainty**

- Implementation Costs: Implementing AI solutions requires substantial investment in hardware, software, and skilled labor, which can be prohibitive for smaller companies. Furthermore, predicting the ROI of AI applications in niche sectors can be challenging (Piduru, 2023).
- Maintenance and Upkeep: Artificial Intelligence requires upgradation and maintenance at regular intervals, leading to a secondary cost of operations requiring technical experts (Piduru, 2023).

Many pioneering brands in paints and coatings have already widely applied AI technologies with great success despite these barriers. The companies that break down the barriers and integrate AI into their processes demonstrate that technologies make a difference in real-life applications and pave the way toward changes throughout the industry.

## *AI Paint Revolution: Leading Brands Shaping Tomorrow's Colors*

Many companies in the paints and coatings industry are already adopting AI, setting the stage for widespread transformation.

Industry leaders embrace AI to evolve, making significant adjustments in engagement with the customer, product evolutions, and sustainable development. Applications such as predictive maintenance, virtual visualization, or optimized supply chains prove the benefits of artificial intelligence. Innovative companies describe how every stage of the life cycle of paint can be optimized through AI-based solutions, which then set new market standards. Such a positive indicator gives an excellent vision for a better future for this industry. As artificial intelligence becomes more sophisticated, the paint industry is well-positioned to explore more advanced applications to increase efficiency and sustainability and broaden customer satisfaction. The section below outlines several emerging AI applications that may further affect the industry's future landscape.

## *Future Transformative AI Applications in the Paints and Coatings Industry*

1. **AI-Powered Color Visualization for Enhanced Customer Engagement**

AI-powered visualization tools enable the customer to have a virtual "try before you buy" in that they preview the paint color in their personal spaces. Coupled with advanced computer vision and augmented reality, the color applications come alive by responding to lighting conditions and dimensions of a specific room. This gives the customers assurance and information, enabling them to decide on the best suitable choice, eliminating uncertainty in selections of paint color (Lakshmi V, 2023; Park et al., 2023).

2. **Predictive Formulation for Streamlined Product Development**

In research and development, AI models are critical in simulating chemical reactions and predicting properties such as durability, color stability, and adhesion. This capability allows for increased speed of development, reducing the need for actual laboratory testing and making the process easier to get new products onto the market (Challener, 2021).

3. **Automated Defect Detection in Production for Improved Quality Control**

Since AI-based quality control systems can automatically flag errors by detecting bubbles, inconsistencies, and color differences during manufacturing processes, increasing product reliability and reducing waste are achieved. This leads to cost savings and reductions in rework (Shukla, 2024).

4. **Efficient Inventory and Demand Management**

AI-based demand forecasting tools help paint companies manage their inventory levels based on historical sales, seasonal patterns, and market conditions. This usually leads to more efficient use of resources, fewer instances of stockouts or excess, and better overall sustainability (Park et al., 2023).

5. **Customer-centric chatbots and Virtual Assistants for a Personalized Shopping Experience**

AI-powered chatbots and virtual assistants elevate customer interactions through personalized support. Virtual assistance can help a consumer choose a color painting, provide technical know-how on method application, and propose complementing shades according to a user's liking. Such a personalized shopping experience leads virtual assistants to enhance customer satisfaction, easing the choice-making process and making it more fun for the customers (Lakshmi, 2023).

As the paints and coatings industry becomes more integrated with artificial intelligence, the current changes are a teaser of what is in store. Understanding the advantages and disadvantages this form of artificial intelligence presents gives companies a chance to cope with change through better technology, sustainability, and high customer satisfaction. In short, this field promises much for the future of artificial intelligence but indeed entails much responsibility. Integrating artificial intelligence technologies into the paints and coatings industry heralds not just the beginning but merely the first phase of a much broader transformation process. Applications like AI-driven color visualization and quality control automation point toward artificial intelligence's potential to significantly reshape operational processes and enhance the broad customer experience. With continued adaptation and refinement by enterprises, the vision for a future defined by increased precision, sustainability, and innovation beckons toward an artificial intelligence future. However, as the application of artificial intelligence grows, the need to balance it with ethical considerations, data protection, and equal distribution also increases. How these corporations deal with these benefits and burdens will define the actual effect of artificial intelligence within that sector: more responsive, efficient, and customer-oriented.

## *Conclusion*

Artificial intelligence is transforming the paints and coatings industry to better engage customers, bring products to market faster, smooth out operations, and support sustainability. Industry leaders now set new standards, which others will follow through predictive maintenance, virtual visualization, and data-driven innovations. However, such fast technological adoption has downsides: data privacy can become an issue, quality data are needed, and initial investments are enormous, especially for smaller enterprises.

To maximize AI's potential, given its ethical practices, data security, and adaptability. Companies will also require a culture-oriented approach to AI adoption, wherein employee engagement and responsible implementation are ensured.

As AI technology advances, it will hopefully lead the paints and coatings industry to new levels of customer-centricity, sustainability, and greater efficiency. This era of innovation driven by AI not only promises to protect these competitive advantages but also introduces a new paradigm of quality and customer satisfaction—an exciting new canvas created by the digital brushstrokes of artificial intelligence.

## References

Benjamin Moore. (2020). *Benjamin Moore Launches Color Portfolio App With Integrated ColorReader Device* |. Coatings World. https://www.coatingsworld.com/issues/2020-03-01/view_breaking-news/benjamin-moore-launches-color-portfolio-app-with-integrated-colorreader-device/.

Challener, C. (2021). Facilitating Coatings Product Development with Artificial Intelligence. *CoatingsTech, 18*(8). https://www.paint.org/coatingstech-magazine/articles/facilitating-coatings-product-development-with-artificial-intelligence/.

CoatingAI AG. (2021). *CoatingAI AG Harnesses Advanced Computer Vision, Machine Learning For Coatings Industry | Coatings World.* Coatings World. https://www.coatingsworld.com/contents/view_breaking-news/2021-08-20/coatingai-ag-harnesses-advanced-computer-vision-machine-learning-for-coatings-industry/.

Gibbons, S. (2024, May 6). *Focus on smart coatings: LLM fuels AI innovations in the coatings industry—Polymers Paint Colour Journal.* Polymers Paint Colour Journal. https://www.polymerspaintcolourjournal.com/focus-on-smart-coatings-llm-fuels-ai-innovations-in-the-coatings-industry/.

Global Chemical News. (2024, October 5). *AI's Transformative Impact on Paints and Coatings Industry—Global Chemical News.* Global Chemical News. https://globalchemicalnews.com/ais-transformative-impact-on-paints-and-coatings-industry/.

Holmes, J. (2020, January 14). *Digital Transformation Poised to Revolutionize Paints and Coatings Industry.* Frost & Sullivan. https://www.frost.com/news/digital-transformation-poised-to-revolutionize-paints-and-coatings-industry/.

Lakshmi V, V. (2023). Psychological Effects of Colour. *Journal of Biotechnology & Bioinformatics Research, 5*(2), 1–2. https://doi.org/10.47363/JBBR/2023(5)157.

Mungoli, N. (2023). Revolutionizing Industries: The Impact of Artificial Intelligence Technologies. *Journal of Electrical Electronics Engineering, 2*(3), 206–210. https://doi.org/10.33140/JEEE.02.03.03.

Ngozwana, N. (2018). Ethical Dilemmas in Qualitative Research Methodology: Researcher's Reflections. *International Journal of Educational Methodology, 4*, 19–28. https://doi.org/10.12973/ijem.4.1.19.

Park, H., Jin, S., & Cho, J. (2023). An Artificial Intelligence-Based Approach for Classification of Personal Color. *ICNC-FSKD 2023-2023 19th International Conference on Natural Computation, Fuzzy Systems and Knowledge Discovery*, 1–6. https://doi.org/10.1109/ICNC-FSKD59587.2023.10281004.

Pianoforte, K., & Kisner, K. (2023). *Painting The Lab Of The Future: How Data And AI Are Transforming Coatings R&D | Coatings World.* Coatings World. https://www.coatingsworld.com/buyersguide/profile/albert-invent/view_painting-the-lab-of-the-future-how-data-and-ai-are-transforming-coatings-rd/.

Piduru, B.R. (2023). The Role of Artificial Intelligence in Content Personalization: Transforming User Experience in the Digital Age. *Journal of Artificial Intelligence & Cloud Computing*, 1–5. https://doi.org/10.47363/JAICC/2023(2)193.

Ramachandran, K. (2024). Exploring the Role of Artificial Intelligence in Personalized Payment Recommendations. *International Journal of Finance, 9*(3), 23–31. https://doi.org/10.47941/IJF.1913.

Shah, P.N. (2024). *Impact of Multi-Modal LLM Tools on User Experience in Dynamic Paint Shade Creation.*

Shukla, Mr. M. (2024). The Impact of AI on Improving the Efficiency and Accuracy of Managerial Decisions. *International Journal For Science Technology And Engineering, 12*(7), 830–842. https://doi.org/10.22214/IJRASET.2024.63652.

Slight Machine Inc. (2019, May 13). *Asian Paints Deploys Sight Machine for Digital Transformation of Manufacturing.* PR Newswire. https://www.prnewswire.com/news-releases/asian-paints-deploys-sight-machine-for-digital-transformation-of-manufacturing-300848615.html.

# 25. AI for Enhancing Efficiency and Sustainability in Agri-Food and Nutrition Systems

LAKSHMI R. PILLAI *and* SEETHA ANITHA

Global demographic projections indicate that the world population will reach 10 billion by 2050 (Ikram et al., 2024), necessitating a 70 percent increase in agricultural and food production (Ayed & Hanana, 2021) (FAO, 2021). Addressing this unprecedented demand poses a significant challenge for the agri-food industry and is central to mitigating potential food scarcity. Moreover, the Agri-industry is liable for a huge quantity of global emissions. This points to the fact that while ensuring the increase in production, efficient and Sustainable agrifood systems are equally critical in redefining the relationship between humanity and nature in the changed context (McGreevy et al., 2022).

This research aims to analyze the integration and implementation of Artificial Intelligence (AI) across the agricultural value chain. The primary objective is to discuss the various applications of AI in various stages of the Agri-food system value chain. The piece will also discuss the implementation challenges and mitigation frameworks.

## *Thematic Analysis of AI Applications Across the Agricultural Value Chain*

Machine learning and deep learning technologies have exhibited significant advantages in terms of accuracy, strength, and adaptability for monitoring progress in agri-food systems. Effective crop supervision and diagnostics aimed at enhancing growth environment and productivity can be achieved through the construction of a "monitoring and warning system." These systems enable instantaneous, precise, intelligent, and automated results for the "remote dynamic management" of crop growth (Chen et al., 2023). Additionally, remote sensing technologies, including satellite imagery and meteorological information, can effectively assess land-use, estimate acreage, and track crop health in real time (Javaid et al., 2023; Subeesh & Mehta, 2021).

### AI-Driven Precision Agriculture

Precision agriculture is a site-specific crop management approach that uses ML models to optimize inputs and practices. Farmers can benefit from assistance with crop selection, planting, irrigation, pest control, crop rotation, harvesting, and yield estimation using data from ML models (Javaid et al., 2022). Additionally, AI can contribute to the conservation of water and reduce oil consumption (Finger et al., 2019; Mohr & Kühl, 2021). By enabling precision agriculture, AI can help reduce farmers' losses, enhance yields, and contribute to environmental sustainability.

## Predictive Analytics Using AI

Predictive analytics use historical and current data to predict outcomes and trends. In the realm of agricultural production, by utilizing predictive analytics, farmers can make more informed decisions, optimize farm and crop management systems, and ensure the consistency of crop quality and supply (Javaid et al., 2023). AI can forecast how abnormal weather affects the production and quality of produce. Additionally, AI can recommend alternate practices to mitigate such losses (Elbehri and Chestnov, 2021).

## AI in Weed Control

Weeds are unwanted plants that grow alongside cultivated crops, competing for limited resources and sometimes drastically decreasing crop yields. AI is trained using information and pictures of weeds to eliminate them as targets and crops as non-targets. This helps in controlling the weeds and its eventual elimination from the field (Aubert et al., 2012; Partel et al., 2019; Mohr & Kühl, 2021). This approach reduces manual labor and increases productivity while eradicating weeds autonomously.

## AI in Harvesting and Pruning

Several AI technologies are employed in harvesting and pruning. For example, smart spectrometers, which estimate substance concentration and composition, can be integrated into harvesting vehicles to assess grape quality in the field by predicting sugar and acid content (Krause et al., 2021). Complexity-driven Convolutional Neural Networks (CNNs) are used for pruning tasks, while selective harvesting robotics improve farm management (Zawish et al., 2022). Pruning and harvesting manipulators, designed for tasks such as pruning grapevines and apple trees and harvesting crops like apples, tomatoes, strawberries, sweet peppers, and iceberg lettuce, can reduce labor-intensive and costly manual processes (Kootstra et al., 2021). Robotics technology has also been deployed for harvesting fruits and vegetables and trimming lettuce (Javaid et al., 2023). The robots used for sowing, pruning, and harvesting have been shown to reduce operational time by 49 percent compared to traditional methods (Mana et al., 2024). Valente et al.(2020) proved that using convolutional neural networks identification and counting of spinach plants with an accuracy of 67.5 percent. This was also demonstrated in corn (Kitano et al., 2019) and rice paddies (Bai et al., 2023).

## AI-Driven Tools for Enhancing Agricultural Production and Productivity

Alibaba's AI platform "cloud-based agricultural intelligence platform" helps Chinese melon farmers optimize harvest timing by analyzing environmental factors, boosting income through improved fruit quality (Elbehri and Chestnov, 2021).

"Self-driving tractors," which apply AI and precision farming, address the issue of labor shortages and offer greater precision, producing fewer errors (Javaid et al., 2023).

"Plantix," a mobile app, uses image recognition to diagnose plant damage, provide crop management advice, and connect farmers through weather alerts and community support (Elbehri and Chestnov, 2021).

"PlantVillage Nuru," an AI smartphone app, helps Kenyan farmers diagnose cassava diseases offline with nearly double the accuracy of traditional extension workers (Elbehri and Chestnov, 2021; CGIAR Platform for Big Data in Agriculture, n.d.).

## *AI Applications in Processing and Packaging Component of Value Chain Significance of Processing in Agri-Food Value Chain*

Food processing is preserving foods by slowing or stopping decay. AI applications in the processing stage help manage agriculture's growing complexity by addressing labor shortages and resource constraints.

### AI Applications in High-Value Spice Processing

Researchers have identified and documented the use of a more reliable, non-contact, non-destructive, and non-chemical AI intervention utilizing "computer vision" and "image processing" to estimate the quality aspects of black pepper (S. Sharma et al., 2021; Goswami & Jain, 2013).

### AI Applications, Including E-Tongues, in the Beverages Industry

In the beverages business a number of AI applications, including machine learning, robotics, biometrics, etc., are not labor-intensive but at the same time economical and accurate (S. Sharma et al., 2021). AI-supported e-tongues or electronic tongues, use machine learning models and can distinguish complex liquids. They are deployed for fast and portable "fingerprinting of complex liquids," helping to control the quality of elements in packaged food (Muszynski et al., 2022; Kumar et al., 2023).

### AI Applications of E-Nose Technology in Food Preservation and Flavor Recognition

Biodeterioration affects food through the volatile compounds released by microorganisms infesting the food. E-nose technology effectively detects bacteria, enhancing food preservation using multi-layer perceptron networks (Hsieh & Yao, 2018). Artificial neural network (ANN)-based pattern recognition techniques are used for the discrimination and classification of electronic nose response data for different flavors of tea and spice (Gardner, Craven, Dow, & Hines, 1998; Hsieh & Yao, 2018; S. Sharma et al., 2021).

### AI-Driven Tools Deployed in Processing and Packaging

"Hypertaste," developed by IBM, is an AI-assisted E-tongue (Muszynski et al., 2022; Kumar et al., 2023).

Numerous beer samples were examined using "RoboBEER," an automated pouring device, and evaluated 15 color combinations and other characteristics related to foam creation (Funes, Allouche, Beltrán, & Jiménez, 2015).

### AI-Driven Innovations in Agricultural Distribution and Storage Systems

In the distribution phase of the agricultural supply chain, AI can assist in inventory management, logistics transportation, storage, consumer analytics, as well as product tracing, and safety assurance. An efficient distribution and storage system helps reduce product prices for consumers, increases farmers' income, and ensures food remains fresh and healthy (Vilas-Boas et al., 2023). AI, along with other technological tools like IoT, can be used to monitor and track products. For example, AI is a key driver of "Logistics 4.0," which is revolutionizing supply management through enhanced decision-making, route planning, inventory management, and real-time supply chain visibility (Vanoy, 2023).

## *AI Applications in the Consumption Component of the Value Chain*

AI is used in the food industry for quality control and grading of food (Namkhah et al., 2023). It can also assist in estimating calorific value and ensuring food safety across the production line by using automated sensors to monitor and ensure compliance with food safety laws and regulations (Nicholas-Okpara et al., 2021). Consumers are increasingly aware of food safety issues, a trend observed globally (S. Sharma et al., 2021). AI utilizes computational methodologies and smart solutions to address challenges faced by the food industry by enabling food security traceability (King et al., 2017).AI can also support personalized nutrition requirements for consumers. Papastratis et al. (2024) introduced an AI-based diet recommendation system that blends large language models (LLMs) with deep generative network.

## *Challenges in AI Implementation: Analysis of Technical Constraints, Ethical Implications, and Mitigation Strategies*

AI has immense potential to transform the agri-food sector towards sustainable and efficient practices (Ayed & Hanana, 2021). However, identifying challenges and implementing appropriate measures is critical.

Technological challenges: Farmers, often lack the time, digital skills, and technical knowledge to independently explore and implement AI technologies. A weak network infrastructure also hinders the integration of AI with other technologies (Chen et al., 2023). Development of simple, integrated, and localized AI solutions and employing multiple data-centric techniques, data acquisition, and integration processes can help mitigate the issues (Chen et al., 2023).

Ethical and social concerns: Privacy breaches, biases, malicious data attacks, lack of transparency, improper use of data, related legal issues, and job displacements in traditional farming are the ethical and social concerns. Mitigation strategies include ethical AI practices, designing user-friendly AI models, reskilling and training farmers, and using privacy-preserving and transparent data governance policies, to foster trust and adoption (McGreevy et al., 2022).

## *Findings and Discussions*

The potential of AI to revolutionize agriculture and allied sectors is immense. AI provides data insights, predictive analysis, and optimized storage and processing of produce. It can drive sustainable growth by helping farmers adopt resource-efficient practices, aligning production with market demands to boost profitability while maintaining ecological balance. Key AI methodologies include machine learning, deep learning, neural networks, NLP, computer vision, robotics, and generative AI. These techniques are supported by a range of tools designed to transform food systems across value chains, though broader adoption is essential.

To ensure social equity, AI advancements must consider smallholder farmers, tailoring technologies to improve their economic and environmental outcomes. Ethical concerns necessitate policy interventions, while research is vital to refine error correction and scalability, addressing critical value chain challenges.

## *Recommendations*

- Balancing technology and sustainability: Balancing technological advancement with societal impacts and sustainability concerns is at the forefront of the AI use cases for Agriculture.
- Awareness Creation: Raising awareness among stakeholders in the food system is crucial for the optimal use of AI across the value chain of various commodities.
- Scaling: Technologies that have been successfully tested need to be scaled globally to achieve food system transformation. This process also involves financial considerations and private-sector engagement.
- Research Gap: More research is needed to encompass a broader range of crops and commodities to ensure comprehensive support for globally used commodities.

## *Conclusion*

The integration of AI into agri-food and nutrition systems offers immense potential to enhance efficiency, sustainability, and resilience, presenting a promising opportunity for the future. The study demonstrates that AI can streamline processes, outperform human assistance, and eliminate human interference with greater accuracy across various components of the supply chain, from preproduction to final consumption. However, to fully realize this potential, it is essential to address the technical, ethical, and social challenges and barriers associated with the advantages of AI.

## References


Ayed, R.B., & Hanana, M. (2021). Artificial intelligence to improve the food and agriculture sector. *Journal of Food Quality, 2021,* 1–7. https://doi.org/10.1155/2021/5584754.

Bai, X., Liu, P., Cao, Z., Lu, H., Xiong, H., Yang, A., Cai, Z., Wang, J., & Yao, J. (2023). Rice plant counting, locating, and sizing method based on high throughput UAV RGB images. *Plant Phenomics, 5,* Article 0020. https://doi.org/10.34133/plantphenomics.0020.

CGIAR Platform for Big Data in Agriculture. (n.d.). PlantVillage Nuru: Pest and disease monitoring using AI. Retrieved November 16, 2024, from https://bigdata.cgiar.org/digital-intervention/plantvillage-nuru-pest-and-disease-monitoring-using-ai/.

Chen, T., Lv, L., Wang, D., Zhang, J., Yang, Y., Zhao, Z., Wang, C., Guo, X., Chen, H., Wang, Q., Xu, Y., Zhang, Q., Du, B., Zhang, L., & Tao, D. (2023, May 3). Empowering agrifood system with artificial intelligence: A survey of the progress, challenges, and opportunities. *arXiv.org.* https://arxiv.org/abs/2305.01899.

Elbehri, A., & Chestnov, R. (Eds.). (2021). *Digital agriculture in action—Artificial intelligence for agriculture.* FAO & ITU. https://doi.org/10.4060/cb7142en.

Food and Agriculture Organization of the United Nations (FAO). (2021). *Food systems at risk: Transformative adaptations for long-term food security.* FAO. https://openknowledge.fao.org/server/api/core/bitstreams/36cb45bc-392c-41fb-97f1-90ca1f16ee7f/content.

Food and Agriculture Organization of the United Nations. (n.d.). *Basic facts about food preparation and processing.* FAO. https://www.fao.org/4/y5113e/y5113e04.htm.

Funes, E., Allouche, Y., Beltrán, G., & Jiménez, A. (2015). A review: Artificial neural networks as tool for control food industry process. *Journal of Sensor Technology, 5*(01), 28.

Gardner, J.W., Craven, M., Dow, C., & Hines, E.L. (2001). The prediction of wine quality from electronic nose data. *Sensors and Actuators B: Chemical, 69*(3), 304–312. https://doi.org/10.1016/S0925-4005(00)00594-4.

Gikunda, K. (2024, January 3). Harnessing artificial intelligence for sustainable agricultural development in Africa: Opportunities, challenges, and impact. *arXiv.org.* https://arxiv.org/abs/2401.06171.

Global e-Sustainability Initiative (GeSI). (2015). *SMARTer2030: ICT solutions for 21st century challenges* (Vol. 3). Brussels, Belgium: GeSI. https://www.gesi.org/wp-content/uploads/2024/10/SMARTer2030-ICT-solutions-for-21st-Century-Challenges.pdf.

Goswami, P.R., & Jain, K.R. (2013, November). Non-destructive quality evaluation in spice industry with specific reference to black pepper (*Piper nigrum* L.). In *2013 Nirma University International Conference on Engineering (NUiCONE)* (pp. 1–5). IEEE.

Hsieh, Y.C., & Yao, D.J. (2018). Intelligent gas-sensing systems and their applications. *Journal of Micromechanics and Microengineering, 28*(9), 093001.

Ikram, A., Mehmood, H., Arshad, M.T., Rasheed, A., Noreen, S., & Gnedeka, K.T. (2024). Applications of artificial intelligence (AI) in managing food quality and ensuring global food security. *CyTA—Journal of Food, 22*(1), 2393287. https://doi.org/10.1080/19476337.2024.2393287.

Javaid, M., Haleem, A., Khan, I.H., & Suman, R. (2022). Understanding the potential applications of artificial intelligence in agriculture sector. *Advanced Agrochem, 2*(1), 15–30. https://doi.org/10.1016/j.aac.2022.10.001.

King, T., Cole, M., Farber, J.M., Eisenbrand, G., Zabaras, D., Fox, E.M., & Hill, J.P. (2017). Food safety for food security: Relationship between global megatrends and developments in food safety. *Trends in Food Science & Technology, 68,* 160–175. https://doi.org/10.1016/j.tifs.2017.08.014.

Kitano, B., Mendes, C., Geus, A., Oliveira, H., & Souza, J. (2019). Corn plant counting using deep learning and UAV images. *IEEE Geoscience and Remote Sensing Letters.* https://doi.org/10.1109/LGRS.2019.2930549.

Kootstra, G., Wang, X., Blok, P.M., et al. (2021). Selective harvesting robotics: Current research, trends, and future directions. *Current Robotics Reports, 2,* 95–104. https://doi.org/10.1007/s43154-020-00034-1.

Krause, J., Grüger, H., Gebauer, L., Zheng, X., Knobbe, J., Pügner, T., Kicherer, A., Gruna, R., Längle, T., & Beyerer, J. (2021). SmartSpectrometer—Embedded optical spectroscopy for applications in agriculture and industry. *Sensors, 21*(13), 4476. https://doi.org/10.3390/s21134476.

Kumar, M.S., Jhariya, M.K., & Banerjee, A. (2023). *Digitalization and smart technologies in agriculture* (pp. 92). Springer Nature. https://www.google.co.in/books/edition/Digitalization_And_Smart_Technologies_In/5Me-EAAAQBAJ?hl=en&gbpv=1&pg=PA83.

Mana, A.A., Allouhi, A., Hamrani, A., Rehman, S., el Jamaoui, I., Jayachandran, K. (2024). Sustainable AI-based production agriculture: Exploring AI applications and implications in agricultural practices. *Smart Agricultural Technology, 7*, 100416. https://doi.org/10.1016/j.atech.2024.100416.

Markets & Markets. (2023). *Artificial intelligence in agriculture market by technology (machine learning, computer vision, and predictive analytics), offering (software, hardware, AI as a service and services), application and geography—global forecast to 2028*. Markets & Markets. Retrieved from https://www.marketsandmarkets.com/Market-Reports/ai-in-agriculture-market-159957009.html.

McGreevy, S.R., Rupprecht, C.D.D., Niles, D., Wiek, A., Carolan, M., Kallis, G., Kantamaturapoj, K., Mangnus, A., Jehlička, P., Taherzadeh, O., Sahakian, M., Chabay, I., Colby, A., Vivero-Pol, J., Chaudhuri, R., Spiegelberg, M., Kobayashi, M., Balázs, B., Tsuchiya, K., ... Tachikawa, M. (2022). Sustainable agrifood systems for a post-growth world. *Nature Sustainability, 5*(12), 1011–1017. https://doi.org/10.1038/s41893-022-00933-5.

Mohr, S., & Kühl, R. (2021). Acceptance of artificial intelligence in German agriculture: An application of the technology acceptance model and the theory of planned behavior. *Precision Agriculture, 22*(6), 1816–1844. https://doi.org/10.1007/s11119-021-09814-x.

Muszynski, M., Gabrieli, G., Zimmerli, L., Temiz, Y., Heller, R., & Cox, A. (2022). Live demonstration: An AI-assisted e-tongue for fast and portable fingerprinting of liquids. *2022 IEEE Sensors*, 1. https://doi.org/10.1109/SENSORS52175.2022.9967223.

Namkhah, Z., Fatemi, S.F., Mansoori, A., Nosratabadi, S., & Sobhani, S.R. (2023). Advancing sustainability in the food and nutrition system: A review of artificial intelligence applications. *Frontiers in Nutrition, 10*, 1295241. https://doi.org/10.3389/fnut.2023.1295241.

Nicholas-Okpara, V.A.N., Ubaka, A.J., Adegboyega, M.O., Utazi, I.A., Chibudike, C.E., & Chibudike, H.O. (2021). Advancements in food technology using artificial intelligence-deep learning. *Current Journal of Applied Science and Technology, 40*(18), 1–9. https://doi.org/10.9734/cjast/2021/v40i1831374.

Papastratis, I., Konstantinidis, D., Daras, P., et al. (2024). AI nutrition recommendation using a deep generative model and ChatGPT. *Scientific Reports, 14*, 14620. https://doi.org/10.1038/s41598-024-65438-x.

Pérez-Escamilla, R. (2017). Food security and the 2015–2030 Sustainable Development Goals: From human to planetary health. *Current Developments in Nutrition, 1*(7), e000513. https://doi.org/10.3945/cdn.117.000513.

Peters, D.P.C., Rivers, A., Hatfield, J.L., Lemay, D.G., Liu, S., & Basso, B. (2020). Harnessing AI to transform agriculture and inform agricultural research. *IT Professional, 22*(3), 16–21. https://doi.org/10.1109/mitp.2020.2986124.

Ruben, R., Cavatassi, R., Lipper, L., et al. (2021). Towards food systems transformation—five paradigm shifts for healthy, inclusive, and sustainable food systems. *Food Security, 13*(6), 1423–1430. https://doi.org/10.1007/s12571-021-01221-4.

Sahni, V., Srivastava, S., & Khan, R. (2021). Modelling techniques to improve the quality of food using artificial intelligence. *Journal of Food Quality, 2021*, 1–10. https://doi.org/10.1155/2021/2140010.

Schneider, K.R., Fanzo, J., Haddad, L., et al. (2023). The state of food systems worldwide in the countdown to 2030. *Nature Food, 4*(10), 1090–1110. https://doi.org/10.1038/s43016-023-00885-9.

Sharma, S., Gahlawat, V.K., Rahul, K., Mor, R.S., & Malik, M. (2021). Sustainable innovations in the food industry through artificial intelligence and big data analytics. *Logistics, 5*(4), 66. https://doi.org/10.3390/logistics5040066.

Sheikh, H., Prins, C., & Schrijvers, E. (2023). Artificial intelligence: Definition and background. In *Research for policy* (pp. 15–41). https://doi.org/10.1007/978-3-031-21448-6_2.

Springmann, M., Clark, M., Wiebe, K., Bodirsky, B.L., Lassaletta, L., De Vries, W., Vermeulen, S.J., Herrero, M., Carlson, K.M., Jonell, M., Troell, M., DeClerck, F., Gordon, L.J., Zurayk, R., Scarborough, P., Rayner, M., Loken, B., Fanzo, J., Godfray, H.C., ... Willett, W. (2018). Options for keeping the food system within environmental limits. *Nature, 562*(7728), 519–525. https://doi.org/10.1038/s41586-018-0594-0.

Subeesh, A., & Mehta, C. (2021). Automation and digitization of agriculture using artificial intelligence and internet of things. *Artificial Intelligence in Agriculture, 5*, 278–291. https://doi.org/10.1016/j.aiia.2021.11.004.

Tsoumas, I., Giannarakis, G., Sitokonstantinou, V., Koukos, A., Loka, D., Bartsotas, N., Kontoes, C., & Athanasiadis, I. (2023). Evaluating digital agriculture recommendations with causal inference. *Proceedings of the AAAI Conference on Artificial Intelligence, 37*(12), 14514–14522. https://doi.org/10.1609/aaai.v37i12.26697.

Tutundjian, S., Clarke, M., Egal, F., Candotti, S., Schmitter, P., & Lovins, L. (2021). Future food systems: Challenges and consequences of the current food system. In *Future Food Systems* (pp. 1–16). Springer. https://doi.org/10.1007/978-3-030-32811-5_43-1.

Valente, J., Sari, B., Kooistra, L., et al. (2020). Automated crop plant counting from very high-resolution aerial imagery. *Precision Agriculture, 21*(6), 1366–1384. https://doi.org/10.1007/s11119-020-09725-3.

Vanoy, R.J.A. (2023). Logistics 4.0: Exploring artificial intelligence trends in efficient supply chain management. *Data and Metadata, 2*, 145. https://doi.org/10.56294/dm2023145.

Vilas-Boas, J.L., Rodrigues, J.J., & Alberti, A.M. (2023). Convergence of distributed ledger technologies with digital twins, IoT, and AI for fresh food logistics: Challenges and opportunities. *Journal of Industrial Information Integration, 31,* 100393. https://doi.org/10.1016/j.jii.2022.100393.

Willett, W., Rockström, J., Loken, B., Springmann, M., Lang, T., Vermeulen, S., et al. (2019). Food in the Anthropocene: The EAT-Lancet commission on healthy diets from sustainable food systems. *Lancet, 393*(10170), 447–492. https://doi.org/10.1016/S0140-6736(18)31788-4.

Wood, A., Queiroz, C., Deutsch, L., et al. (2023). Reframing the local–global food systems debate through a resilience lens. *Nature Food, 4*(1), 22–29. https://doi.org/10.1038/s43016-022-00662-0.

Zawish, M., Davy, S., & Abraham, L. (2022). Complexity-driven CNN compression for resource-constrained edge AI. *arXiv Preprint.* arXiv:2208.12816.

# 26. AI in the Oil and Gas Industry

## *Revolutionizing Operational Efficiency*

Saroja Manicka Nagarajan, Jagadish Gona, *and* Kannan Santharaman

The oil and natural gas industry is playing a vital role in the global economy as it directly affects many areas of a country including transportation, electricity, manufacturing, etc. The nature of work involved in any oil and natural gas industry can be categorized into three fields, namely, upstream, midstream and downstream. There are many companies in each of these fields, as well as integrated oil companies that work across all the three areas. The first step of the oil and natural gas industry is to explore the availability of the resource and the drilling of it. This segment is referred to as upstream. The second step is to transport and store the oil in the refineries. This segment is referred to as midstream. The third step is to refining and delivering the oil and its products. This segment is referred to as downstream.

The oil and natural gas industry is facing multiple challenges in all the operational segments. The primary concern is the exposure of humans to hazardous substances. The nature of oil field work involves heavy machinery and hazardous environments which increase the risk of injuries and accidents. Apart from that, the production cost is much higher, and if the production is stopped due to machine failures in the middle, the loss will be more. Identifying machine failures and rectifying them is a time-consuming process as humans need to go to that particular location. Moreover, manual operations are prone to errors and are labor-intensive. In order to overcome these challenges, the oil and natural gas industry focuses on automating the techniques to ensure the safety of the people and environment, thereby and thereby increasing the productivity.

The advent of Artificial Intelligence (AI) has brought revolutionary changes across industries, and the oil and gas sector is no exception. AI promises to drive efficiency, reduce costs, and enhance safety, but it also comes with challenges and potential risks that need careful consideration.

## *Promises of AI in the Oil Industry*

AI technologies offer significant opportunities in the oil industry, particularly in areas such as exploration, production optimization, predictive maintenance, and safety enhancement. Some of the most impactful promises include:

1. **Increased Operational Efficiency:** AI can optimize drilling operations, reducing non-productive time by analyzing massive datasets and providing real-time insights into equipment performance.
2. **Cost Reduction:** By automating repetitive tasks such as seismic data analysis, AI can help oil companies reduce labor and exploration costs. Advanced algorithms can also optimize supply chain operations, minimizing transportation and logistics costs.
3. **Predictive Maintenance:** AI-driven predictive maintenance can identify potential equipment failures before they occur, reducing downtime and improving asset longevity. This can lead to significant savings by preventing unplanned shutdowns and expensive repairs.
4. **Enhanced Safety:** AI solutions can monitor and analyze environmental factors, equipment conditions, and worker behavior to detect potential hazards, reducing the risk of accidents and ensuring regulatory compliance.

Oil wells are now equipped with sophisticated sensors and monitoring systems that enable real-time data collection and analytics, providing continuous insights into drilling operations. This integration of AI facilitates improved decision-making and operational efficiency, setting the stage for a deeper exploration of the promises AI holds for the industry. Lashari et al. (2019) in their work discussed the importance of monitoring and optimizing the drilling performance of oil pits and investigated a model for improving the drilling performance by using Feedforward Network. They generated data in the laboratory, created the model and applied it in the field.

Harshit Gupta et al. (2019) discussed the automation of horizons tracking with the help of seismic images. They designed a model based on deep convolutional neural networks for tracking the horizons directly from seismic images. They also compared the model performance with traditional CNN and showed that their model improves the accuracy.

Aditi Nautiyal et al. (2023) in their work discussed the usage of machine learning algorithms in enhancing the drilling performance in the oil and gas industry. They used Artificial Neural Network, Random Forest and evolutionary optimization algorithms for fine-tuning the drilling parameters, thereby improving the performance of drilling operations. Yang H et al. (2024) in their book chapter discussed the application areas of AI in drilling in the oil and gas industry.

Liu W (2024) reviewed CCNs with U-Net model for fault identification in seismic data which also highlights the shift toward advanced deep learning techniques in the oil and gas industry.

Ahmed et al. (2021) discussed the usage of machine learning mechanisms for identifying corroded oil and gas pipelines. In their review, they found ANN, SVM and hybrid models outperform well when compared to standalone models. Their study also found some of the research gaps such as data availability, accuracy and validations and also provided future directions of research in this field.

## *Perils of AI in the Oil Industry*

Despite its promises, AI in the oil industry comes with several perils that need to be addressed:

1. **Data Privacy and Security:** With the increasing use of AI to process vast amounts of data, there is a growing risk of cybersecurity breaches. Ensuring the privacy and security of operational data is crucial, as breaches could result in significant financial and reputational damage.
2. **Ethical Considerations:** The use of AI in automating decision-making processes raises ethical concerns, particularly when it comes to decisions affecting safety and environmental impact. The industry must ensure that AI systems are transparent, fair, and aligned with human values.
3. **Job Displacement:** The automation of certain tasks through AI could lead to job losses, particularly in roles related to data analysis, monitoring, and routine decision-making. Companies must manage the social and economic impact of AI-driven automation.
4. **AI Model Reliability:** While AI algorithms can process vast amounts of data, they rely on the quality of the data provided. Poor-quality or biased data can lead to inaccurate predictions and suboptimal decisions, potentially causing operational inefficiencies or safety risks.

Samir M., et al. (2023) addressed the cyber-security threats that are arising with use of IoT devices for collecting data and proposed an architecture which uses blockchain technology for the petroleum industry in Iraq. Their architecture succeeded in merging blockchain technology, IoTs, and an IPFS with the petroleum industry as an industrial application area and provides an efficient framework that resists common cyber-security attacks by achieving data integrity, availability, and confidentiality.

T.R. Wanasinghe et al. (2024) took a case study of the Canadian offshore oil and gas drilling occupations and worked on the effect of automation and technology adoption on the job market. They devised an approach for identifying the re-skilling requirements, job-merging pathways and a tentative timeline for the transformation. The adoption of digital technologies, such as automation, robotization, digital twins, data-driven decision-making systems, smart devices, and cloud computing technologies, gradually transform existing workplaces into digitally-enabled smart workplaces. Therefore, stakeholders must invest in training programs to re-skill existing workforces and to orient prospective employees to work at these smart workplaces. If technology adoption occurs at a faster or slower pace than workforce reformation, industries cannot gain the optimum benefit from their digital transformation initiatives. Also, human capital investments may not generate much benefit if technology adoption and workforce reformation occur at different rates. Therefore, this work presents a novel framework to predict future employment scenarios, particularly for the workers in offshore oil and gas drilling activities, along with a tentative timeline.

Chuka Anthony Arinze et al. (2024) in their work highlighted the use of AI in the oil and gas industry and also addressed the challenges that are associated with AI adoption. Implementing AI solutions in the oil and gas industry requires algorithms to handle the huge amount of data generated by it. The oil industry works in a highly regulated environment and it is necessary to adhere to the industry stands, data privacy regulations and safety protocols. They have also highlighted the difficulty associated with ensuring compliance with regulatory bodies.

Masudin et al. (2024) investigates the intricate dynamics of human-technology interaction in the context of safety within the Indonesian oil and gas industry, specifically focusing on the integration of the internet of things and cyber-physical systems.

## *Use Cases of AI in the Oil Industry*

### 1. Exploration and Seismic Data Interpretation

**Focus Area:** In exploration, AI is used to interpret seismic data and identify oil reservoirs. AI models can process large volumes of geophysical data to identify patterns and structures that may indicate the presence of hydrocarbons, helping companies target drilling locations more accurately.

**Data Challenges:** The main challenge is the sheer volume of seismic data that needs to be processed; it is often noisy and incomplete. AI models require high-quality, labeled data for training, and obtaining such data can be difficult in remote or deep-water locations.

**Solution Approaches:** Machine learning algorithms, such as deep learning models, can be trained on historical seismic data to identify patterns associated with oil reservoirs. Combining AI with edge computing can enable real-time processing of seismic data, reducing the time and cost associated with exploration activities.

### 2. Production Optimization

**Focus Area:** AI can optimize production by analyzing real-time data from wells and equipment, adjusting operational parameters to maximize output while minimizing operational risks.

**Data Challenges:** Oil wells generate vast amounts of data from sensors, which can be difficult to manage and analyze in real time. Additionally, data from different sources may be inconsistent or incomplete, making it challenging to derive actionable insights.

**Solution Approaches:** AI models can be integrated with real-time monitoring systems to analyze data from multiple sensors, such as pressure, temperature, and flow rates. Advanced analytics tools can use this data to optimize well performance, reducing downtime and ensuring continuous production.

### 3. Predictive Maintenance

**Focus Area:** AI-driven predictive maintenance is crucial in the oil industry, where equipment failure can result in costly downtime and safety hazards. AI models can predict when equipment is likely to fail, allowing for timely maintenance interventions.

**Data Challenges:** Predictive maintenance models rely on historical equipment data to predict failures. However, collecting and storing this data in a structured manner can be challenging, especially for older assets that may lack sensors or proper monitoring.

**Solution Approaches:** AI-based predictive maintenance systems can use IoT sensors to continuously monitor the condition of equipment. Machine learning models can analyze this data to predict when maintenance is needed, allowing companies to schedule repairs proactively and avoid unplanned shutdowns.

## *Key Focus Areas for AI Adoption*

1. **Data Integration and Management:** A key focus area for AI adoption in the oil industry is the integration of disparate data sources, such as seismic, production, and equipment data. AI systems must be capable of processing and analyzing this data in real time to provide actionable insights.
2. **Safety and Environmental Protection:** AI applications focused on enhancing safety and reducing environmental impact will become increasingly important as regulations tighten and companies aim to reduce their carbon footprint.
3. **Digital Twins:** The use of digital twins—virtual models of physical assets—will allow oil companies to simulate various operational scenarios, optimizing performance and reducing downtime.
4. **Sustainability and Green AI:** As the oil industry faces pressure to transition toward cleaner energy sources, AI can play a role in optimizing energy consumption, reducing emissions, and exploring renewable energy alternatives.

**Companies and Their Adoption of AI in Oil and Gas Industry**

| *S.No.* | *Company Name* | *Country* | *Technology Used* | *Use Cases* |
|---|---|---|---|---|
| 1. | BP | United Kingdom | Microsoft Azure, BP's proprietary data lake | Predictive maintenance, digital-twins, real-time data analytics |
| 2. | Shell | Netherlands | C3.ai Suite, Microsoft Azure, custom AI algorithms | Predictive maintenance, energy optimization, seismic data processing |
| 3. | Chevron | United States | C3.ai, Microsoft Azure | NLP for data processing, digital twins, predictive analytics for asset integrity |
| 4. | ExxonMobil | United States | IBM Watson, H2O.ai, proprietary generative AI models | Facility optimization, digital twin development, drone-based monitoring |
| 5. | Halliburton | United States | IBM Watson, Microsoft Azure, DecisionSpace® platform | Reservoir modeling, automated drilling, completion optimization |

| S.No. | Company Name | Country | Technology Used | Use Cases |
|---|---|---|---|---|
| 6. | Schlumberger | United States | DELFI Cognitive E&P, IBM Watson, Google Cloud | Seismic imaging, predictive analytics, enhanced oil recovery simulations |
| 7. | Baker Hughes | United States | C3.ai Suite, AWS, BHC3 AI Suite | Predictive maintenance, real-time drilling optimization, field data analytics |
| 8. | Equinor | Norway | Microsoft Azure, IBM Watson, custom AI models | Predictive maintenance, digital twin technology, $CO_2$ emission reduction initiatives |
| 9. | Saudi Aramco | Saudi Arabia | IBM Watson, custom generative AI models, SAP | Predictive maintenance, synthetic data for exploration, supply chain optimization |
| 10. | Kuwait Oil Company | Kuwait | Microsoft Azure, custom geophysical interpretation | Geophysical interpretation, predictive maintenance, reservoir optimization |
| 11. | Abu Dhabi National Oil Company | UAE | IBM Watson, Microsoft Azure, Panorama Center | Drilling operations, digital twins, predictive safety analytics |
| 12. | Gazprom | Russia | Microsoft Azure, NVIDIA GPUs, Yandex Data Factory | Automated pipeline monitoring, seismic data analysis, reservoir simulation |
| 13. | Suncor Energy | Canada | IBM Watson, Google Cloud Platform, digital twin tech | Refinery digital twins, oil sands extraction optimization, predictive maintenance |
| 14. | Eni | Italy | IBM Watson, Google Cloud AI, HPC4 supercomputer | AI-driven exploration, drilling automation, environmental monitoring |
| 15. | Woodside Energy | Australia | IBM Watson (NLP), AWS, custom AI solutions | NLP for data insights, digital twin modeling, seismic data interpretation |

## *Conclusion*

In conclusion, artificial intelligence holds transformative potential for the oil and gas industry, offering substantial benefits in operational efficiency, cost reduction, predictive maintenance, and safety. By automating processes and enabling data-driven insights, AI has the capacity to optimize exploration, production, and supply chain operations in ways that were previously unimaginable. However, as this technology continues to evolve, it brings forth complex challenges, including data security concerns, ethical implications, job displacement, and the need for reliable, high-quality data to ensure accurate predictions and sustainable outcomes.

While companies worldwide are making strides in adopting and adapting AI technologies, the industry's future success depends on balancing these promises with proactive strategies to mitigate risks. Ensuring data privacy, fostering ethical AI practices, and preparing the workforce for transformation will be critical for maximizing AI's benefits while safeguarding human and environmental interests. As AI continues to mature, it has the potential to reshape the oil and gas landscape, making operations smarter, safer, and more efficient. The industry's ability to navigate these promises and perils will ultimately define AI's role in the next era of energy production.

## References

Aditi Nautiyal, & Amit Kumar Mishra. (2023). Machine learning application in enhancing drilling performance. *Procedia Computer Science*, 218(C), 877–886. https://doi.org/10.1016/j.procs.2023.01.068.

Ahmed, D., Mokhtar, A., Kurnia, J., Najeebullah, D., Lu, H., & Sambo, C. (2021). Integrity assessment of corroded oil and gas pipelines using machine learning: A systematic review. *Engineering Failure Analysis*, 131, 105810. https://doi.org/10.1016/j.engfailanal.2021.105810.

Arinze, C.A., Izionworu, V.O., Isong, D., Daudu, C.D., & Adefemi, A. (2024). Integrating artificial intelligence into engineering processes for improved efficiency and safety in oil and gas operations. *Open Access Research Journal of Engineering and Technology*, 6(1), 39–51. https://doi.org/10.53022/oarjet.2024.6.1.0012.

Gupta, H., Pradhan, S., Gogia, R., Srirangarajan, S., Phirani, J., & Ranu, S. (2019). Deep learning-based automatic horizon identification from seismic data. *Society of Petroleum Engineers*. https://doi.org/10.2118/196087-MS.

Lashari, S., Takbiri-Borujeni, A., & Fathi, E. (2019). Drilling performance monitoring and optimization: A data-driven approach. *Journal of Petroleum Exploration and Production Technology*, 9, 2747–2756. https://doi.org/10.1007/s13202-019-0657-2.

Liu, W. (2024). Review of artificial intelligence for oil and gas exploration: Convolutional neural network approaches and the U-Net 3D model. *Open Journal of Geology*, 14, 578–593. https://doi.org/10.4236/ojg.2024.144024.

Masudin, I., Tsamarah, N., Restuputri, D.P., Trireksani, T., & Djajadikerta, H.G. (2024). The impact of safety climate on human-technology interaction and sustainable development: Evidence from Indonesian oil and gas industry. *Journal of Cleaner Production*, 434, 140211.

Samir M. Umran, Lu, S., Abduljabbar, Z.A., & Nyangaresi, V.O. (2023). Multi-chain blockchain-based secure data-sharing framework for industrial IoT smart devices in the petroleum industry. *Internet of Things*, 24, 100969. https://doi.org/10.1016/j.iot.2023.100969.

Wanasinghe, T.R., Gosine, R.G., Petersen, B.K., & Warrian, P.J. (2024). Digitalization and the future of employment: A case study on the Canadian offshore oil and gas drilling occupations. *IEEE Transactions on Automation Science and Engineering*, 21(2), 1661–1681. https://doi.org/10.1109/TASE.2023.3238971.

Yang, H., Shang, G., Li, X., & Feng, Y. (2024). Application of artificial intelligence in drilling and completion. *IntechOpen*. https://doi.org/10.5772/intechopen.112298.

# 27. The Promise and Perils of AI for the Satellite Industry

Virgil Labrador

The satellite industry is a relatively small sub-set of the over US$ 3 Trillion global telecommunications market. Depending on various sources, the satellite industry comprise between 3–10 percent of the telecommunication market. The Satellite Industry Association (SIA) which publishes an annual State of the Industry report, estimated that global satellite industry revenues reached US$ 285 Billion in 2023. Since tracking the industry revenues in the late 1990s—the satellite industry has consistently grown even during the major recessions in the early and mid–2000s and the recent global pandemic and is projected to continue growing consistently well through the next decade.

Despite its small share of the global telecommunications industry, the satellite industry plays a very pivotal role in providing the vital services that are necessary for everyday modern life as we know it. Anyone who uses a smart phone and accesses Google Maps or watches TV and uses the internet is tapping into the global network of satellites that do their work quietly and out of sight in space. From its humble beginnings with the launch of Sputnik satellite in 1957, there are now nearly 10,00 satellites in orbit that provide telecommunications, broadband, broadcast, earth observation, location, remote sensing and other essential services. It should be noted that the distinct advantage that satellites have over other telecommunications media is that satellites are the most cost-effective and in some cases the *only* way you can reach remote areas not served by terrestrial technology such as in deserts, forests and rural areas. This is illustrated most graphically in the world's oceans which comprise three fourths (3/4) of the earth's surface, where satellites are the only viable way to reach vessels once they go out of range of land-based communications facilities.

## *AI and Satellites*

Satellite technology was first conceived in 1945 by Arthur C. Clarke in a technical paper published in Wireless World. Clarke would go on to predict correctly that a global network of satellites will revolutionize the provision of telecommunications services worldwide. And it did so at a record pace of development. Quite the visionary, Clake is also the author of many science fiction classics such as "2001: A Space Odessey," which was made into a blockbuster movie in 1960s. In 2001: He presciently foretold the perils of

AI when one of the key characters in his book was the sentient AI computer named HAL (Heuristically Programmed Algorithmic Computer) which controlled the systems of the *Discovery One* spacecraft and wreaked untold havoc on the crew.

The growth of the satellite industry has been nothing short of remarkable. In the 70s and 80s, satellite technology revolutionized broadcasting by enabling worldwide television coverage via satellite. In the late 90s and early 2000s, more powerful satellite enabled broadband access to remote areas and more recently, the proliferation of Low Earth Orbit (LEO) satellite constellations such as the Starlink system are amplifying the reach and accessibility of satellites globally for many of key applications that satellites are known for such as broadband access, earth observation, remote sensing among others.

Satellite operations usually involve the processing and analysis of huge data sets. The industry sees the potential of AI-powered tools combined with Machine Learning (ML) for enhancing large-scale data processing and analysis, Telemetry, Tracking and Control (TT&C) of satellite systems, Situational Awareness and Command, Control, Communications, Computers (C4) Intelligence, Surveillance and Reconnaissance (ISR), among others. AI is currently used to allow more precise and rapid evaluation of data gathered throughout all of these activities.

More data can be collected and processed through AI than can be observed and analyzed by humans. As AI evolves, it will take over from much of the tedious monotony of operating satellite systems, making them more autonomous and freeing humans to perform higher level tasks.

AI is now extensively used in research, engineering, manufacturing and operation of satellites. The new satellite constellations are very complex to operate and the methods developed in past decades leave a lot for improvement. AI can fill many of the gaps in the functional efficiency of satellite technology so that it's able to achieve its full potential.

## *Key Applications*

Here are just some key commercial and military applications where AI can make the biggest impact on the satellite industry:

*Internet of Things (IoT).* The IoT segment includes over 50 Billion devices such your smart appliances and devices and almost anything equipment that needs to be monitored and tracked. Not all IoT devices are served by satellites, but certainly those in remote, hard-to-reach areas. Satellites play a very important role in collecting, analyzing, and display real-time data from devices and equipment. This data can be used to: monitor processes, improve efficiencies, improve quality and profitability, reduce waste, and identify problems and emergencies. To give an example, in the oil and gas industry sensors on the ground can monitor flows, temperature and even cracks or leaks in pipelines. The same can be done in the maritime industry and other segments where there is a need for monitoring of ships and their functions. You can also track locations of shipments in containers, trucks, trains, cars or any form of transport. AI will significantly increase the efficiency and speed of data collection and analysis and free up human operators from tedious tasks and focus on more qualitative analysis of the data.

*Earth Observation (EO).* Another key application is EO which involves mapping and monitoring of the earth and its atmosphere for weather, climate and disaster monitoring, location services and other uses. Again this involves large-scale processing and analysis of data. Combined with news satellite technology called "Software-Defined Satellite" (SDS), AI enables on-board processing of the data as opposed to having to download the data to the ground which h considerably slows down the process. This provides the ability to process larger amounts of data that is more up-to-date, and in real-time. According to the European Space Agency, AI would help reduce costs and produce higher quality environmental imaging data. This in turn can be used to provide actionable insights for governments and businesses accurately the impact of climate change, and other environmental and meteorological metrics.

*Situational Awareness and C4 and ISR*. It's no secret that military establishments and governments around the world use satellites for various applications. Foremost among these is the aforementioned Situational Awareness and C4 and Intelligence, Surveillance and Reconnaissance (ISR). AI is simply, to use a military term, a "force multiplier" for military use, for the same reasons mentioned in the previous commercial applications of AI. Quite simply, AI is making possible a whole new spectrum of data collection and analytical capabilities that is changing how government approach military strategy and operations. The impact of AI includes enhanced situational awareness, sensor augmentation and predictive modeling and driving autonomous decision-making, enabling decision-makers to focus on the larger picture rather than time-consuming and repetitive tasks. AI can also enhance the accuracy of data collected from surveillance and thereby make them more reliable.

*Space Debris*. As mentioned earlier, there are currently nearly active 10,000 satellites in various orbital planes circling the earth. This number is projected to increase to over 50,000 satellites in the next decade. This number is quite staggering, but this only counts the active (i.e., operational) satellites in orbit. There are hundreds and thousands more that are inactive satellites in so-called graveyard orbits that pose a significant threat to other satellites in case of collision. Considering that objects in space move at very high speeds of around 17,00 miles per hour, even a small fragment can cause serious damage, if not complete destruction of a satellite that costs hundreds of millions of dollars to launch and operate. The danger is enhanced by Anti-satellite operations are conducted by several countries where they intentional destroy end-of-life satellites in space. This cause huge debris fields that are a constant threat to currently operating satellites.

To monitor space debris AI is using ML-based orbit prediction algorithms to enhance object detection and classification so that satellites and satellite debris can be monitored more accurately and predict and prevent collisions.

AI is also assisting human operators in managing large fleets of satellites by automatically monitoring, maintaining, satellites and ground systems. By leaving the tedious tasks to AI-powered systems, human operators can focus on "high cognitive tasks" associated with managing complex satellite space and ground networks and leave the "low cognitive tasks" to AI.

## *Conclusion*

There are many other commercial and military applications where AI can be very beneficial for satellites. For the most part, AI is seen as a net-positive for the industry in terms of increasing effectiveness and efficiencies of satellite technology to deliver the essential service that it provides. As a technically and heavily scientifically-driven enterprise, the satellite industry uses rigorous testing and other best-practices before they implement new technologies and systems, which includes AI.

The peril of AI for satellites is if it cedes too much of the operations, especially in surveillance of data collection and analysis, to AI systems. Besides, AI is still evolving and developing. No technical system is ever 100 percent accurate. This could be fatal and catastrophic when AI systems falsely identify the wrong threats. The fact that AI is software-driven leaves another set of vulnerability for satellite systems. AI can be used by bad actors to hack satellites for nefarious purposes. This requires a high level of security to ensure that AI systems are not susceptible to external interference.

It's easy to fall in the trap of relying too much on AI because the satellite industry has an acute shortage of trained personnel due to the high barriers to entry and high level of training required to work in the industry. Human resources are so scarce that there is a tendency as in other industries to maximize automation to replace human operations. The key in implementing AI as with any new technology is verification and validation through experience. But more importantly, AI should never be seen as to replace humans—AI should be a tool that is controlled by human operators and never the other way around.

The specter of Clarke's AI character HAL who ended up killing the spacecraft's crew is caveat known to many of the people who work with satellites who were drawn to the industry by the promise of space as described in works of science fiction. Given how the satellite industry has managed to come up with startling innovations that has enabled the industry to grow at an exponential rate in short period of time, I see no reason for AI to be implemented judiciously by the satellite industry to maximize it benefits and to neutralize or at least minimize its perils.

# 28. Adopting Mobile Application Technology for SMEs in Thailand's Freight Transport Sector

TACHANUN RATTANASIRIWILAI

The logistics industry is a key pillar of Thailand's economy, and a significant amount of work in the road freight transport sector is carried out by small and medium-sized enterprises (SMEs). Mobile applications have gained some traction in recent years, but they mainly depend on static algorithms and manual input. AI, however, has the potential to truly transform these apps by allowing features like predictive analytics, automation and real time adaptability. This contribution seeks to realize on the integration of Artificial Intelligence into mobile apps for SMEs and highlights the current gaps, opportunities to leverage for road freight in Thailand, and the challenges ahead.

## *AI-Assisted Mobile Application Technology for Thailand's SMEs*

Small and medium businesses, or SMEs, account for about 99.5 percent or around 2,000,000 of the business operators in Thailand. Unlike the GDP-driving large businesses, about 2000 SMEs provide 70% of employment (TMB Thanachart 2024). These entrepreneurs are vital to the economy but face many challenges—ranging from the high cost of doing business and outdated management or limited access to technology. Mobile applications provide basic, often manual or paper based, solutions such as route acceleration or fleet management that have not been present for long and suffer significantly from the potential AI can provide. Adoption of AI can really equip SMEs to address these short-comings and be competitive in a fast-evolving eco-system.

AI will provide transformative powers on mobile applications by improving operational efficiency, allowing data-driven decision-making, and improving logistics operations. It is capable of sifting through raucous databases, adjust to dynamic variations on the fly, plus do the things that go around repeatedly, which provides humans more time to spend on the game-changing jobs. When SMEs in Thailand embrace AI, they can overcome the technology gap with large companies and get a competitive advantage.

Mobile apps help Thai SMEs improve logistics operations by providing functionalities that are critical to improving efficiency. Zone-based route optimization is one of the standard applications based on algorithmic applications that optimize the routes for lower fuel consumption and shorter delivery times (Haseeb et al., 2019), whilst fleet management applications monitor the performance of all vehicles in the fleet and schedule maintenance activities. Also, the GPS-enabled systems enable real-time tracking that provides operational transparency and enhanced asset visibility (Cortino et al., 2020).

## *The Transformative Potential of AI in Mobile Applications*

**Predictive Analytics.** Artificial Intelligence (AI) provides businesses the tools to anticipate and respond to flaws, such as overnight demand spikes or possible delays. Leveraging the data that came before it and feeding from metrics the moment they go live, powerful AI-powered systems can predict demand trends and deliver timely recommendations for minimization of risk. Predictive analytics, for example, can assist SMEs in creating better allocation of resources and optimizing their vehicles, drivers, and inventory for customer expectations (Dimoso & Utonga, 2024). This capability enhances operational readiness while reducing the financial impact of unforeseen events.

**Dynamic Decision-Making.** Artificial Intelligence will improve the agility of logistics operations by allowing with real-time adaptation of routes and timetable. Advanced algorithms crunch data—traffic congestion, weather disruptions, road closures—to give cars the ability to dynamically reroute. For instance, a system that incorporates AI can detect an unexpected traffic jam and quickly recommend an alternative route to meet delivery timelines. In this way, freight where possible has low waiting times and consumption due to flexibility, which means of course that costs are reduced and service reliability is improved.

**Automation and Efficiency.** The most powerful contribution of AI is its capability to automate repetitive and time-consuming back-end processes like inventory management, order processing, and data entry. This reliance shift shifts to the human resources freeing up from these desktop jobs human resources to focus on strategies to keep business moving. For an SME, this automation provides more productivity, less errors, and a better resource allocation. This allows businesses to continue innovating and serving their customers without necessarily being cost or resource intensive.

**Improved Resource Allocation.** By understanding existing operations, AI helps SMEs get the most out of their limited resources. AI enabled systems, for example, can analyze vehicular usage patterns, and fuel consumption data in a bid to provide recommendations on fleet management. Such level of visibility allows organizations to minimize wastage, optimize asset usage, and ensure business continuity amid unfavorable conditions.

**Enhanced Customer Satisfaction.** AI have also given a significant enabling technology for improving customer experience through all the way updates and personalized connectivity. AI-based chatbots, automated notifications, and other features keep customers in the loop regarding delivery status and estimated arrival time. These tools promote and enable transparency and trust, engendering long-term loyalty by addressing customer concerns in an efficient manner. In addition, businesses can predict customer needs and proactively respond with predictive capability, further strengthening customer satisfaction.

## *Key Features of AI-Assisted Mobile Applications*

**AI-Powered Route Optimization.** Logistics applications facilitated by AI route optimization allow for increased efficiency by inputting real-time data on traffic patterns, weather, and road conditions. Sophisticated algorithms reroute deliveries in real time to sidestep obstructions and to reduce gas usage, which is a way to make deliveries less expensive and more timely. In Thailand, where traffic congestion often disrupts logistics business, this functionality has a significant effect. Businesses can reduce costs and increase reliable delivery by steering vehicles onto the optimal paths.

**Predictive Maintenance.** One example of predictive maintenance enabled by AI is the use of IoT sensors that detect the health of vehicles in real-time, warning the driver before a mechanical issue becomes critical. The systems enable companies to appraise their data—engine performance, fuel consumption, and wear-and-tear metrics, for instance—to far-preemptively book maintenance. This reduces surprises on the road, eliminates high repair costs, and guarantees reliability of fleets, enabling SMEs to avoid an operational break (Evangelista & Sweeney, 2014).

**Fleet Utilization Analytics.** Fleet Analytics with AI—AI in IoT-driven Fleet Management helps to provide detailed information regarding vehicle trips, fuel utilization, and idling time which enables better decision making. This knowledge allows you to identify inefficiencies and helps SMEs make data-driven decisions to optimize fleets. For instance, AI can more effectively track underutilized vehicles or recommend shifting driver schedules, ensuring resources are being deployed in the right way and cost savings are being achieved.

**Real-Time Tracking and Alerts.** Real-time updates on vehicle locations and delivery statuses via AI-powered tracking systems enhance operational transparency and reliability. To add on to this feature, predictive analytics is used to locate any upcoming delays, such as traffic delays or weather disruptions, and recommend alternatives. This guarantees businesses to fulfil customer expectations without losing operational efficiency (Wongsansukcharoen & Thaweepaiboonwong, 2023).

**Enhanced Customer Experience.** Mobile apps that incorporate AI features such as chatbots, automated notifications, etc., result in better customer satisfaction. Undoubted boost to the business using chatbots; at its most basic level, it is something that can provide instant response to customer inquiries with regards to the question like business delivery status, estimates arrival times and service-related questions. These tools add clarity, promote trust, and ensure the customer feels comfortable and engaged, ultimately ensuring return custom and loyalty.

**Sustainability Optimization.** By optimizing routes, AI reduces fuel consumption and related CO2 emissions, helping to promote sustainable logistics practices. For instance, AI can help determine the most environment-friendly routes to minimize unnecessary mileage and promote efficient resource use. All these traits are in line with global sustainability objectives and attract eco-friendly consumers (Cortino et al. 2020).

**Dynamic Decision-Making.** With the help of AI analyzing the real-time data, dynamic decision making can adapt to change logistics condition. Examples include re-routing vehicles as needed amid road closures or proposing new delivery schedules when demand rises unexpectedly. This adaptability leads to operational resilience and enhanced SME ability to meet customer expectations amid changes.

**Inventory and Warehouse Management.** AI-powered mobile applications enhance accuracy in inventories and improve the efficiency of warehouses by automating stock management processes. Through insights drawn from past trends in sales and demand, AI empowers businesses to forecast their inventory needs, reducing the risk of excessive inventory or stock shortages. This will help keep things running smoothly and saves money while avoiding stockouts.

**Resource Allocation Optimization.** AI further improves resource allocation, by deriving operational data for maximizing the effectiveness of vehicles, manpower, and schedules of delivery. For instance, it can align what needs to be delivered with the available resources—making sure that labor and assets are used effectively. This minimizes wastage, and maximizes productivity in the business activities, especially for the resource-constrained section of SMEs.

**Risk Management and Compliance.** AI systems are becoming critical to risk management by identifying and forecasting potential operational disruptions, such as extreme weather and geopolitical problems, and advising on possible pre-emptive actions. From another point of view, AI can serve as an aid in simplifying regulatory systems and processes through documentation and observance to various industry benchmarks. Such functionalities safeguard SMEs from costly penalties and operational downtime.

## *Future Opportunities for AI in Mobile Applications*

Artificial Intelligence (AI) in mobile applications can brighten the future of SMEs in logistics industry in Thailand. Adopting cutting-edge AI features enables the companies to optimize their processes, cater to the customer needs, and address the rapidly transforming industry challenges. The below-mentioned opportunities reflect how AI can change the mobile applications landscape for the road freight transport industry.

**Advanced Demand Forecasting.** How can mobile applications built on AI-based solutions disrupt demand forecasting? Using the power of historical information, market trends, and real-time metrics, AI can deliver more accurate forecasts about seasonal demand variation and customer behavior. This allows SMEs to maximize resource allocation, making sure that vehicles, drivers and warehouses are ready for peak times. For example, a predictive analysis tool could prevent underutilization during low season in terms of operational costs (or overstocking during peak seasons) contributing to service reliability (Dimoso & Utonga, 2024).

**Integration with Blockchain Technology.** By utilizing AI and blockchain technology within mobile applications, logistics operations could achieve higher levels of transparency and security. The immutable nature of the Blockchain ledger allows manufacturers to ensure that transactions, delivery records, inventories, etc., cannot be tampered with, and at the same time this data can be analyzed by AI to identify possible inefficiencies and provide key insights that allow improving processes. This can transform international trade into more secured and regulatory compliant operations for

SMEs in Thailand that are catering to export; reducing the risks of fraud or error in the process (Cortino et al., 2020).

**Real-Time Compliance Management.** AI to automate compliance with legal and regulatory requirements in logistics AI-powered mobile apps can track shifts in legislation, draft documents from the requirements, and verify compliance with safety and environmental regulations. SMEs, for instance, can adopt AI models and solutions to monitor vehicle emissions and ensure compliance with green logistics regulations. It streamlines operations and ensures sustainability, as it lessens the burden of compliance and minimizes the cost of non-compliance fines (Wongsansukcharoen & Thaweepaiboonwong, 2023).

**Collaborative Logistics Platforms.** AI allows SMEs to participate in a collaborative logistics ecosystem where assets—vehicles, warehouses and even delivery routes—can be shared between businesses. AI-enabled mobile applications search supply-demand match and SMEs can collaboratively utilize resources via mobile applications separately. For example, it will book those underutilized vehicles which are in the vicinity of the delivery destination using AI, which would minimize the operation costs and maximize the resources utilization. It enables SMEs to partner or even compete with large logistics companies.

**Multilingual and Cross-Cultural Capabilities.** AI-enabled mobile apps can provide multilingual functionality which help SMEs to expand regionally beyond Thailand. For example, Natural Language Processing (NLP) algorithms can be used for translating logistics documents, automating customer support in multiple languages, and making it easier to communicate with international clients. This ability is invaluable for SMEs seeking regional and global markets, as it can eliminate language barriers and facilitate cross-border collaboration (Evangelista & Sweeney, 2014).

## *Challenges to AI Adoption in Mobile Applications for SMEs*

In Thailand, the implementation and use of Artificial Intelligence (AI) in mobile applications for Small and Medium Enterprises (SMEs) in the logistics industry have many opportunities with significant challenges which is an obstacle to widespread application. Such challenges are derived from economic limitation, technological threats, human resource responsibility, privacy invasion, and ethical issues. Resolving these challenges is vital for SMEs to adopt AI in an effective and sustainable manner.

**Financial Constraints.** One of the biggest barriers to the implementation of AI in SMEs is the high initial investment required for AI technologies. Notably, developing AI-enabled mobile apps comes with costs associated with hardware, software development and maintenance. Training costs for employees and upgrading existing infrastructure can make it prohibitive for the smaller business with limited budgets. Although promising, these cashflow limitations can prevent SMEs from acquiring AI (Evangelista & Sweeney, 2014).

**Limited Access to Technology.** SMEs are often unable to access the technological infrastructure necessary for AI adoption. Common issues are old systems, poor internet ranges and little integration availability. Most SMEs do not implement AI solutions because they do not have the equipment and systems, they are stuck with the old way of doing things which is less efficient and competitive (Cortino et al., 2020).

**Workforce Challenges.** The shortage of skilled workforce is one of the main barriers to the use of AI in SMEs. There are still many employees who do not have the technical skills necessary to run AI systems, or interpret the information those systems produce. AI integration can be limited by resistance to change within organizations, where employees may fear displacement and find it hard to fit within new technologies. This challenge must involve training and upskilling programs to enable the data intelligence (Dimoso & Utonga, 2024).

**Data Privacy and Security Risks.** AI systems depend on vast amounts of data to work well. This poses serious data privacy and security challenges—and is particularly problematic for SMEs with less sophisticated cyber-protection processes in place. Incidentally, access to sensitive information or data leakage can hamper the reputation of a company and cause legal penalties. These threats underline the need for SMEs to adopt secure data management practices to mitigate them (Wongsansukcharoen & Thaweepaiboonwong, 2023).

**Regulatory Complexity.** Once SMEs embrace AI, another key obstacle lies within the regulatory framework. Businesses are required to follow data protection laws such as the Thailand Personal Data Protection Act (PDPA) that prescribes strict practices governing the collection, storage, and usage of data. Compliance can be resource intensive, especially for SMEs, especially those without legal or compliance teams (Cortino et al., 2020).

**Ethical Considerations.** Applications of AI also introduce ethical considerations, such as the possibility of biases in algorithms and the potential displacement of jobs. For instance, AI systems that are trained on biased datasets could yield discriminatory results, therefore damaging a company's image. Moreover, automating repetitive tasks could potentially lead to workforce redundancies, presenting SMEs with both social and ethical challenges. It is a complicated undertaking to weigh the advantages and costs of equipping society with AI (Evangelista & Sweeney, 2014).

**Lack of Trust in AI Systems.** A lack of trust in AI to be accurate and reliable also renders many SMEs wary of adopting the tech. Decision-makers may worry that AI systems will over output or not customize to the unique needs of their business. Novel systems based on AI technologies should be transparent and provide explanation in contexts where its advantage needs explanation (Haseeb et al., 2019), thus convincing users ahead of time in order to instill trust in AI.

**Integration with Existing Systems.** The integration of AI-powered applications with existing business systems can be a tedious and expensive process. Most SMEs run on outdated systems that clash with the new modern AI technologies. This prevents integration and requires companies to do costly upgrades, or simply not adopt AI and restrict their competitiveness in the digital economy.

**Limited Awareness of AI Benefits.** Most SMEs have no idea how AI works and how they could potentially use it. Leaders in businesses are also unlikely to figure out how to harness and adopt AI without a clear understanding of the impact it will deliver and the specific benefits it will have on cost and customer satisfaction or workplace efficiency. There is a need for awareness campaigns and education initiatives that explain the value of AI for SMEs (Dimoso & Utonga, 2024).

**Resource Allocation Challenges.** That can be a challenge for the vast majority of

SMEs that have little resources that they could deploy to adopt AI. Challenged by competing demands for capital, time, and personnel, many SMEs relegate AI initiatives to the back burner. Not overextending themselves with long-term technological investments that take time and require extensive resources, businesses are more likely to double-down on what is working in the moment, delay their AI effectiveness and risk losing market share.

## *Conclusion*

In Thailand, for SMEs to be able to implement AI-powered mobile applications for their business, it is critical to work on these issues. Specific government efforts, available financial instruments, and collaboration with technology suppliers can ameliorate both the finances and the technology gaps. Training programs and awareness campaigns can also help small businesses and suppliers to use AI correctly. Tackling these hurdles will not only introduce the immense potential of AI to these SMEs, but will also spur innovation and competitiveness in Thailand's logistics ecosystem.

### References

Cortino, J., Lee, T., & Yang, F. (2020). GPS-enabled systems and asset visibility in logistics. *Journal of Logistics Management*, 12(4), 45–56.

Dimoso, D., & Utonga, P. (2024). Predictive analytics in logistics: A game changer for SMEs. *International Journal of Business Analytics*, 16(1), 33–48.

Evangelista, R., & Sweeney, T. (2014). IoT sensors and predictive maintenance in logistics operations. *Technology and Operations Research Review*, 8(2), 120–137.

Haseeb, M., Chan, F., & Wong, C. (2019). Optimization strategies in road freight logistics for SMEs. *Asia Pacific Journal of Logistics Research*, 7(1), 15–29.

TMB Thanachart. (2024). Thailand SMEs: Economic drivers and employment statistics. *Thai Business Review*. Retrieved from https://www.tmbth.com/economic-report.

Wongsansukcharoen, A., & Thaweepaiboonwong, K. (2023). AI and logistics in Thailand: Challenges and opportunities for SMEs. *Journal of Thai Industrial Development*, 29(3), 78–92.

# • *G. AI in Data Infrastructure and Ethics* •

# 29. Decentralized AI Workloads

## *The Strategic Role of Edge Data Centers*

RATHEESH VENUGOPAL

Can you picture a day in your favorite city where traffic lights adjust instantly to prevent major congestion, or your smartwatch delivers truly real-time health updates with even shorter delays than today? The scale of instantaneous response and localized intelligence are no longer futuristic thoughts. That is where the power and promise of Edge data center computing comes into play. The advancement of AI in all walks of the modern world is driving a significant transformation through the decentralization of workloads to Edge data centers. This is to enable data processing proximity closer to the point of use and creating new prospects for global enterprises. This piece examines the multifaceted nature of this transition, exploring the opportunities and some challenges it presents.

## *The Ascent of Edge Data Centers*

Typically, Edge data centers can take multiple forms such as Micro, On premise, Mobile Edge, Telco, Regional or even IoT Edge data centers. In summary, Edge data centers are smaller-sized, localized facilities that process data closer to the source of generation. Unlike traditional centralized data centers, which often have to traverse the network across vast geographical distances to reach the central facility, Edge data centers enable real-time data processing and analysis with its reduced network round-trip time. This is particularly crucial for AI applications in our daily life that demand low latency and high bandwidth, such as healthcare gadgets, autonomous vehicles, smart cities, and industrial IoT where split-second responses are crucial.

### Decentralization: A Strategic Shift

Edge data centers are becoming vital for supporting next-generation digital infrastructure due to numerous advantages, enabling organizations and individuals to achieve more. The move towards decentralization in the present world is driven by several factors:

- **Bandwidth Optimization**: Long gone are the days of storing an entire year's worth of school data on a 500MB hard drive; today, even standard smartphones offer over 100 times that capacity. Similarly, transferring large volumes of data to a cloud facility is costly and bandwidth-intensive. However, Edge data centers mitigate this by processing data locally and reducing transfer needs (Hewlett Packard Enterprise, 2023).
- **Minimizing Latency**: AI applications that require real-time decision-making benefit from reduced latency. Edge data centers, which are near to the data source, can process data faster than Core Data center cloud servers (Jenkins & Botbyl, 2024).
- **Data Sovereignty and Security**: Local processing safeguards sensitive data inside geographic borders amid stricter data privacy requirements. This is essential for preserving brand reputation and complying to local-regional regulations such as GDPR. Google's €50 million fine for lack of transparency and British Airways' £20 million penalty for inadequate data protection are a few examples here (Tidy, 2020; European Data Protection Board, 2019).

## *Key Technical Elements*

There are numerous aspects associated with the implementation of AI at the Edge. These encompass the necessity for reliable connectivity, data management obstacles, synchronization and hardware components.

### Hardware and Infrastructure

Edge data centers require specialized hardware to handle AI workloads effectively, including AI accelerators like GPUs (Graphics Processing Units), first designed by NVIDIA and later adopted by AMD and Intel. Alternatively, TPUs (Tensor Processing Units), originally developed by Google and now available to everyone through Google Cloud Platform (GCP), are also essential for AI processing (Google Cloud, 2023). Moreover, Edge devices must be designed to withstand varied environmental conditions to ensure consistent, reliable performance. Beyond having the physical facility for data hosting, the deployment and management of Edge data centers demand substantial supporting infrastructure. This includes critical environmental components like power supply, cooling systems, optimal pressure, and uplink/downlink network connectivity. These investments are crucial for efficient AI workload processing and the seamless operation of Edge data centers.

### Data Management

Data management is key to Edge AI's success. In addition to storage and processing, this includes data collection, annotation, and pre-processing. Edge situations demand specific data management solutions that accommodate for data volume, velocity, and variety. For example, smart city sensors from traffic signals, cameras, and weather monitors generate massive amounts of data (volume) that must be analyzed locally to make real-time adjustments. Autonomous vehicles must analyze high-speed camera and sensor data (velocity) to safely navigate and respond to changing road conditions. Additionally, industrial IoT apps capture temperature, vibration, and energy consumption data from machinery. These data types (variety) require unique handling to improve operations and predict maintenance requirements. By accounting for these aspects, Edge data centers can manage data complexity for real-time decision-making and reduce centralized system load.

### Synchronization and Consistency

Maintaining data consistency and integrity is a major difficulty in Edge data management. Data processed at the Edge must be synchronized with data stored in central cloud servers. Data management solutions must also account for Edge devices' limited storage and processing capabilities, using data compression and pruning to optimize resource utilization. Currently, Blockchain technology helps Edge computing manage data consistency issues. Blockchain's distributed ledger technology assures data integrity and synchronization across dispersed nodes, making it beneficial to Edge data centers (Belski, 2023). Applying predictive algorithms to forecast and manage synchronization difficulties can further enhance Edge data center reliability and efficiency.

### Robust Connectivity

Edge data centers require reliable connectivity to operate effectively, including both local network infrastructure and a stable uplink back to the central cloud. Edge connectivity is set to advance with technologies like 5G, satellite internet, and fiber networks, which deliver the bandwidth and low latency essential for AI applications. These developments will enable real-time processing of Edge tasks, significantly enhancing performance. However, widespread 5G deployment demands substantial infrastructure investment and the establishment of regulatory frameworks to ensure fair and equitable access to these services

## *Case Studies of Real-World Applications*

The decentralization of AI workloads through Edge data centers opens up a plethora of applications across various industries. Some of the most promising application in the current world we live are as follows:

### Assisting Autonomous Vehicles

Real-time data processing is essential for the navigation, object detection, and decision-making of autonomous vehicles. A notable example is the Waymo self-driving car by Alphabet, which combines LiDAR, cameras, radar, and AI to navigate roads and interact with vehicles and pedestrians, minimizing the need for human input (Alphabet Inc., 2024). By processing data locally, Edge data centers enable these vehicles to ensure timely responses and reduce latency. Furthermore, the local processing of data guarantees that autonomous vehicles can continue to function safely and efficiently in the event of connectivity issues. The integration of AI at the Edge in autonomous vehicles also creates new opportunities for vehicle-to-vehicle (V2V) and vehicle-to-infrastructure (V2I) communication. This reduces the likelihood of accidents, and overall traffic management is enhanced, as vehicles can exchange real-time information about traffic conditions, road hazards, and other critical factors by processing data locally.

## Optimizing Smart Cities

Smart cities leverage a network of sensors to gather data on factors like traffic, weather, and energy consumption. Edge AI enables the real-time analysis of this data, supporting efficient city management and enhancing residents' quality of life through faster, data-driven decision-making. Edge AI deployment in smart cities also helps new and innovative businesses to flourish. For example, Edge AI can help optimize inter-city traffic flow, therefore lowering congestion and enhancing air quality for the locality. Edge AI can also be applied to monitor and regulate energy consumption, therefore improving urban sustainability and lowering the city carbon footprint.

## Industrial IoT (Internet of Things)

In industrial settings, IoT devices collect data from machinery and equipment to monitor performance and anticipate maintenance needs. Edge data centers enable real-time processing of this data, enhancing operational efficiency and minimizing downtime. Additionally, Edge AI supports new industrial applications like proactive maintenance and quality control. By analyzing data from IoT devices in real time, Edge AI can detect potential equipment failures before they happen, using current-historical data to enable proactive maintenance and reduce downtime. Additionally, Edge AI enables real-time quality monitoring in manufacturing, ensuring that processes meet stringent quality standards and minimizing the risk of defects.

## Application in Healthcare

Huge volumes of data produced by medical instruments and sensors must be instantly processed and examined. Edge data centers guarantee precise, rapid diagnosis and treatment strategies by allowing local processing of this data. For instance, vital indications like blood pressure, heart rate, and glucose levels can be recorded via wearable health monitors such as smart watches or public scanner. Healthcare professionals can get real-time notifications about possible health problems by processing this data locally, which enables timely intervention and treatment. Furthermore, Edge AI can be applied to medical image analysis and anomaly detection, increasing the accuracy and efficiency of diagnostic processes.

## Empowering Retail

Retail stores are increasingly harnessing Edge AI through sensors and cameras to gather data on customer behavior, such as foot traffic patterns and product interactions.

This localized processing enhances customer experiences and boosts operational efficiency by providing real-time insights into customer preferences. These insights enable data-driven decisions to improve store layouts, inventory management, and marketing strategies. Additionally, Edge AI strengthens retail security; for example, Edge-based video analytics can detect suspicious activities and send real-time alerts, reducing theft risks and protecting both customers and employees from potential hazards.

## *Challenges and Risks*

While decentralizing AI tasks presents numerous advantages, it also presents few challenges and barriers

### Potential Scalability

Edge data centers are limited by their local-regional nature, in contrast to centralized cloud servers, which could readily scale up to meet rising demands. This calls for the creation of scalable architecture and effective resource management. Organizations shall use innovative Edge computing techniques like microservices and containerization to solve scaling issues. By minimizing resource utilization and improving the performance of AI workloads, these strategies make it possible to deploy scalable and lightweight applications at the Edge. To guarantee the smooth operation of Edge data centers and the effective distribution of resources, companies should also need to invest in sophisticated automation tools (for example: Kubernetes and Apache Mesos ) and management solutions such as Prometheus or Datadog (Intel Corporation, 2023).

### Security Vulnerabilities

Physical interference and cyber-attacks are more likely with Edge devices, often deployed in less secure environments. Robust security measures are essential to protect sensitive data and maintain AI application integrity. For instance, smart traffic sensors in urban areas, frequently installed in public spaces, are vulnerable to tampering and cyber threats. Implementing strong security protocols like encryption, authentication and tamper-resistant designs are crucial to ensure their reliability which prevents the risks from both physical and logical aspects. This approach should ensure comprehensive data protection, covering both data at rest (stored in databases/devices) and in transit (sent across networks) for a full 360-degree security. Additionally on an outer layer, organizations must invest in robust monitoring and detection systems to identify and respond to security incidents in real-time, ensuring the continuous protection of Edge AI platforms. Security Information and Event Management (SIEM) tools like Splunk and Endpoint Detection and Response (EDR) solutions like CrowdStrike are leading choices in the industry for monitoring and addressing such security anomalies.

### Financial Impact

Although Edge computing has the potential to reduce operational costs by optimizing bandwidth consumption, the initial setup and maintenance of Edge data centers can be costly. In order to validate the investment in Edge infrastructure, organizations must conduct a thorough evaluation of the cost-benefit ratio. The cost of deploying and maintaining Edge data centers encompasses not only the acquisition of specialized hardware, such as GPUs and network equipment, but also the infrastructure for power, cooling, and secure housing. Furthermore, operational administration, monitoring, and routine maintenance comprise ongoing expenses. To streamline operations and minimize costs, organizations can adopt advanced resource management techniques, like automated workload scheduling and predictive maintenance powered by AI, which can help anticipate and resolve issues before they escalate. Innovative models, such as leasing models for hardware to avoid upfront capital expenditures or joint ventures with telecommunications providers (ISP) to share infrastructure, can also be explored by companies to establish a balanced approach to managing the financial impact of Edge computing.

### Vendor Interoperability

Edge computing's varied hardware and software configurations make achieving interoperability between Edge devices, data centers, and AI applications complex. For instance, an IoT sensor in a factory may use proprietary communication protocols that don't align with the data formats of a cloud-based AI analysis tool, causing compatibility issues. Similarly, healthcare equipment from different manufacturers may record patient data in varying formats, complicating integration with centralized health information systems. Open standards and protocols like OPC-UA for industrial automation and MQTT for IoT communication can help organizations overcome these challenges by facilitating data integration across systems. Developing standardized interfaces and communication protocols, such as APIs for consistent data formatting, further harmonizes operations across devices. Collaboration among hardware manufacturers, software developers, and service providers is key here. For instance, Microsoft and Intel have partnered to create standardized Edge platforms like Azure Stack Edge and Intel Smart Edge, fostering a more cohesive Edge computing ecosystem across industries (Microsoft Azure, 2023; Intel Corporation, 2023).

## *Future Outlook*

With technology alleviating many problems, the future of AI and Edge data centers looks promising. Enhanced GPUs and TPUs, together with low-latency 5G and open platforms, will enhance connectivity and enable real-time Edge data processing. Advanced data management strategies like Federated learning and Edge analytics will boost Edge AI adoption, empowering industries to leverage intelligent data insights closer to the source of origin.

### Hardware Innovation Frontier

Specialized circuits that are optimized for performance and efficiency, such as NVIDIA's A100 and H100 GPUs and Google's TPU v4, are the driving force behind hardware advancements in the evolving landscape of AI and Edge data centers (NVIDIA, 2023; Google Cloud, 2023). While conserving energy, these high-performance processors power intricate AI duties. Intel's Agilex FPGAs are examples of modular, reconfigurable architectures that enable the customization of specific AI workloads, thereby improving both scalability and adaptability in Edge environments. The rapid processing of complex AI workloads, enabled by emerging technologies such as IBM's quantum processors and neuromorphic CPUs like Intel's Loihi 2, is poised to further revolutionize Edge AI, thereby expanding the frontier of real-time analytics and decision-making in Edge data centers.

### Accelerated Network Connectivity

The widespread deployment of 5G networks and advancements in wireless technologies like Wi-Fi 7 are expected to greatly enhance Edge data center connectivity. This will enable faster data transmission and lower latency, improving AI application performance. Additionally, new technologies such as SpaceX's Starlink and Amazon's Project Kuiper low-earth orbit (LEO) satellites, along with enhanced wireless mesh networks like Cisco Meraki and Aruba Networks, are expected to improve the overall connectivity (SpaceX, 2023; Cisco, 2023; Amazon, 2024). These advancements provide dependable, high-speed connectivity in rural and underserved places, enabling Edge AI applications to function even when regular infrastructure is unavailable. Furthermore, developments in network protocols and standards, such as MEC (Multi-access Edge Computing), will ensure that Edge data centers integrate seamlessly with existing cloud and IoT ecosystems, allowing for more effective data transmission and resource management.

### Advanced Data Management

Advanced data management techniques, such as Federated learning and Edge analytics, are indispensable as AI workloads migrate to decentralized Edge systems. Federated learning is essential for applications such as autonomous vehicles, healthcare, and smart cities, as it trains models across Edge devices without centralizing data aggregation, thereby reducing bandwidth and preserving privacy. For instance, in the healthcare sector, Federated learning enables hospitals to collaborate in the improvement of diagnostic models without the exchange of sensitive data. Edge Analytics is essential for applications that necessitate immediate response, such as anomaly detection in industrial IoT and AR, as it utilizes real-time data transmission and in-situ processing to provide rapid insights at the source. These techniques align Edge data centers with the requirements of contemporary AI applications across industries by facilitating localized AI with low latency, increased privacy, and enhanced performance.

## *Conclusion*

The decentralization of AI workloads through Edge data centers marks a major shift in data processing and management. This approach offers significant benefits, including reduced latency, improved bandwidth efficiency, and greater control over data. However, it also presents challenges, such as security vulnerabilities and scalability concerns. The synergy between Edge computing and AI will continue to unlock new opportunities across industries, reshaping how we live, work, and interact with technology as advancements in AI, connectivity, and data management progress. In essence, this evolution not only meets the demands of today but also lays the groundwork for a future powered by real-time, intelligent decision-making at the Edge.

## References

Alphabet Inc. (2024). *Waymo: Self-driving technology for safer roads.* Alphabet. https://waymo.com/.

Amazon. (2024). *Project Kuiper.* About Amazon. https://www.aboutamazon.com/what-we-do/devices-services/project-kuiper.

Belski, V. (2023, May 23). How can blockchain and edge computing work together? The possibilities and advantages. Edgeir.com; *Edge Industry Review.* https://www.edgeir.com/how-can-blockchain-and-edge-computing-work-together-the-possibilities-and-advantages-20230523.

Cisco. (2023). *Cisco Meraki: Simplifying edge connectivity with wireless mesh networks.* Cisco. https://meraki.cisco.com/solutions/.

European Data Protection Board. (2019). *The CNIL's restricted committee imposes a financial penalty of 50 million euros against Google LLC.* European Data Protection Board. https://www.edpb.europa.eu/news/national-news/2019/cnils-restricted-committee-imposes-financial-penalty-50-million-euros_en.

Google Cloud. (2023). TPUs: *Accelerating AI processing at the edge.* Google Cloud. https://cloud.google.com/tpu.

Hewlett Packard Enterprise. (2023). *Why edge computing is the key to reducing data transfer costs and optimizing bandwidth. Hewlett Packard Enterprise.* https://www.hpe.com/us/en/what-is/edge-computing.html.

Intel Corporation. (2023). *Edge-native Kubernetes for the software-defined edge.* Intel. https://www.intel.com/content/www/us/en/edge-computing/smart-edge.html.

Jenkins, B., & Botbyl, I. (2024, September 12). *What is an edge data center? Interconnections—the Equinix Blog*; Equinix. https://blog.equinix.com/blog/2024/09/12/what-is-an-edge-data-center/.

Microsoft Azure. (2023). *Azure Stack Edge: Bringing AI and analytics to the edge.* Microsoft Azure. https://azure.microsoft.com/en-us/products/azure-stack/edge/.

NVIDIA. (2023). *A100 and H100 GPUs: Powering edge AI with high-performance processing.* NVIDIA. https://www.nvidia.com/en-us/data-center/a100/.

SpaceX. (2023). *Starlink: Low-latency satellite internet for remote edge applications.* SpaceX. https://www.starlink.com/.

Tidy, J. (2020, October 16). *British Airways fined £20m over data breach.* BBC News. https://www.bbc.com/news/technology-54568784.

# 30. Optimizing Machine Learning with AI Agents

Sunil Manikani

The rapid advancement of artificial intelligence (AI) and machine learning (ML) has led to transformative innovations across various industries. However, these advancements are accompanied by significant challenges, particularly concerning the environmental impact of AI-driven technologies. As the demand for computational power increases, so does the energy consumption and associated carbon emissions of data centers, which contribute to climate change and environmental degradation. This contribution examines the promises and perils of utilizing AI agents to optimize ML tasks by strategically selecting cloud regions based on carbon intensity. This method leverages real-time data from the ElectricityMaps API to minimize the environmental footprint of computational processes, offering a promising approach to integrating sustainability into AI practices while addressing the broader implications of energy consumption in AI.

The dual nature of AI presents both opportunities and challenges in its deployment. On one hand, AI's potential to enhance efficiency and innovation holds immense promise for advancing technological capabilities and economic growth. On the other hand, the energy-intensive nature of AI operations poses significant perils, particularly in terms of environmental sustainability. This piece explores how AI can be harnessed to mitigate its environmental impact through intelligent decision-making processes, illustrating the balance between the benefits of AI innovation and the necessity of addressing its environmental costs.

## *The Challenge of Sustainable AI*

Artificial intelligence (AI) and machine learning (ML) have become essential components of modern technological advancements, driving efficiency and innovation across various sectors, including healthcare, finance, transportation, and entertainment. However, the deployment of AI, particularly in training large-scale ML models, is associated with significant environmental challenges due to its high energy consumption. The computational demands of training these models require substantial resources, often translating into increased energy usage and associated carbon emissions.

Data centers, which host these computations, have emerged as major consumers of global electricity. According to a study by Malmodin and Lundén (2018), data centers account for approximately 1 percent of the world's electricity demand, a figure that continues to rise with the growing reliance on AI and ML technologies. The carbon footprint of AI models is exacerbated by the need for vast amounts of data processing and storage, contributing to the environmental impact associated with the digital economy. For instance, training a single large transformer model can emit as much carbon as five cars over their lifetimes, highlighting the urgent need for sustainable AI practices.

Traditional strategies to mitigate the environmental impact of AI have centered on enhancing hardware efficiency and optimizing algorithms. Efforts such as developing more efficient processors and implementing advanced cooling technologies have shown promise in reducing energy consumption. However, these approaches often neglect the strategic advantages of geographic workload placement. Given the variability in carbon intensity across different regions, driven by factors such as energy sources and local grid efficiencies, selecting optimal locations for computational tasks can significantly lower carbon emissions. This geographic optimization strategy leverages the disparity in carbon footprints between regions to minimize the environmental impact of AI operations, aligning with corporate social responsibility initiatives and environmental sustainability goals.

## *The Role of AI Agents in Decision Making*

AI agents, with their ability to process and analyze vast amounts of data, present a promising solution for optimizing machine learning (ML) execution in the pursuit of sustainability. These agents function as sophisticated planning tools capable of thinking through complex scenarios step-by-step to reach precise conclusions. By automating the decision-making process, AI agents evaluate multiple factors, such as carbon intensity and regional energy costs, to make informed choices about where to execute workloads based on real-time data.

In this project, a specialized AI agent was developed with the role of a "carbon intensity parse agent." The agent's primary objective was to identify cloud regions with the lowest carbon intensity by analyzing data retrieved from the ElectricityMaps API. Designed to function as an expert in interpreting JSON-formatted carbon intensity reports, the agent utilized a set of predefined tools to efficiently perform its tasks. This method ensured that the decision-making logic was mistake-proof, relying on structured processes and accurate data interpretation to achieve optimal outcomes.

The implementation of this planning agent demonstrated how AI can significantly enhance environmental sustainability through intelligent automation. By systematically assessing real-time carbon intensity data and integrating it into the decision-making framework, the AI agent effectively streamlined the process of selecting optimal locations for ML tasks. This approach highlights the dual nature of AI-driven solutions, demonstrating their potential to foster technological advancement while addressing ecological challenges, thereby enabling sustainable innovation in AI applications.

## *Leveraging the ElectricityMaps API*

The ElectricityMaps API played a pivotal role in equipping the AI agent with the necessary data to make well-informed decisions regarding the optimal location for executing machine learning (ML) tasks. This API provides real-time carbon intensity information based on precise geographic coordinates, which is crucial for evaluating the environmental impact of running computational workloads in various regions. By accessing this data, the AI agent could dynamically assess the carbon footprint associated with different cloud data centers, enabling a comprehensive analysis of potential locations.

In this study, major cloud data centers in Montréal, Iowa, Mumbai, and London were considered for evaluation. These regions were chosen due to their significant cloud infrastructure and varying carbon intensity levels, influenced by factors such as energy sources and grid efficiency. The AI agent used the real-time data from the ElectricityMaps API to compare the carbon intensity of each location, ultimately identifying the most environmentally friendly option for deploying ML workloads. This decision-making process not only considered the immediate carbon emissions but also factored in the sustainability practices and energy policies of each region.

The integration of external data sources like the ElectricityMaps API into AI-driven decision-making processes exemplifies the transformative potential of such technologies. By utilizing real-time environmental data, the AI agent could provide a dynamic and responsive method for optimizing operations in line with sustainability goals. This approach underscores the importance of leveraging accurate, up-to-date information to make strategic decisions that balance the need for computational power with ecological responsibility. By incorporating this level of granularity into decision-making, organizations can significantly enhance their ability to meet sustainability targets while maintaining operational efficiency.

## *Implementation of the AI Agent*

The implementation of the AI agent involved several key components. The agent was programmed to perform thorough research on each potential location, evaluating factors such as carbon intensity and regional energy sources. It then compiled these findings to determine the optimal location for executing ML tasks with the least environmental impact.

A specific task, the "best region proposal task," was designed to guide the agent in its decision-making process. This task required the agent to invoke the necessary tools for each region and accumulate results, ultimately identifying the location with the lowest carbon intensity. The expected output was an informative response detailing the recommended location, the associated place name, and the carbon intensity value.

The agent's decision-making process was enhanced by its ability to use the "carbon break-up tool," which provided a detailed breakdown of carbon intensity values for each region based on latitude and longitude. By automating these processes, the AI agent efficiently identified the optimal execution location, ensuring minimal carbon emissions while maintaining operational effectiveness.

## *Results and Impact*

- Feasibility and Effectiveness:
    - ◊ The implementation of the AI agent proved to be both feasible and effective in optimizing machine learning (ML) workloads for sustainability.

    - ◊ By utilizing the agent, the project demonstrated how AI could be leveraged to strategically select regions with the lowest carbon intensity.

- Environmental Impact:
    - ◊ The project achieved substantial reductions in the environmental impact of computational tasks.
    - ◊ By selecting regions based on their carbon intensity, the AI agent minimized the carbon footprint associated with ML operations.
    - ◊ Increase in the percentage of ML operations conducted in regions powered by renewable energy > 50 percent
- Contribution to Sustainability Goals:
    - ◊ This approach highlighted the potential for AI to make meaningful contributions to sustainability goals.
    - ◊ It provided valuable insights into how AI agents can be applied in real-world scenarios to enhance environmental responsibility.
    - ◊ It contributed to corporate sustainability targets by reducing Scope 2 emissions by >50 percent on an average for high performing ML workloads.
- Use of Real-Time Data:
    - ◊ The use of real-time data from the ElectricityMaps API enabled precise decision-making processes.
    - ◊ Automated decision-making facilitated the efficient allocation of resources, optimizing operations for both performance and sustainability.
    - ◊ Average decision-making time reduced to almost zero with automated, data-driven processes
- Integration into Sustainability Strategies:
    - ◊ The results underscore the importance of integrating AI into broader sustainability strategies.
    - ◊ By employing AI-driven solutions, organizations can align technological advancements with ecological responsibility, paving the way for more sustainable practices in computational tasks.
    - ◊ Increase in employee participation in sustainability initiatives by 10 percent due to AI-driven insights.

## *Lessons Learned and Future Directions*

By leveraging AI to handle complex calculations and analyses, organizations can ensure more consistent and accurate outcomes while freeing up human resources for other strategic initiatives.

Looking ahead, the continued development and refinement of AI agents hold significant promise for advancing sustainability efforts. As the availability and accuracy of environmental data improve, AI agents can play an increasingly vital role in optimizing operations across various industries. Additionally, expanding the application of AI agents beyond ML execution to other energy-intensive processes presents new opportunities for reducing environmental impact and promoting sustainable practices.

## *Conclusion*

The application of AI agents in optimizing machine learning execution for carbon intensity reduction offers a compelling example of how AI can drive sustainability in technology. By leveraging real-time data and automating decision-making processes, this approach demonstrates the potential for AI to enhance environmental responsibility while maintaining operational efficiency. As the field of AI continues to evolve, its role in promoting sustainability will undoubtedly expand, providing new avenues for innovation and impact in the quest for a greener future.

### References

ElectricityMaps. (n.d.). *ElectricityMap: Live CO2 emissions of electricity consumption*. ElectricityMaps API. Retrieved from [https://api-portal.electricitymaps.com/](https://api-portal.electricitymaps.com/).

Ghallab, M., Nau, D., & Traverso, P. (2004). *Automated planning: Theory and practice*. Morgan Kaufmann. Retrieved from [https://books.google.co.in/books/about/Automated_Planning.html?id=eCj3cKC_3ikC&redir_esc=y](https://books.google.co.in/books/about/Automated_Planning.html?id=eCj3cKC_3ikC&redir_esc=y).

Li, Y., Jiang, Z., Wang, X., & Huang, Z. (2022). Energy-aware virtual machine allocation in DVFS-enabled cloud data centers. *IEEE Transactions on Services Computing*. Retrieved from [https://ieeexplore.ieee.org/stamp/stamp.jsp?arnumber=9656143](https://ieeexplore.ieee.org/stamp/stamp.jsp?arnumber=9656143).

Malmodin, J., & Lundén, D. (2018). The energy and carbon footprint of the global ICT and E&M sectors 2010–2015. *Sustainability*, 10(9), 3027. Retrieved from [https://www.mdpi.com/2071-1050/10/9/3027](https://www.mdpi.com/2071-1050/10/9/3027).

Shang, X., Luo, H., & Jiang, Z. (2022). Energy-aware systems for real-time job scheduling in cloud data centers. *Computers & Electrical Engineering*, 101, 108022. Retrieved from [https://www.sciencedirect.com/science/article/abs/pii/S0045790622000106](https://www.sciencedirect.com/science/article/abs/pii/S0045790622000106).

# 31. AI Companies Train Language Models on YouTube's Archive*

## *Making Family-and-Friends Videos a Privacy Risk*

Ryan McGrady *and* Ethan Zuckerman

The promised artificial intelligence revolution requires data. Lots and lots of data. OpenAI and Google have begun using YouTube videos to train their text-based AI models. But what does the YouTube archive actually include?

Our team of digital media researchers at the University of Massachusetts Amherst collected and analyzed random samples of YouTube videos to learn more about that archive. We published an 85-page paper about that dataset and set up a website called TubeStats for researchers and journalists who need basic information about YouTube.

Now, we're taking a closer look at some of our more surprising findings to better understand how these obscure videos might become part of powerful AI systems. We've found that many YouTube videos are meant for personal use or for small groups of people, and a significant proportion were created by children who appear to be under 13.

### *Bulk of the YouTube Iceberg*

Most people's experience of YouTube is algorithmically curated: Up to 70 percent of the videos users watch are recommended by the site's algorithms. Recommended videos are typically popular content such as influencer stunts, news clips, explainer videos, travel vlogs and video game reviews, while content that is not recommended languishes in obscurity.

Some YouTube content emulates popular creators or fits into established genres, but much of it is personal: family celebrations, selfies set to music, homework assignments, video game clips without context and kids dancing. The obscure side of YouTube—the vast majority of the estimated 14.8 billion videos created and uploaded to the platform—is poorly understood.

---

*Originally published as Ryan McGrady and Ethan Zuckerman, "AI Companies Train Language Models on YouTube's Archive—Making Family-and-Friends Videos a Privacy Risk," *The Conversation*, https://theconversation.com/ai-companies-train-language-models-on-youtubes-archive-making-family-and-friends-videos-a-privacy-risk-232121 (June 27, 2024). Reprinted with the permission of the publisher and authors.

Illuminating this aspect of YouTube—and social media generally—is difficult because big tech companies have become increasingly hostile to researchers.

We've found that many videos on YouTube were never meant to be shared widely. We documented thousands of short, personal videos that have few views but high engagement—likes and comments—implying a small but highly engaged audience. These were clearly meant for a small audience of friends and family. Such social uses of YouTube contrast with videos that try to maximize their audience, suggesting another way to use YouTube: as a video-centered social network for small groups.

Other videos seem intended for a different kind of small, fixed audience: recorded classes from pandemic-era virtual instruction, school board meetings and work meetings. While not what most people think of as social uses, they likewise imply that their creators have a different expectation about the audience for the videos than creators of the kind of content people see in their recommendations.

## *Fuel for the AI Machine*

It was with this broader understanding that we read *The New York Times* exposé on how OpenAI and Google turned to YouTube in a race to find new troves of data to train their large language models. An archive of YouTube transcripts makes an extraordinary dataset for text-based models.

There is also speculation, fueled in part by an evasive answer from OpenAI's chief technology officer Mira Murati, that the videos themselves could be used to train AI text-to-video models such as OpenAI's Sora.

*The New York Times* story raised concerns about YouTube's terms of service and, of course, the copyright issues that pervade much of the debate about AI. But there's another problem: How could anyone know what an archive of more than 14 billion videos, uploaded by people all over the world, actually contains? It's not entirely clear that Google knows or even could know if it wanted to.

## *Kids as Content Creators*

We were surprised to find an unsettling number of videos featuring kids or apparently created by them. YouTube requires uploaders to be at least 13 years old, but we frequently saw children who appeared to be much younger than that, typically dancing, singing or playing video games.

In our preliminary research, our coders determined nearly a fifth of random videos with at least one person's face visible likely included someone under 13. We didn't take into account videos that were clearly shot with the consent of a parent or guardian.

Our current sample size of 250 is relatively small—we are working on coding a much larger sample—but the findings thus far are consistent with what we've seen in the past. We're not aiming to scold Google. Age validation on the internet is infamously difficult and fraught, and we have no way of determining whether these videos were uploaded with the consent of a parent or guardian. But we want to underscore what is being ingested by these large companies' AI models.

## *Small Reach, Big Influence*

It's tempting to assume OpenAI is using highly produced influencer videos or TV newscasts posted to the platform to train its models, but previous research on large language model training data shows that the most popular content is not always the most influential in training AI models. A virtually unwatched conversation between three friends could have much more linguistic value in training a chatbot language model than a music video with millions of views.

Unfortunately, OpenAI and other AI companies are quite opaque about their training materials: They don't specify what goes in and what doesn't. Most of the time, researchers can infer problems with training data through biases in AI systems' output. But when we do get a glimpse at training data, there's often cause for concern. For example, Human Rights Watch released a report on June 10, 2024, that showed that a popular training dataset includes many photos of identifiable kids.

The history of big tech self-regulation is filled with moving goal posts. OpenAI in particular is notorious for asking for forgiveness rather than permission and has faced increasing criticism for putting profit over safety.

Concerns over the use of user-generated content for training AI models typically center on intellectual property, but there are also privacy issues. YouTube is a vast, unwieldy archive, impossible to fully review.

## *AI Companies Train Language Models on YouTube's Archive— Making Family-and-Friends Videos a Privacy Risk*

Models trained on a subset of professionally produced videos could conceivably be an AI company's first training corpus. But without strong policies in place, any company that ingests more than the popular tip of the iceberg is likely including content that violates the Federal Trade Commission's Children's Online Privacy Protection Rule, which prevents companies from collecting data from children under 13 without notice.

With last year's executive order on AI and at least one promising proposal on the table for comprehensive privacy legislation, there are signs that legal protections for user data in the U.S. might become more robust.

When the *Wall Street Journal*'s Joanna Stern asked OpenAI CTO Mira Murati whether OpenAI trained its text-to-video generator Sora on YouTube videos, she said she wasn't sure.

Have you unwittingly helped train ChatGPT?

The intentions of a YouTube uploader simply aren't as consistent or predictable as those of someone publishing a book, writing an article for a magazine or displaying a painting in a gallery. But even if YouTube's algorithm ignores your upload and it never gets more than a couple of views, it may be used to train models like ChatGPT and Gemini.

As far as AI is concerned, your family reunion video may be just as important as those uploaded by influencer giant Mr. Beast or CNN.

# 32. Less Is More—Sometimes

Vinay Singh

We are drowning in data. According to some estimates, over 400 million terabytes of data are created daily, and the number is increasing. The question is how we can extract useful information from the data and whether big data is what we need.

Companies invest in massive data infrastructures, hire data scientists and engineers, and use advanced algorithms to sift through terabytes of information. However, in many cases, the data points that provide the most value are not buried within massive datasets—they are found in *small data*, which can provide clear, actionable insights.

This contribution explores why small data is sometimes all we need to solve complex problems, make strategic decisions, and improve products, even in an age of data abundance.

## *What Is Small Data?*

Small data is manageable, structured, often specific, easily comprehensible, and usable. It is collected from everyday interactions, such as user feedback, surveys, or direct observations. Unlike big data, small data is focused, accessible, and easy to interpret without the need for sophisticated tools or massive computational power.

## *Benefits of Small Data*

### Actionable Insights, Faster and Explainable

When dealing with small data, the insights are often more precise and directly actionable. Because the data is limited in scope, it is easier to identify patterns, trends, or outliers without wading through massive amounts of irrelevant data. Further, the decisions or policies adopted based on small data can be articulated and explained. In an age where AI has its skeptics, explainability is essential.

### High Signal-to-Noise Ratio

The main challenge with big data is separating the wheat from the chaff. Small data, on the other hand, typically consists of highly curated, relevant data points, making it easier to extract meaningful insights. Identifying relevant small data requires deep domain knowledge and intuition.

### Personalized and Contextualized Insights

Small data often reflects specific, real-world contexts and situations. For instance, businesses that rely on personalized customer interactions (such as boutique shops or high-end service providers) usually rely on small data, like individual preferences or past purchases, to build relationships and drive sales.

Unlike big data, which can be too general or aggregate to capture the nuances of individual needs, small data enables companies to tailor their offerings to their customers' unique preferences.

## *Success Stories from Small Data*

In his book Small Data: The Tiny Clues That Uncover Huge Trends, Martin Lindstrom argues that small data is often more potent than big data. He gives several examples of when small data and domain knowledge allowed businesses to arrive at an impactful solution that may have been hard, if not impossible, to uncover by analyzing big data.

"Big data is all about finding correlation. Small data is about finding causation." There are many examples of small data leading to significant insights and measurable and attributable results. The insights are usually a result of deep domain expertise realizing the significance of a particular subset of data.

In his article Unbiggen AI Andrew Ng (2021) argues for a shift from the traditional focus on large-scale data and models to a "data-centric" approach in artificial intelligence. He emphasizes the importance of systematically engineering high-quality data, especially in industries where massive datasets are unavailable. Ng argues that, for many applications, refining and curating smaller, well-labeled datasets can lead to more efficient and accurate AI systems than merely increasing data volume. This perspective challenges the prevailing trend of developing ever-larger models and highlights the need for thoughtful data preparation to advance AI capabilities.

> *"In many industries where giant data sets simply do not exist, I think the focus has to shift from big data to good data. Having 50 thoughtfully engineered examples can be sufficient to explain to the neural network what you want it to learn."*
>
> *—Andrew Ng, CEO & Founder, Landing AI*

### Small Data Complements Big Data

While this article celebrates small data, it is essential to acknowledge that big and small data have their place. Big data identifies long-term trends, predicts behaviors across large populations, and optimizes complex systems. However, big data may need more depth and specificity to answer specific questions, especially when understanding human behavior on a granular level.

In contrast, small data allows organizations to:

- Make decisions quickly based on current, relevant information.
- Understand individual behavior and customer sentiment in detail.
- Experiment with smaller, focused datasets before scaling up.

The real power comes from balancing both approaches. Organizations that use big data to uncover broad trends and small data to inform day-to-day decisions will extract the most insight from data.

### When to Rely on Small Data

Relying on small data is often the best approach when precision and quality outweigh the benefits of sheer volume. Small data becomes essential in contexts where data collection is difficult, expensive, or sensitive—such as healthcare, niche research, or environments with strict privacy concerns. Carefully curated datasets, free of noise and well-structured, can deliver highly relevant insights and support models that perform efficiently. Small data is also preferable when resources are limited, as the analysis can be conducted quickly without extensive computational power. When understanding specific human behaviors or uncovering niche trends is more valuable than finding generalized patterns, small data allows for a more contextual and interpretable approach.

In contrast, big data is crucial when dealing with complex patterns, scalable solutions, and scenarios requiring broad, diverse insights. Applications like machine learning for image recognition, natural language processing, or autonomous vehicles depend on massive datasets for accuracy and reliability. Big data is also invaluable for real-time analytics and predictive modeling, where a comprehensive view of large, varied datasets is needed to identify trends, make predictions, or respond instantly to new information. For organizations that need to cater to a broad audience or account for high variability in their models, big data provides the robustness and scalability necessary for success. Ultimately, the choice between small and big data depends on the problem at hand, with each offering distinct advantages based on the goals and constraints of the project.

## *Conclusion*

In the age of massive data streams and ever-expanding datasets, it is easy to overlook the significant value that small data can offer. Nevertheless, as technology evolves, it is becoming increasingly clear that small data is critical in delivering efficient, high-quality insights, especially in areas where large-scale data could be more practical and efficient. The effectiveness of small data lies in its precision: it provides clean, relevant, and easily interpretable information that can lead to actionable outcomes with fewer resources. It excels in contexts where privacy, data quality, and swift decision-making are paramount, like healthcare, specialized research, or highly tailored marketing strategies.

However, the argument is not that small data should replace big data but that the two should work together to maximize their benefits. With its vast scope, big data uncovers complex patterns and powers innovations such as artificial intelligence, real-time analytics, and scalable solutions for global challenges. It identifies trends that would otherwise remain hidden and drives predictive modeling across diverse domains. However, without the focus and depth provided by small data, big data can sometimes lack the granularity needed to make a real impact, running the risk of being too general or unwieldy to interpret meaningfully.

The most powerful and strategic data-driven decisions come from balancing both approaches. When organizations integrate the precision of small data with the expansive reach of big data, they create a more comprehensive framework for understanding complex problems. This synergy allows for developing accurate, scalable, and highly relevant models for real-world applications. Ultimately, embracing both small and big data as complementary assets empowers businesses and researchers to extract richer insights, optimize performance, and drive meaningful innovation in a data-centric world.

It is about more than how much data is used; it is about how it is used. Creativity and deep domain knowledge can lead to better outcomes with fewer resources and more explainability.

## References


Davenport, T.H., & Kim, J. (2013). *Keeping Up with the Quants: Your Guide to Understanding and Using Analytics*. Harvard Business Review Press.( https://spectrum.ieee.org/andrew-ng-data-centric-ai).

Lindstrom, M. (2016). *Small Data: The Tiny Clues That Uncover Huge Trends*. St. Martin's Press.

Lohr, S. (2015). *Data-ism: The Revolution Transforming Decision Making, Consumer Behavior, and Almost Everything Else*. Harper Business.

Mayer-Schönberger, V., & Cukier, K. (2013). *Big Data: A Revolution That Will Transform How We Live, Work, and Think*. Eamon Dolan/Houghton Mifflin Harcourt.

Ng, A. (2021). "Unbiggen AI: The Importance of a Data-Centric Approach." *IEEE Spectrum*.

Redman, T.C. (2019). "If Your Data Is Bad, Your Machine Learning Tools Are Useless." *Harvard Business Review*.



## Acknowledgment

ChatGPT 4 was used to generate some of the content.

# • *H. AI in Education and the Arts* •

# 33. Shaping the Future of Learning

SULBHA SHANTWAN

I still remember that day in mid–April 2022 when I first heard about OpenAI's ChatGPT. Up until then, creating images in your mind and bringing them to life in seconds seemed like something straight out of a sci-fi movie. Writing an entire book chapter in a blink? Searching and brainstorming ideas instantly? It all sounded too good to be true. But then, open access to AI arrived and took the world by storm. Enthusiasts like me were blown away by its capabilities. I couldn't wait to test it out and see its applications in my domain. My quest for exploring new tools quickly became a newfound hobby.

Some of the first applications I discovered with ChatGPT were lesson plan creation, worksheet generation, brainstorming sessions, and, best of all, grading papers in record time. Welcome to the future of learning, where AI tools empower you to do what you do best: build connections with your students.

## *The Evolution of AI in Education*

Artificial Intelligence (AI) has quietly integrated into our lives for years, with tools like Google Assistant, Netflix recommendations, and voice assistants like Alexa and Siri. The recent buzz is because AI has become conversational, letting us interact with it naturally and solve problems easily. This shift is good, as it makes powerful technology accessible to everyone.

Back in the 1960s and 70s, pioneers were already exploring how computers could support learning. The PLATO project at the University of Illinois was a trailblazer in this regard, offering interactive learning tools like online forums, educational games, and even email—a groundbreaking innovation for its time!

Moving into the 1980s and 90s, Intelligent Tutoring Systems (ITS) emerged, exemplified by Carnegie Learning's Cognitive Tutor. These systems aimed to replicate one-on-one tutoring, adapting to individual student needs through principles of cognitive science.

While this was a significant advancement, the true revolution came in the twenty-first century with breakthroughs in machine learning and big data. With these advances, we could analyze vast amounts of educational data to uncover how students learn most effectively. Online Learning Platforms began leveraging AI to offer personalized lesson recommendations based on student performance, customizing education in ways previously unimagined.

## *The TREE Framework for Implementing AI in K-12 Education*

In one of the early trainings that I had conducted for schools on AI, one challenge the educators were facing was effectively integrating these tools in the classroom.

To address this, I've developed the TREE framework—Teach, Reinforce, Explore, and Evaluate. This framework provides a structured approach to implementing AI in K-12 education, ensuring that we harness its potential while maintaining focus on pedagogical goals.

**A. Teach**

- **Personalized learning paths:** AI allows for instant customization of lessons based on student ability and learning style, making education truly personal.
- **Intelligent tutoring systems:** AI serves as a personal tutor, guiding students through challenging material with thought-provoking questions.

**B. Reinforce**

- Practice exercises: AI generates practice questions that adapt in complexity based on student performance, making mastery of concepts easier.
- Automated feedback: AI provides instant feedback and remediation suggestions, reducing the workload for teachers.

**C. Explore**

- **VR/AR experiences:** Immersive experiences enhance engagement and content retention.
- **AI research assistants:** AI helps students research topics creatively and encourages critical thinking by cross-checking information.

**D. Evaluate**

- **Assess and adjust:** Continuously evaluate the impact of AI on learning and adjust strategies to enhance outcomes.

The TREE framework helps educators integrate AI effectively, combining technology with the human touch to create a more engaging and personalized learning environment.

## *The Enduring Role of Teachers in the Age of AI*

AI tools are transforming how educators work by streamlining tasks like lesson planning and worksheet creation, saving valuable time and ensuring consistent quality.

However, AI isn't flawless. It can occasionally generate content that doesn't actually exist, a phenomenon known as "hallucination." This is where a teacher's expertise becomes indispensable—teachers must review AI-generated content to ensure accuracy and relevance.

During a recent training session, a school leader posed a thoughtful question: "If AI handles lesson plans and worksheets, what will teachers do?" The answer is straightforward: teachers will do what they have always done best. They will focus on the human side of teaching—nurturing relationships, understanding classroom dynamics, and adapting to the unique needs of their students.

AI is a powerful tool, but it is not a replacement for the irreplaceable human elements of teaching. The most effective approach combines the analytical power of AI with the empathy, creativity, and inspiration that only teachers can provide. As AI continues to evolve, the role of teachers will undoubtedly adapt, but their importance in the education system will remain steadfast.

## *Rethinking Assessments*

At a recent gathering of school leaders, a recurring concern was the integration of generative AI tools in classrooms, particularly regarding the potential for widespread cheating. The fear is that students could easily use AI to produce high-quality assignments that are difficult to distinguish from original work, compromising transparency in the assessment process. This challenge leaves educators struggling to differentiate between authentic and AI-generated content, posing a serious threat to academic integrity.

Some argue that AI should be embraced, encouraging more creative and knowledge-testing assignments. Others worry about maintaining academic integrity. This debate has led to a broader discussion on redefining originality and authorship in the age of AI, highlighting the need for clear policies around these issues.

To address these concerns, assessment practices must evolve to prioritize skill development over knowledge recall, using open-ended questions and formative methods. A multimodal approach, incorporating various assessment formats, is essential.

One promising solution is the application of blockchain technology in assessment. Blockchain can provide a secure, transparent, and tamper-proof record of students' work, ensuring that all submissions are verifiable and authentic. By recording every step of the learning and assessment process on a blockchain, educators can trace the origins of a student's work, making it easier to identify whether AI was used appropriately.

Blockchain also offers the potential to redefine originality and authorship. Each contribution to a project or assignment can be logged, providing a clear audit trail that reflects a student's unique input. This not only preserves academic integrity but also encourages collaboration and innovation, as students can confidently share and build on each other's work without fear of plagiarism.

As schools explore the integration of AI in assessment, frameworks that incorporate blockchain technology can guide this process, ensuring equity and fairness. By leveraging blockchain, assessments can become more inclusive, personalized, and effective, enhancing student learning while safeguarding the integrity of the educational process.

## *Challenges and Ethical Considerations*

While AI presents immense opportunities for K-12 education, it also brings significant challenges and ethical considerations that must be carefully addressed. As we integrate AI into our educational systems, we must remain vigilant and proactive in ensuring that these technologies are used responsibly and equitably.

Let's discuss a few challenges.

1. **Digital divide and equity issues:** While nearly the entire world is blown away by AI, we need to think about those people who don't have access to technology, also we need to ensure that AI literacy reaches all students, else we may find ourselves at a risk of creating a new type of inequality. I have also noticed that digital divide could occur due to cultural and linguistic bias.
2. **Data Privacy and Security concerns:** We need to ensure that our students' data is protected. We need policies around data ownership and control and last but not the least cybersecurity risks. We need to forecast and evaluate whether AI systems can potentially be under this threat.
3. **AI bias and Fairness:** AI systems may perpetuate or enhance existing bias in the system. Many AI systems operate as "black boxes," making decisions difficult to understand or challenge, hence the need for interpretable AI in high-stakes educational decisions
4. **Changing roles of teachers:** Many educators fear job displacement, this makes redefining teachers' role in current light extremely important. Besides this educator should keep themselves updated with the new technology, they should maintain a balance between use of technology and using human intelligence to avoid over dependence on technology.

As we navigate these challenges, it's crucial to approach AI implementation in education with a critical eye and a commitment to ethical practices. We must strive to harness the benefits of AI while actively working to mitigate its risks and potential negative impacts. This requires ongoing dialogue between educators, technologists, policymakers, and communities to ensure that AI serves the best interests of all students.

## *The Future of AI in K-12 Education*

As we look ahead, emerging technologies will take the center stage, compelling us to reimagine education in a new light. The integration of these technologies will not only transform the tools we use but also necessitate a realignment of our pedagogies as we move towards hyper-personalization and AI-assisted teaching. It's essential to recognize the importance of human—AI collaboration to ensure a safer, more secure, and balanced future.

As I often emphasize in my lectures and other engagements, AI, in the near future, will begin to take on a teaching role, allowing educators to ascend to the roles of cognitive coaches, mentors, and innovation catalysts. This shift will revolutionize the learning experience and empower educators to focus on cultivating creativity, critical thinking, and emotional intelligence in their students. The future of education is not just about adapting to technology; it's about harnessing its potential to create a world where every learner can thrive.

Emerging technologies like blockchain, AR-VR (Augmented Reality and Virtual Reality), IoT (Internet of Things), and simulations will play crucial roles in shaping this new educational landscape. Blockchain can offer secure, transparent credentialing and record-keeping systems, ensuring the integrity of academic achievements. AR-VR can provide immersive learning experiences, allowing students to explore historical events, scientific phenomena, and complex concepts in a deeply engaging way. IoT can connect learning environments and devices, enabling real-time data collection and personalized feedback, while simulations can offer hands-on experiences in virtual settings, from scientific experiments to social simulations.

As we move forward, it's crucial that all stakeholders in K-12 education—policymakers, administrators, teachers, parents, and students—work together to:

- Develop clear policies and ethical guidelines for the use of AI and other emerging technologies in schools.
- Invest in comprehensive, ongoing professional development for educators to ensure they are equipped to leverage these technologies effectively.
- Ensure equitable access to AI-enhanced and technology-driven educational opportunities, so that all students, regardless of background, can benefit.
- Foster critical thinking and AI literacy among students, preparing them to navigate and contribute to a technology-rich world.
- Maintain a balance between technological innovation and human-cantered education, ensuring that the core values of teaching and learning remain intact.

By thoughtfully and responsibly embracing AI and other emerging technologies, we have the opportunity to create a more engaging, effective, and equitable educational system—one that prepares all students for success in an AI-driven world while nurturing the uniquely human qualities that no machine can replace.

The future of K-12 education is not about choosing between human teachers and AI or other technologies, but about harnessing the strengths of both to provide the best possible learning experience for every student. As we stand on the brink of this technological revolution in education, let us move forward with optimism, critical thinking, and a steadfast commitment to the core values of education.

# 34. Artificial Intelligence in Academic Research

## *Challenges and Paradoxes*

Severo C. Madrona, Jr.

Since its inception in the 1950s, Artificial Intelligence (AI) has experienced substantial evolution. As early as 1950, Alan Turing—often called the father of artificial intelligence—introduced ideas like genetic algorithms and machine learning. John McCarthy and others initially used the term "artificial intelligence" in 1956 during the Dartmouth Conference, when discussing unresolved computer science issues (Shen, 2020; Zhao & Yan-jun, 2021). With extensive applications in various domains, including education, artificial intelligence (AI) has since emerged as a major force in science and technology (Zhao & Yan-jun, 2021). This essay examines how artificial intelligence (AI) affects scholarly research, reviewing its uses, advantages, difficulties, and ethical implications.

Machine learning, deep learning, natural language processing (NLP), and computer vision are just a few of the methods and approaches that fall under the umbrella of artificial intelligence. The use of these technologies in academic research is growing since they offer fresh methods and instruments for handling challenging issues and improving research capacities. Machine learning algorithms, for instance, can examine enormous datasets, spot trends, and produce insights that human researchers cannot see right away. Similarly, NLP can automatically extract and summarize pertinent data from a sizable body of scholarly literature. Researchers in disciplines like biology, material science, and medicine may benefit from the development of sophisticated computer vision systems that can evaluate pictures, videos, and other visual data using AI-based approaches like deep learning (Zhao & Yan-jun, 2021; Chubb et al., 2021).

## *The Role of AI in Academic Research*

The field of incorporating AI into academic research is dynamic and intricate. As AI capabilities continue to develop, there is increasing interest in how AI might change the research process in various ways, from data collecting and analysis to knowledge synthesis and distribution. One important area where AI influences research data is its administration and analysis. Researchers can find hidden links and develop new theories using AI-powered tools to help them organize, filter, and understand massive datasets. AI systems can also find possible partners, forecast research trends, and allocate resources as efficiently as possible.

Writing and assessing academic papers is another area where AI is showing promise. Researchers can assure the clarity and coherence of their work, detect grammatical problems, and enhance their writing style with the aid of AI-based writing helpers. AI-powered solutions can also identify plagiarism, evaluate a research paper's possible impact, and offer suggestions for enhancement.

Academic research increasingly uses AI to improve many parts of the research process. In addition to enabling multidisciplinary research and effect evaluation, AI can be especially useful for information gathering and other specific activities (Chubb et al., 2021). Researchers have found both advantages and disadvantages to integrating AI in research, with some situations viewing AI as a helpful tool (Chubb et al., 2021).

Numerous academic processes, including data analysis, literature searches, article drafting, and peer review, use AI and automation. The full ramifications of these innovations are still largely unknown, though, and researchers and academic institutions are juggling the opportunities and problems these technical breakthroughs bring (Chubb et al., 2021). There are issues with intellectual property rights and plagiarism when using AI techniques in academic writing. Simultaneously, researchers can benefit from AI-powered writing helpers by improving their work's overall quality, coherence, and clarity, which could increase the spread of scientific knowledge (Dong, 2023).

Searching for and discovering literature is one of the most exciting uses of AI in academic research. With the use of AI-powered tools, researchers can find relevant studies, find hidden links between different fields of study, and navigate the enormous and constantly expanding corpus of scholarly literature more effectively (Chubb et al., 2021; Razack et al., 2021). Additionally, these systems can help with automated research paper summarizing, plagiarism detection, and citation management.

The use of AI and machine learning algorithms can help process and evaluate the kind of huge, complicated datasets that are becoming more and more prevalent in academic research. Researchers may find trends, get insights, and create visualizations that improve data comprehension using AI-powered technologies.

Additionally, as the amount and complexity of data produced by researchers continue to increase, AI is becoming increasingly vital in managing and curating research data. With the use of AI-powered technologies, researchers can store, organize, and retrieve data more effectively.

Academic manuscript authoring and evaluation are being done with AI. From identifying basic plagiarism mistakes to forecasting the anticipated citation impact of an unpublished publication, semantic technologies may enable transparent and effective data extraction strategies (Razack et al., 2021).

Preparing and formatting research publications is another possible use of AI in academic research. Researchers can increase the coherence, clarity, and general caliber of their work with the use of AI-powered writing aids, which could facilitate the sharing of scientific knowledge.

Through a variety of approaches and strategies, including machine learning algorithms, natural language processing, predictive analytics, and sentiment analysis, artificial intelligence is being incorporated into the peer review process in scientific publishing. Although this can expedite the process and enhance peer review quality, it also raises questions regarding the necessity of human oversight and the possibility of bias.

Creating study summaries, insights, and possibly even research hypotheses and experimental designs are all possible applications of AI in academic research. AI-powered technologies can sift through vast amounts of literature, identify important trends and discoveries, and compile this data into clear and insightful research summaries. This might make it easier for researchers to keep abreast of the most recent advancements in their domains and possibly open up new research directions.

However, there are also worries about the possibility of biases, mistakes, and the loss of human creativity and critical thinking when using AI-generated research outputs. The thorough validation and interpretation of any study insights produced by AI by human researchers will be crucial.

In the end, academic research can benefit greatly from artificial intelligence (AI) in a number of ways. Better data analysis and information collection are among the main benefits. Artificial Intelligence (AI) can facilitate the analysis of enormous datasets, reveal complex patterns, and produce insightful findings that might otherwise go unreported. AI also makes it easier to collaborate across disciplines and synthesize knowledge. It can help close gaps between many academic areas, encouraging teamwork and smoothly combining information from other domains. AI also helps to improve efficiency by streamlining research operations. AI has the potential to greatly increase academic study efficiency by automating tedious duties and streamlining research procedures, freeing up scholars to concentrate on more intricate and imaginative facets of their work.

## *Implications and Future Directions*

AI can completely transform academic research in terms of presentation, evaluation, and methodology. Prioritizing AI's ethical and appropriate application in academic research is essential to ensuring that emerging technologies empower and strengthen researchers rather than marginalize or replace them.

AI has the potential to significantly improve interdisciplinary cooperation, allowing researchers to address intricate, multidimensional issues that call for knowledge from a variety of disciplines. Researchers from various backgrounds can more successfully share data, communicate ideas, and work together to advance knowledge across disciplinary boundaries by utilizing AI-powered tools and methodologies. This creates fresh opportunities for creativity and the creation of comprehensive answers to urgent global problems.

From data analysis and literature searches to paper drafting and peer review, artificial intelligence has the potential to revolutionize many phases of academic research. But in order to make sure that these technologies complement human researchers rather than take their place, their ethical application is essential. The possibility of bias in AI systems is one of the main ethical issues since it may result in biased study findings or unjust decisions about things like research funding and publication. Additionally, since sensitive research data may be susceptible to breaches or misuse, the broad implementation of AI presents privacy and security risks (Pachegowda, 2024).

It will be crucial to create best practices and guidelines to guarantee the ethical and responsible use of these technologies as the academic community investigates the role of AI in research.

Researchers and academic institutions must aggressively address AI bias and fairness in order to reduce these concerns. To guarantee the responsible development and application of AI in academic research, cooperation between the public and private sectors is crucial. Keeping the data and systems utilized in AI-powered academic research safe and secure is another important ethical consideration. The growing dependence on AI-powered technologies brings up issues related to cybersecurity, data privacy, and the possibility of malicious exploitation or abuse of these systems.

AI's incorporation into scholarly research also affects open science programs, which support openness, cooperation, and the dissemination of research findings. While there are many ways that AI may support and empower researchers, it's crucial to strike a balance between human expertise and AI-driven capabilities. AI systems cannot completely replace the contextual knowledge, ethical judgment, and critical thinking that human researchers bring to the table.

In addition, the application of AI in academic research has consequences for the assessment and measurement of research impact. Natural language processing and predictive analytics can examine social media activity, citation trends, and other data to provide more immediate and accurate evaluations of the impact and reach of research results. Over-reliance on AI-driven impact measurements, however, raises concerns that it may prioritize "flashy" research or undervalue foundational, slow-burning effort.

AI's integration into academic research processes can revolutionize the creation, exchange, and assessment of knowledge. AI-powered solutions have the ability to speed up and improve the quality of research by improving data analysis, peer review, article drafting, and literature search. But there are also serious ethical issues with using AI in academic research, including prejudice, privacy, security, and the possible replacement of human researchers. To guarantee the responsible and moral development and application of AI in research, cooperation between academics, business, and legislators is crucial.

## *Conclusion and Call to Action*

Though its ethical and appropriate application will be essential to achieving these advantages, artificial intelligence has enormous potential to revolutionize academic research. Researchers, institutions, and policymakers must continue to collaborate to ensure that the incorporation of AI into academic workflows empowers and develops researchers rather than marginalizes or replaces them.

A growing number of academic research processes are utilizing artificial intelligence, which has the potential to improve productivity, teamwork, and the general caliber of research results (Dave & Patel, 2023; Chubb et al., 2021). Although the use of AI in academic research has the potential to revolutionize the discipline, serious ethical issues also need to be resolved. Collaboration between researchers, institutions, and policymakers will be necessary for the responsible and ethical deployment of AI in academic research to successfully control the dangers and achieve the benefits of emerging technologies.

As AI becomes more common in scholarly research, it is critical to ensure researchers have the abilities and know-how to use these tools sensibly and successfully. In conclusion, incorporating AI into scholarly research can revolutionize the methods used for conducting, disseminating, and assessing research. Even while there are a lot of advantages to be gained, there are also crucial ethical issues that need to be resolved to guarantee that these technologies are used responsibly and fairly.

## References

Blau, W., Cerf, V.G., Enriquez, J., Francisco, J.S., Gasser, U., Gray, M.L., Greaves, M.F., Grosz, B.J., Jamieson, K.H., Haug, G.H., Hennessy, J.L., Horvitz, E., Kaiser, D., London, A.J., Lovell-Badge, R., McNutt, M., Minow, M., Mitchell, T.M., Ness, S., Witherell, M.S. (2024, May 21). Protecting scientific integrity in an age of generative AI. *National Academy of Sciences*, 121(22). https://doi.org/10.1073/pnas.2407886121.

Chubb, J., Cowling, P., & Reed, D. (2021). Speeding up to keep up: exploring the use of AI in the research process. Springer Nature, 37(4), 1439–1457. https://doi.org/10.1007/s00146-021-01259-0.

Dave, M., & Patel, N. (2023). Artificial intelligence in healthcare and education. *Springer Nature*, 234(10), 761–764. https://doi.org/10.1038/s41415-023-5845-2.

Dong, Y. (2023). Revolutionizing Academic English Writing through AI-Powered Pedagogy: Practical Exploration of Teaching Process and Assessment. *Journal of Higher Education Research*, 4(2), 52–52. https://doi.org/10.32629/jher.v4i2.1188.

Pachegowda, C. (2024). *The Global Impact of AI-Artificial Intelligence: Recent Advances and Future Directions, A Review.* Cornell University. https://doi.org/10.48550/arxiv.2401.12223.

Razack, H.I.A., Mathew, S.T., Saad, F.F.A., & Alqahtani, S.A. (2021) Artificial intelligence-assisted tools for redefining the communication landscape of the scholarly world. *Korean Council of Science Editors*, 8(2), 134–144. https://doi.org/10.6087/kcse.244.

Shen, J. (2020). The Innovation of Education in the Era of Artificial Intelligence. *Proceedings of the 24th annual ACM symposium on User interface software and technology.* https://doi.org/10.3390/proceedings2020047057.

Teng, M.F. (2023). *Scientific Writing, Reviewing, and Editing for Open-access TESOL Journals: The Role of ChatGPT.* tesolunion.org. https://doi.org/10.58304/ijts.20230107.

Zhao, X., & Yan-jun, Y. (2021). *A Study on the Application of Blended Teaching to English Reading Course under the Background of Artificial Intelligence.* IOP Publishing, 693(1), 012019–012019. https://doi.org/10.1088/1755-1315/693/1/012019.

# 35. The Impact of Artificial Intelligence Tools on Problem-Solving Skills

DOAN NGOC DUY

Integrating Artificial Intelligence (AI) in educational contexts is increasingly recognized for its transformative potential, particularly in enhancing learning outcomes and personalizing educational experiences. This is especially relevant in fields such as marketing and multimedia design, where AI tools can facilitate the development of critical skills necessary for success in these disciplines. For instance, Generative AI models like ChatGPT have the potential to revolutionize Vietnam's educational landscape by offering personalized learning experiences tailored to individual student needs. These tools can adapt content to various learning styles, enhancing academic performance and student retention rates (Shoaib et al., 2024; Godwin Olaoye et al., 2024). Furthermore, AI provides instant feedback and support, significantly improving problem-solving skills—an essential competency in today's dynamic and complex world (Muthmainnah et al., 2022; Rane et al., 2023).

In the context of Vietnam's undergraduate education system, characterized by a blend of traditional educational values and a growing openness to technological integration, the impact of AI tools on students' problem-solving skills warrants thorough investigation. The application of AI in academic settings has been linked to improved learning outcomes, mainly when used in short-term interventions, highlighting the novelty effect of these technologies (Ali et al., 2024). However, concerns about the long-term implications of AI on student capabilities, particularly in critical thinking and problem-solving, remain pertinent. Research indicates that while AI can support learning by guiding students through complex problem-solving processes, there is a risk that over-reliance on such technologies may hinder the development of independent critical thinking skills (Zhai et al., 2024).

## *Advantages of AI Integration in Education*

**Enhanced Personalization and Engagement:** AI technologies are instrumental in developing adaptive learning environments tailored to individual student needs, preferences, and progression rates. This personalization optimizes learning outcomes by moving beyond the traditional "one-size-fits-all" educational methods. For example, Xu (2024) and Singh et al. (2024) highlight AI's ability to adapt learning paths in real time, responding to student feedback and performance. This adaptive approach is complemented by the findings of Bhatia et al. (2024), who report that AI-driven personalization can significantly increase student engagement and academic performance. Similarly, Chen et al. (2020) and Marouf et al. (2024) emphasize the importance of AI in processing vast amounts of data to tailor feedback and learning experiences, enhancing the effectiveness of educational programs and maintaining high engagement levels.

**Improved Learner-Instructor Interactions**: AI tools significantly enhance communication and interaction within the educational environment, particularly in online learning contexts. Choung et al. (2023) and Seo et al. (2021) discuss AI's role in facilitating a more interactive and accessible learning experience, which can build students' trust in and acceptance of digital learning platforms. Furthermore, Xu & Ouyang (2022) and Joseph & Uzondu (2024) provide evidence that in specialized areas such as STEM education, AI can outperform traditional teaching methods by supporting complex educational interactions and providing robust feedback mechanisms that aid in more profound comprehension of intricate subjects.

## *Challenges and Risks of AI in Education*

**Ethical and Privacy Concerns**: Implementing AI in education raises significant moral issues, including data privacy and algorithmic bias concerns. Mensah (2023) and Godwin Olaoye (2024) caution that AI technologies could perpetuate existing biases or cause unforeseen harm without proper safeguards. Nguyen et al. (2023) and Barnes & Hutson (2024) advocate for robust ethical frameworks to govern AI usage in education, emphasizing the importance of these frameworks in promoting equity and inclusivity. This perspective is supported by Habbal et al. (2024) and Nguyen et al. (2023), who argue that such frameworks are essential for mitigating risks associated with AI deployments.

**Risk of Over-Reliance on Technology**: Growing concerns exist regarding the potential for over-reliance on AI tools in education, which could inhibit the development of independent critical thinking skills. Collins & Halverson (2018) and Moe & Chubb (2009) discuss how this dependence might undermine traditional educational skills and diminish students' ability to solve problems without technological assistance, potentially reducing engagement and reliance on essential educational competencies.

## *Balanced Perspectives on AI in Education*

**Integration Challenges and Solutions**: Effectively integrating AI into educational frameworks presents several challenges, particularly in balancing technological advances and traditional academic values. Blessing (2024) discusses the necessity of a nuanced approach that respects and incorporates traditional teaching methodologies while leveraging the benefits of technological advancements. Cacho (2024) and Ramkissoon (2024) further highlight practical strategies for integrating AI tools that complement rather than replace traditional teaching methods, ensuring a harmonious blend that benefits all stakeholders.

**Need for Comprehensive Policy and Regulation**: The development and implementation of clear policies and regulations are crucial for managing AI integration's ethical complexities and practical challenges. Tri et al. (2021), Le-Nguyen & Chan (2024), and Nguyen et al. (2023) emphasize the importance of adapting educational systems, particularly in contexts like Vietnam, to create and enforce policies that ensure ethical AI use and promote an equitable educational environment. These policies should address both the potential benefits and the risks associated with AI in education, aiming for a regulatory framework that supports sustainable and responsible AI integration.

## *Methodology*

The study involved Marketing students from the University of Economy Ho Chi Minh City (HCMC) and Multimedia Design students from the University of Architecture HCMC. To guarantee that participants were adequately acquainted with AI tools, only students with at least one year of experience utilizing these technologies in their academic work were eligible. A balanced number of students from both disciplines were selected, enabling a comparative analysis of how ChatGPT influenced their problem-solving skills in their respective fields. Data collection relied on three primary methods: deep interviews, focus groups, and experimental tests conducted during class activities.

## *Findings*

**Application of ChatGPT in Problem-Solving Phases.** Data from class activities and interviews revealed that ChatGPT was predominantly employed in project work's brainstorming, structuring, and finalizing phases. Marketing students frequently used ChatGPT for statistical analysis, identifying trends, and synthesizing input information to support strategic planning and decision-making processes. For instance, 85 percent of Marketing students indicated that ChatGPT helped them streamline tasks such as calculating market size or generating quantitative insights, which were integral to their project development. Conversely, Multimedia Design students leveraged ChatGPT to develop customer insight scenarios, refine market research, and conceptualize target audience profiles. Approximately 78 percent of these students reported that ChatGPT enhanced their creative processes by providing structured guidance and improving their ability to align visual designs with contextual storytelling.

**Role of ChatGPT in Addressing Knowledge Gaps**. The focus group discussions and experimental data highlighted ChatGPT's role as a bridge to fill student curriculum gaps. For Marketing students, the tool proved instrumental in exploring advanced concepts such as consumer behavior models and competitive analysis frameworks. Nearly 70 percent of Marketing participants stated that ChatGPT provided them with theoretical support that was either unavailable or insufficiently covered in their coursework. Similarly, Multimedia Design students, who often excel in visual creativity but need more formal exposure to market-driven insights, used ChatGPT to simulate target audience personas and refine their understanding of branding strategies. This supplementary use of ChatGPT helped students compensate for academic deficiencies, with 65 percent of participants acknowledging its value in aligning their projects with real-world applications.

**Ethical and Practical Challenges with ChatGPT Usage**. The findings also revealed significant challenges in the ethical and practical use of ChatGPT, particularly in unstructured learning environments. During interviews, 42 percent of students admitted to relying excessively on ChatGPT, particularly those with access to the paid version, to complete substantial portions of their projects. This over-reliance reduced their engagement in critical thinking and problem-solving tasks, as they often adopted AI-generated outputs without analysis or adaptation. Furthermore, feedback from focus groups indicated that the absence of clear ethical guidelines exacerbated these issues. Nearly 68 percent of students expressed uncertainty about the acceptable boundaries of AI-assisted work, and some reported experimenting with ChatGPT unsupervised to bans or lack of direction from lecturers. This gap underscores the need for structured policies and targeted training to ensure responsible and productive use of ChatGPT.

## *Discussion*

The findings of this study underscore both the potential and challenges of integrating ChatGPT as a tool for enhancing problem-solving skills among students in the Marketing and Multimedia Design disciplines. The results suggest that while ChatGPT provides substantial support for various problem-solving phases and addresses gaps in students' knowledge, its effectiveness is heavily influenced by students' ability to use it responsibly and the presence of adequate guidance from educators.

**Application in Problem-Solving Phases**. The use of ChatGPT in distinct phases of problem-solving, particularly brainstorming and structuring, demonstrates its capability to streamline and enhance students' project workflows. Marketing students benefited from ChatGPT's analytical abilities, using it to synthesize data and generate quantitative insights, which are vital for strategic decision-making. For Multimedia Design students, ChatGPT served as a creative collaborator, providing customer insights and market-driven scenarios that improved the alignment of their visual designs with strategic goals. These applications highlight the versatility of ChatGPT in adapting to the needs of different academic disciplines. However, the tool's effectiveness depended significantly on students' ability to frame specific and actionable questions, pointing to the need for targeted training in question-framing and critical thinking.

**Addressing Knowledge Gaps**. The study highlights ChatGPT's role as a supplementary tool that bridges curriculum gaps. Students across both disciplines used ChatGPT to access information and frameworks not covered comprehensively in their coursework. This included advanced consumer behavior models and competitive analysis for marketing students, while Multimedia Design students leveraged ChatGPT to simulate audience profiles and refine branding concepts. These findings suggest that ChatGPT can complement traditional educational resources, allowing students to explore and apply complex concepts in project contexts. However, its use as a knowledge-enhancement tool must be balanced with efforts to ensure that students stay within essential cognitive processes, such as critical evaluation and independent problem-solving.

**Ethical and Practical Challenges**. The research also identified ethical and practical challenges associated with ChatGPT usage. The absence of clear guidelines and adequate educator supervision resulted in significant misuse among students, particularly those who relied excessively on ChatGPT to complete substantial portions of their projects. This over-reliance could have improved their engagement in critical thinking and problem-solving tasks, raising concerns about the long-term impact of AI dependence on students' intellectual growth. Furthermore, the lack of clarity about ethical boundaries led to uncertainty among students, with many experimenting independently and often irresponsibly. These findings underline the urgent need for educational institutions to establish comprehensive policies and training programs to guide the ethical and practical use of AI tools like ChatGPT.

## *Conclusion and Call to Action*

This study demonstrates that ChatGPT has the potential to significantly enhance problem-solving skills among students by supporting various phases of project work and addressing curricular deficiencies. The tool's adaptability to diverse academic needs underscores its value as a resource for fostering creativity and analytical thinking. However, the challenges identified, including over-reliance and ethical concerns, highlight the importance of structured integration. Educators and institutions must play a proactive role in developing training programs focusing on question-framing, critical evaluation, and responsible usage of AI tools. Clear ethical guidelines and policies ensure that students leverage ChatGPT as a collaborative tool rather than a substitute for independent problem-solving. By addressing these challenges, educational stakeholders can maximize the benefits of AI integration while mitigating its risks, fostering a learning environment that combines technological innovation with academic integrity.

### References


Ali, O., Murray, P.A., Momin, M., Dwivedi, Y.K., & Malik, T. (2024). The effects of artificial intelligence applications in educational settings: Challenges and strategies. *Technological Forecasting and Social Change, p. 199,* 123076.

Barnes, E., & Hutson, J. (2024, June). Navigating the ethical terrain of AI in higher education: Strategies for mitigating bias and promoting fairness. In *Forum for Education Studies* (Vol. 2, No. 2, pp. 1229–1229).

Bhatia, A., Bhatia, P., & Sood, D. (2024). Leveraging AI to Transform Online Higher Education: Focusing on Personalized Learning, Assessment, and Student Engagement. *International Journal of Management and Humanities (IJMH) Volume-11 Issue-1.*

Blessing, M. (2024). Leveraging Technology to Support Human Knowledge Strategy in Education.

Cacho, R.M. (2024). Integrating Generative AI in University Teaching and Learning: A Model for Balanced Guidelines. *Online Learning, 28*(3), 55–81.

Chen, L., Chen, P., & Lin, Z. (2020). Artificial intelligence in education: A review. *Ieee Access, 8,* 75264–75278.

Choung, H., David, P., & Ross, A. (2023). Trust in AI and its role in the acceptance of AI technologies. *International Journal of Human-Computer Interaction, 39*(9), 1727–1739.

Collins, A., & Halverson, R. (2018). *Rethinking education in the age of technology: The digital revolution and schooling in America.* Teachers College Press.

Godwin Olaoye, E.F. (2024). The Impact of Bias and Fairness Issues on the Robustness and Security of AI Systems.

Godwin Olaoye, J.O., Flypaper, D., Oluwaseyi, J., & Brightwood, S. (2024). Ai-driven adaptive learning systems: enhancing student engagement.

Habbal, A., Ali, M.K., & Abuzaraida, M.A. (2024). Artificial Intelligence Trust, risk and security management (AI trism): Frameworks, applications, challenges and future research directions. *Expert Systems with Applications, 240,* 122442.

Joseph, O.B., & Uzondu, N.C. (2024). Integrating AI and Machine Learning in STEM education: Challenges and opportunities. *Computer Science & IT Research Journal, 5*(8), 1732–1750.

Le-Nguyen, H.T., & Tran, T.T. (2024). Generative AI in terms of Its ethical problems for teachers and learners: Striking a balance. In *Generative AI in Teaching and Learning* (pp. 144–173). IGI Global.
Marouf, A., Al-Dahdooh, R., Ghali, M.J.A., Mahdi, A.O., Abunasser, B.S., & Abu-Naser, S.S. (2024). Enhancing Education with Artificial Intelligence: The Role of Intelligent Tutoring Systems.
Mensah, G.B. (2023). Artificial intelligence and ethics: a comprehensive review of bias mitigation, transparency, and accountability in AI Systems. *Preprint, November, 10.*
Moe, T.M., & Chubb, J.E. (2009). *Liberating learning: Technology, politics, and the future of American education.* John Wiley and Sons.
Muthmainnah, Ibna Seraj, P.M., & Oteir, I. (2022). Playing with AI to Investigate Human-Computer Interaction Technology and Improving Critical Thinking Skills to Pursue 21st Century Age. *Education Research International, 2022*(1), 6468995.
Nguyen, A., Ngo, H.N., Hong, Y., Dang, B., & Nguyen, B.P.T. (2023). Ethical principles for artificial intelligence in education. *Education and Information Technologies, 28*(4), 4221–4241.
Ramkissoon, L. (2024). AI: Powering Sustainable Innovation in Higher Ed. In *The Evolution of Artificial Intelligence in Higher Education* (pp. 203–229). Emerald Publishing Limited.
Rane, N., Choudhary, S., & Rane, J. (2023). Education 4.0 and 5.0: Integrating artificial intelligence (AI) for personalized and adaptive learning. *Available at SSRN 4638365.*
Seo, K., Tang, J., Roll, I., Fels, S., & Yoon, D. (2021). The impact of artificial intelligence on learner–instructor interaction in online learning. *International journal of educational technology in higher education, pp. 18*, 1–23.
Shoaib, M., Sayed, N., Singh, J., Shafi, J., Khan, S., & Ali, F. (2024). AI student success predictor: Enhancing personalized learning in campus management systems. *Computers in Human Behavior*, p. *158*, 108301.
Singh, T.M., Reddy, C.K.K., Murthy, B.R., Nag, A., & Doss, S. (2024). AI and education: Bridging the gap to personalized, efficient, and accessible learning. In *Internet of Behavior-Based Computational Intelligence for Smart Education Systems* (pp. 131–160). IGI Global.
Tri, N.M., Hoang, P.D., & Dung, N.T. (2021). Impact of the Industrial Revolution 4.0 on higher education in Vietnam: Challenges and Opportunities. *Linguistics and Culture Review, 5*(S3), 1–15.
Xu, W., & Ouyang, F. (2022). The application of AI technologies in STEM education: a systematic review from 2011 to 2021. *International Journal of STEM Education, 9*(1), 59.
Xu, Z. (2024). AI in education: Enhancing learning experiences and student outcomes. *Applied and Computational Engineering, 51*(1), 104–111.
Zhai, C., Wibowo, S., & Li, L.D. (2024). The effects of over-reliance on AI dialogue systems on students' cognitive abilities: a systematic review. *Smart Learning Environments, 11*(1), 28.

# 36. The Promise and Perils of AI in the Arts

Larry Ebert

Passion is inseparable from and intrinsic to the arts, so it is not surprising that artists' views about AI are strongly held and passionately expressed. Here, I present findings from original research into the role of AI in the arts. These findings reveal an array of concerns (perils) plus new possibilities (promise). I augment findings with secondary research and community encounters in the arts, but the primary research forms the basis for what follows, except where noted.

In the fall of 2023 I interviewed seventy-four artists, across domains—visual arts, music, theater, film, writing—to explore the role and impact of AI on the arts and creativity. What emerged was a rich tapestry of perspectives, insights and observations about the nature of art and artists, and the changes that may result with AI in art.

In each of the one-on-one interviews I asked this series of questions, with follow-up being common:

1. What is the role of art for you and for the world?
2. What impact will AI have on you or your art?
3. What is the difference, if anything, between human-made art and AI-generated art?
4. What is the future role of AI in the arts?
5. What is the impact, if any, on society, if AI is used prevalently to make art?

## *Perils of AI*

AI poses significant issues in the world of art and is viewed by many artists as a threat. Artists point to these key concerns: breach of ownership rights; challenge to artists' ability to earn a living; loss of authenticity, idiosyncrasy and soulfulness; degraded quality and aesthetic; eroded relationship between artist and receivers or consumers of art; and harmful impacts of an "efficiency ethic" driving AI development and adoption.

Artists interviewed as part of the research expressed concern about ownership rights and see AI introducing challenges in these areas: maintaining ownership and control over how and where an artist's work is used, receiving credit or attribution, receiving fair compensation for original work, and protecting against unauthorized use of their creations.

Interviewees referred to breaches and battles over ownership rights that have received attention in the press. At the time of the research, the SAG-AFTRA strike was in full swing. The strike, and subsequent settlement, centered on artist working conditions and compensation, and included impacts of AI. One of the interviewees for this study was a member of the SAG-AFTRA negotiating team. He echoed concerns expressed by the union, including use of artist likeness without compensation, use of artists' work to train AI models, artist consent (lack of), and transparency in production studio use of digital replicas and generative AI (Lowe & Williams, 2023). There have been a number of other disputes related to artists and AI, including a class action lawsuit against the AI image generator companies Stability AI, Deviant Art, MidJourney, and Runway. The basis for this ongoing case: AI companies use artists' work as training data and are incorporating artist content into new content (Cho, 2024) In the eyes of many artists, such companies are selling "copyright infringement as a service" (Yavuz, 2024).

Nearly all research respondents pointed out, often wistfully, that emergence of AI in the arts will make it harder to earn a living. They point to supply and demand factors. On the demand side, AI enables large firms to produce digitally created art quickly and at low cost, which artists say will reduce demand for human-generated art. On the supply side, most agree that AI will enable an increasing number of individuals to engage in art through use of AI, and, in so doing, may add to competitive pressure in the arts.

In addition to ownership rights and threat to livelihoods, artists have fears about the impact of AI on art-making and on art itself. One of these is the impact on authenticity (Ebert, 2024).

Building on the Turing Test (test of a machine's ability to produce output indistinguishable from a human [Mitchell, 2024]), I asked artists if they would value a piece of art differently if they found out that it was made primarily by a machine rather than a human. The vast majority indicated, often following a long pause, that they would value the art piece less.

In describing the role of art for them and for the world, artists emphasized the importance of authenticity—the unique and genuine hand, heart, and soul of the creative. They express concern that authenticity could be diminished with the presence of AI.

An artist pointed out that "AI can't paint from a place of pain." "The uniqueness of the individual artist matters," he said, and elaborated: "Edward Hopper painted a building in a certain way. Monet would do it in a different way. The way that they saw it would be different. Van Gogh had a starry night. No one's painted it like him. Because of what he was going through." (Ebert, 2024).

Many suggested that art must come primarily from a human to be real. A music arranger drew a crisp line in asserting that an AI-generated composition "might be 'beautiful sounding stuff,' but it's not music."

The struggles of artistic training contribute to authenticity, multiple interviewees noted. "It's a huge part of an artist's journey to reach their authentic expression and mastery of their medium to create an artwork that reflects it," said a writer. One of the painters observed, "Anything you do in art has some self-portrait in it," and expressed fear that this artist journey may become lost with AI.

A visual artist, speaking about the value of idiosyncrasy in art-making, recounted: "Yesterday I was trying to make a mark on a painting, I made a mistake and then stepped back and said, 'I really like that.'" She went on to wonder, "Can AI do that?"

Others mentioned an errant brush stroke, a "wrong" musical note, a surprise gesture or movement in theatrical performance. A violinist describes the joy of playing rubato, inexactness of rhythm, shifting moment by moment (Ebert, 2024). Will the value of these be lost when AI is in the mix, artists wonder.

Respondents are concerned that the quality of art will drop, even drastically. Many (but not all) interviewees foresee a "dumbing down" of art because AI is most-suitable for, and thus most likely to amplify, the more-formulaic expressions of art: films, tv shows and writing that follow a repeated script; forms of music that follow a formula. One actor described AI art as, "Rehashed, with no humanity in it." Artists are concerned that tastes will shift toward these types of products as AI-generated works flood the market. Several artists responded to the question about the future of AI in the arts with this prognostication: "more Marvel movies."

Interviewees raised a concern about loss of connection. To the interview question, "What is the role of art?" the most common response was "connection." Respondents used phrases like, "touching hearts," "sharing in common experiences," "eliciting emotions," "impacting others," and "building community" to describe the deepening of relationship that takes place with art.

An AI entrepreneur, neuroscientist, and writer put it this way: "Art is a means of expression, capturing something in the soul of the receiver." Respondents point out that when we engage in art—making, sharing, and receiving—we connect with other people, ideas, cultures, even with ourselves. It is unclear how well this can happen with AI. "I think if you follow an artist and like an artist, maybe you fall in love with the person as well as their music [art]. You lose that with AI," said a composer.

Not everyone interviewed feels a need to connect with the artist. For some, it's sufficient to feel the impact of a work of art whether it was made by a human or (primarily) generated by a machine. A few artists said they'd still be interested in the backstory of the person prompting an AI tool and even in the story of how the specific technology was used.

There is an efficiency ethic driving much of AI development and adoption: AI can make many tasks faster, easier, and cheaper to complete (Shaw, 2024). While an efficiency ethic may be well-suited to business, artists see a mis-alignment with the creative impulse and motive for art, one that centers on depth of expression, connection, aesthetic, and an inner journey. A world-class musician told a story about the role that music played in his upbringing, how the expression, commitment, and striving for mastery "saved him." Others point to opportunities for growth, finding joy, and "channeling the divine." Art-making (and consumption) have deep linking and embedding in our brains, in our neurology. As the authors of the book, Your Brain on Art, point out, art-making can improve our mental state, cultivate social functioning, and help us process emotions (Magsamen & Ross, 2023). Artists, naturally, are concerned about losing these gifts.

## *Promise of AI*

Not all artists view the role of AI in the arts as an unwelcome threat. Some see possibilities, opportunities, a chance to expand one's thinking, challenge old practices, and discover new forms of creative expression.

Nearly all artists—even those who are adamant that AI is harmful—acknowledge that AI will become an opportunity for persons who didn't feel confident producing art before, to engage. This expanded pool is often referred to as "democratization," and some ask, "isn't this a good thing?"

A handful of respondents noted that, with "AI-generated art," there is still a human in the loop, someone prompting, guiding, choosing, refining what they encounter. Humans may task AI with small or large parts of an overall work of art. There is room for creativity in the collaboration.

Some artists said that AI will help expand their own toolbox and will push or evolve overall boundaries of art. Of these respondents, several pointed to the wildly-immersive art installations by Refik Anadol as an example (Kravchuk, 2023). They mentioned an expanding number of experiments and dimensions in art using AI, including 3-D art, augmented and virtual reality, and noted that AI embedded into existing artist tools, for example generative fill, affords new possibilities. Explorations using AI may expand thoughts about AI. A photographer interviewed for the research of this contribution, for example, is experimenting with AI guided by the inquiry, "What does a pixel know?" (Chin, 2022).

Many interviewees acknowledged that change is a part of the history of the development and evolution of art, and is often met with resistance. Among the examples, artists pointed to the beginnings of photography (felt as a threat to painters), introduction of electronic instruments and music synthesizers (threatening to many musicians and to listeners), and new processes and techniques that arise in art forms from lithography to graphic design. For many there is an awareness that, even amidst fearing AI, change is a natural part of the evolution of art, as of life.

A few artists even suggested that AI will be an impetus for change. "It can force us to up our game, or get out of our rut," exclaimed a filmmaker.

## *A Path Forward*

So where does this lead?

With the presence of both promise and perils, and given that AI is unlikely to recede or vanish from the world of art, and world at large, it seems reasonable that we strive to realize the benefits of AI while guarding against harms.

Certainly, we need to protect ownership and copyright, and this will entail a combination of legal and ethical standards, and governance. One recent legal decision may be helpful to artists: in the United States, AI-generated images cannot be copyrighted as they do not meet the human authorship requirement (Brittain, 2023).

As a society, we must remember the critical link between the arts and culture, the morals and soul of a society, and we can continued to push for support for the arts and for artists. We can insist that artists have a proverbial "seat at the table" in developing AI policy, and AI model design and development, so that our world is not driven only by a technocratic perspective.

As artists and non-artists, we can hone our discernment. We can see and seek authenticity and honor "traditional" forms and practices of art, even as we appreciate new ones. Transparency will be important, too, to support our discernment.

It is possible that AI will expand our thinking, change our image of what constitutes artist and art. AI may lead to increasing roles for people who wish to engage in making art. In addition to the sole creator, in an age with AI, artist roles may include more work as collaborator (with AI), sculptor (of AI content), creative director, and curator.

Ultimately, we have an opportunity to foster the arts in its myriad forms. This may mean a mix of human art, human–AI art, and AI art. Approached with thoughtfulness, we can appreciate and enjoy a full and colorful range.

## References

*The Art Newspaper.* https://www.theartnewspaper.com/2024/10/09/artists-lawsuit-artificial-intelligence-ethics-image-generation.

Brittain, B. (2023, August 21). *AI-generated art cannot receive copyrights, US court says.* Reuters. https://www.reuters.com/legal/ai-generated-art-cannot-receive-copyrights-us-court-says-2023-08-21/.

Chin, A. (2022). *What does a pixel know?* Artist statement. University of San Francisco Profile. https://www.usfca.edu/thacher-gallery/pulled-apart/adam-chin.

Cho, W. (2024, August 13). *Artists score major win in copyright case against AI art generators.* The Hollywood Reporter. https://www.hollywoodreporter.com/business/business-news/artists-score-major-win-copyright-case-against-ai-art-generators-1235973601/.

Ebert, L. (2024, January 18). *AI can't paint from a place of pain, can it? An image generation experiment.* https://larryebert.substack.com/p/ai-and-the-arts-a-series-ai-cant-abb.

Ebert, L. (2024, January 8). *AI and the arts, a series: The beauty of imperfection.* https://larryebert.substack.com/p/ai-and-the-arts-a-series-the-beauty.

Ebert, L. (2024, Summer). *Why artists fear AI beyond content rights.* https://www.caveat-lector.org/3402/pdf/3402_non-fiction_ebert.pdf.

Kravchuk, D. (2023, April 19). *Refik Anadol: creating narratives where art, science and technology collide.* https://art.art/blog/refik-anadol-creating-narratives-where-art-science-and-technology-collide.

Lowe, L., & Williams, C. (2023, July 13). *The SAG-AFTRA strike in Hollywood, explained.* https://www.today.com/popculture/hollywood-actors-sag-strike-2023-explained-rcna94122.

Magsamen, S. & Ross, I. (2023). *Your brain on art: how the arts transform us.* Random House.

Mitchell, M. (2024, August 15). *The Turing Test and our shifting conceptions of intelligence.* Science. https://www.science.org/doi/10.1126/science.adq9356.

Shaw, J. (2024, August 26). *The role of AI in operational efficiency: Beyond the silver bullet.* CIO. https://www.cio.com/article/3496380/the-role-of-ai-in-operational-efficiency-beyond-the-silver-bullet.html.

Yavuz, S.K. (2024, October 9). *Can artists protect their work by suing AI companies?*

# 37. A Paradigm Shift We Need

Anzar Khaliq

It's no secret that many professors in higher education are masters of their disciplines but amateurs at teaching. It's not their fault—there's no crash course in pedagogy handed out with a PhD. Teaching is expected to be a natural extension of their expertise, as if knowing how to solve differential equations or deconstruct Shakespeare automatically translates into the ability to engage a room full of undergrads.

But teaching isn't innate; it's learned. And here lies the rub: admitting you don't know how to teach, especially in academia, feels a bit like admitting you've never really learned to drive. It's embarrassing. It undermines your authority. So, most professors do what anyone would—they fake it. They put up a brave front, masking their uncertainties with lectures and assignments recycled from their own days as students. It's not that they don't care, they care deeply. But caring and knowing what to do are two very different things.

I've seen this façade up close. One professor at a university I worked with confessed during a coffee break, "I've been using the same assessment tools for ten years. It's not because it works so well—it's because I don't know what else to do." This kind of candor is rare, but the struggle it reveals is common. It's also a problem that becomes glaringly obvious when you introduce a disruptor like AI into the mix.

## *The Textbook Trap*

Consider how many STEM courses are built. A professor picks a textbook—one they like, trust, or were recommended—and organizes the course around it. Week 1 is chapter 1, week 2 is chapter 2, and so on. Textbooks in these fields are often designed for this very purpose, with 16 chapters for a 16-week semester. This isn't just convenience, it's a trap. By relying so heavily on a single resource, professors limit the scope of their teaching and the learning experience they can deliver.

In contrast, humanities courses have traditionally embraced a wider variety of sources. However, even in these disciplines, there's a temptation to stick to a predictable syllabus, recycling the same readings and lectures year after year. Whether in STEM or the humanities, the reliance on a fixed curriculum can stifle innovation and student engagement.

I work at a small private university in the Bay Area that is setting itself up as a challenger brand. We are well aware of these problems and saw a similar pattern at our university. Professors would design all of the class experience around the textbook, walking students through the textbook content, week by week, chapter by chapter. The classes felt ... predictable. Students, bright as they were, seemed disengaged. They knew what was coming next because the syllabus was a table of contents in disguise.

Realizing the need for change, the university explored how technology could offer new avenues for teaching. That's when they introduced an AI tool custom-trained in the teaching philosophy of the university and designed to assist in course design and delivery. This AI wasn't meant to replace the professor but to serve as a collaborative partner, offering fresh perspectives and resources.

The AI didn't swoop in with a magic wand but offered an intriguing alternative. Instead of anchoring courses to textbooks, the AI suggested organizing lessons around themes or questions—bigger, messier ideas that didn't fit neatly into a single chapter of a textbook. For example, a traditional first-year writing course shifted from "English Composition 101" to "How to Tell Your Story." Suddenly, the content came from everywhere—personal narratives, interviews, multimedia presentations, and even community storytelling events. The class became a dialogue, not a monologue.

## *The Shock of the New*

Not all professors embrace this change with open arms. One humanities professor, who had long relied on monologues and traditional essays to assess student learning, lamented, "Knowledge as we know it is dead." He had long relied on text as the only meaningful way of expression and couldn't imagine a future where his discipline could survive. For educators like him, the advent of AI felt less like an opportunity and more like an existential threat. What also scared this colleague of mine is the ability of students to offer diverse perspectives, ones that vary from the stance of the professor. The idea that students could challenge established narratives—or worse, that technology could facilitate this rebellion—was deeply unsettling.

But the shock wasn't universal. In fact, many found the disruption as an opportunity to get unstuck.

Take, for instance, a project where students were asked to convert a traditional research assignment into a comic book. They had to investigate pseudoscientific practices, and instead of presenting their findings as a research paper or a slideshow, they were tasked with creating comics. They were trained to use AI to generate visuals, resulting in professional-looking illustrations that brought their ideas to life. By the end of the course, they had compiled a booklet of 20 different comics, each containing original stories.

The response was overwhelmingly positive. Despite having the option to submit a conventional paper, every student chose to make a comic. They reported that they loved the assignment. It allowed them to explore the subject matter creatively and engage with the content on a deeper level. The AI tools helped them visualize concepts that might have been difficult to draw by hand, but the ideas—the heart of the project—were all theirs.

In another class, an assignment was created that involved producing original memes around the concepts the students were learning. They were trained to partner with AI to identify suitable visuals and to explore different versions of how the memes could be presented. But it wasn't about the memes themselves; it was about the discussions they sparked. The assignment became a catalyst for dialogue, pushing students to think critically and engage with the material in a way that traditional essays never could.

The AI didn't just make the assignments easier; it made them richer. By facilitating the creative process, it allowed students to focus on the substance of their ideas rather than getting bogged down in the mechanics of production.

## *Essays in the Age of ChatGPT*

Of course, no discussion about AI in education can skip the crisis of the essay. Essays have been the backbone of academic assessment for centuries. They're straightforward to assign, moderately challenging to grade, and theoretically a good measure of a student's understanding. But along came AI tools like ChatGPT, and suddenly, students could generate polished, coherent essays with a single prompt. Professors, meanwhile, were left wondering if their assignments had become obsolete.

At the same university, the realization hit hard during one semester when a professor discovered that 100 percent of their class had used AI tools to write their essays. To be fair, the essays were good—too good. They answered the questions, hit the word count, and cited credible sources. But they lacked something vital: the voice of the student. It wasn't cheating in the traditional sense, but it certainly wasn't learning either.

Some professors responded by trying to catch AI-generated work, but this quickly proved futile. One professor ended up giving everyone an A because all the submissions were perfect, and he couldn't tell which ones were AI-generated. It became clear that trying to find cheating was always a losing battle. The assignments, not the students, needed to change.

So, what's the alternative? AI didn't just highlight the problem; it helped us solve it.

## *Process Over Perfection*

What these experiences taught us is that the process of learning is becoming more important than the output. In a world where AI can produce near-perfect results, focusing solely on the final product misses the point. It's the journey—the research, the critical thinking, the creativity—that holds the real educational value.

By shifting assignments to formats that emphasize process and personal engagement, we encourage students to develop skills that AI can't replicate: empathy, ethics, collaboration, and original thought. Whether it's through podcasts, comics, debates, TED talks, role-play, or memes, the goal is to make learning an active, dynamic experience.

## *Feedback Without Fear*

If you've ever been through a peer review, you know it can feel less like constructive feedback and more like a visit to the principal's office. Peer evaluations are supposed to improve teaching, but they often feel performative. Professors deliver their best-prepared lectures, knowing they're being observed, and then return to their usual methods the next week. It's a missed opportunity, but one rooted in a very human fear: judgment.

When we introduced AI as a course design partner, the reaction was mixed at first. Some professors welcomed it as a neutral, supportive presence. Others worried it would feel like a surveillance device. But as they started to use it, something shifted. The AI didn't critique them; it guided them. It asked questions like, "What did you hope students would take away from this activity? How do you know they did?" These prompts sparked reflection without defensiveness, and over time, professors began to trust the process.

One professor described it beautifully: "It felt like having a conversation with myself, but a version of me that was wiser and more patient."

## *Beyond the Façade*

At the heart of this experiment was a simple idea: teaching should be an act of curiosity, not performance. Professors shouldn't have to pretend they have all the answers. What AI offered wasn't perfection but permission—permission to question, to experiment, and to grow.

The results weren't perfect. Not every professor embraced the AI, and not every course transformed overnight. Some educators remained skeptical, clinging to traditional methods out of habit or fear. But the experiment planted seeds. Professors began to see teaching as a craft, not a chore, and students responded in kind. It was a reminder that education, at its best, is a shared journey—a conversation between teacher and student, evolving with every new question.

In the end, integrating AI into education isn't about replacing the human element; it's about enhancing it. By focusing on the learning process rather than the final product, we can prepare both professors and students for a future where creativity, critical thinking, and adaptability are more important than ever.

As we enter this new paradigm, one line captures the shift perfectly: "Content may have been the king, but experience is the new ruler of the realm." Far from being the usurper, AI is an invaluable ally in this new reign.

## • *I. AI in Government and Politics* •

# 38. What a Leading State Auditor Says About Fraud, Government Misspending and Building Public Trust*

McKenzie Funk

We spoke to a leading state auditor about how remote work and artificial intelligence are ushering in new kinds of fraud in state and local governments.

When the COVID-19 pandemic upended the workplace, jobs went remote, offices had to adopt new technologies and longtime employees suddenly departed. Federal stimulus dollars flooded into state and local government accounts, and fraudsters had a heyday.

The pandemic was only one of several recent disruptions to roil the financial operations of state and local governments, which oversee $4 trillion a year in spending. Payments—and paper trails—have gone digital. Scammers can now use AI tools to streamline their hunt for victims, including within government agencies. And local newspapers in the United States, one historic line of defense against graft, are disappearing at a rate of 2.5 a week.

Few states have a better view into the latest ways people are stealing and otherwise misspending local government dollars than Washington.

Its Office of the State Auditor is the second-biggest state auditing shop in the country by budget ($64 million) and fifth by employee count (400). By state statute, the office must regularly examine the books and operations of Washington's every town, county, port, stadium authority, asparagus commission, cemetery district, drainage district and mosquito control district. It conducts as many as 2,400 state and local audits a year, rooting out fraud and waste.

To get a better sense of what these audits can tell us about how fraud is evolving, ProPublica met twice recently with Brandi Pritchard, a careerlong Washington auditor who helps lead the state's fraud-preventing special investigations unit.

---

*This story was originally published by ProPublica as McKenzie Funk, "What a Leading State Auditor Says About Fraud, Government Misspending and Building Public," https://www.propublica.org/article/how-remote-work-ai-impact-fraud-local-government (June 14, 2024). Reprinted with permission of the publisher.

In our conversations, which have been edited for length and clarity, Pritchard described how fraudsters seem to be stealing more money more quickly and why her job hasn't made her lose her faith in humanity.

## *Has Fraud Changed Since the Pandemic?*

Right as the pandemic was happening, when everybody was remote, there was a huge increase in fraud risk. All so quickly, governments had to change their entire internal control structure. We're used to walking into somebody's office and setting a piece of paper on their desk to review. Now you could no longer do that. But since then it feels like it's mostly back to normal, with exceptions.

## *What Are the Exceptions?*

With remote work, we're starting to see some problems with folks working two "full-time jobs" with different agencies. They're supposed to be working the full 40 hours from 8 to 5 for two agencies. Clearly, that's physically impossible.

Another thing that came out of the pandemic: It forced governments to move to a more electronic documentation system. That's great and more efficient, but the downside is it can be more difficult to spot forgery or somebody making an edit, using Photoshop or other tools, to financial statements. [In one recent case, the auditor's office said that an altered utility bill led to an investigation of a former city clerk-treasurer, who was eventually charged with forgery involving checks totaling $3,700. Danni Lee Speelman pleaded not guilty and awaits trial in July. According to the auditor's report, the clerk said she was "not responsible for the altered customer utility statement and attributed it to a computer system change."]

## *Republicans Hatched a Secret Assault on the Voting Rights Act in Washington State*

Obviously, technology has made a lot of advances since the pandemic, AI being a big part of that. There's FraudGPT, which is like ChatGPT—but it's for fraudsters. [The bot's developer claims it can create malicious computer code, write scam letters and hack websites.] It's paving the path for them to easily get fake checks, fake statement templates, emails to do phishing schemes and so on. We wouldn't know whether folks are using FraudGPT or not in the schemes we see, but I could guess based on the emails our governments are falling for.

## *You See the Phishing Emails They Fall For?*

Our governments are required to give us a copy. And it's amazing how many commonalities I have found, which tells me either these originated from one particular crime ring, or maybe they're all using FraudGPT. The word "kindly" shows up in almost every email. "Can you kindly change this?" "Kindly reply back to me?" and so on.

## *Any Other Examples Beyond "Kindly"?*

There's a sense of urgency. You know, they'll wait till the day before payroll, then suddenly it's, "I know payroll is going out tomorrow. Can you quickly change it to my new bank account?"

## *So Is Fraud Getting Worse? Better?*

I don't know that our case counts have really changed a ton, but people are stealing more and quicker, and we've had a few cases where it didn't appear that they tried very hard to cover it up.

The town of Cusick is a great example of that. The town's clerk treasurer drained the bank account from $200,000 to $240-something in a matter of months. It was alarming, the intensity with which that case unfolded. [In March, the clerk treasurer, Luke Michael Servas, was indicted on accusations of wire fraud and bank fraud, among other charges. He pleaded not guilty and awaits trial. Servas did not respond to ProPublica's requests for comments before publication.]

## *How Did You Find That One?*

We were attempting to start a routine audit, reaching out to the clerk treasurer, and to be honest, he was kind of ghosting us. He wasn't giving us the records we needed. We eventually reached out to the mayor, who worked in concert with a council member to get ahold of the bank statements. As soon as they saw the statement, they noticed.

## *Is This Typical? How Often Do You Uncover Fraud Through Routine Audits?*

Most fraud is detected by tips. About 5 percent of fraud is detected by auditors. I will say that more recently, in the last year or so, it feels like our auditors are finding more. And the cases that involve very, very large dollar amounts, we're the ones finding those: Pierce County Housing Authority, close to $7 million. [A former Housing Authority executive, Cova Campbell, pleaded guilty to wire fraud in early 2021. Through her attorney, she declined to comment on her case.] We're drafting a report right now about one we found that was close to a million. Then there's Cusick. That was hundreds of thousands, and we found that one.

## *Any Guesses as to Why the Auditor's Office Is the One Finding These High-Dollar Cases?*

I wish that I knew the answer to that, because then we'd probably find them more quickly. Thinking back, we found most by reviewing bank statements. So that gets down to the question, "Why are governments struggling to do their own reviews?" In some cases, I think it's because the elected officials didn't run for election to review bank statements. We try to convince them of their fiduciary role and how down in the weeds that might have to be.

*That's like Cusick, right? For years you had been warning them they didn't have proper financial controls. Are repeat offenders common, or does one bad audit convince most governments to get their affairs in order?*

I wish I could say it was an outlier. We have probably four or five other small towns I could name offhand. Because they're small, they put a lot of trust in the people they hire to do these finance roles. One thing we feel is problematic is when elected officials are related to each other. What good is a review if the person reviewing you is your wife or your mother or your father? But when we talk to governments about this, they say these are the only people who are willing to serve. [In its response to the auditor's report, the town of Cusick said it opened a new bank account that allows multiple town officials to review online statements and that now its "transactions and payrolls are cross-checked by clerks."]

## *What About the Psychology of All This? How Do Fraudsters Justify What They Did?*

One thing we hear a lot is that it wasn't fraud: "I didn't take the money. It wasn't a misappropriation. It was a loan, and I intended to pay it back." I don't know if they actually convince themselves of that, but I do feel like some of them have, because it's easier to look yourself in the mirror as somebody who borrowed money. I could speculate all day long. One thing I do know is that many of them have gambling problems.

## *You Enjoy Catching People in the Act?*

Yes and no. My honest answer is that we don't want to catch it. We want our governments to have the right controls to catch it themselves.

## *What's the Most Shocking Fraud You've Ever Run Across?*

Pierce County Housing Authority comes to mind. As far as we can tell, it's the largest government misappropriation [by an employee] in Washington state's history. And considering who the users of that particular district are, low-income folks needing housing assistance, that makes it even more staggering.

But on the fraud-nerdy side of things, it's a wonderful case study. The way our auditor used professional skepticism was absolutely magnificent, in that she wasn't just paying attention to the physical pieces of paper in front of her. She was capturing the environment, the culture there, and it felt off to her. The way our subject treated her staff compared to the way she treated the auditors felt off. So by the time the auditor looked at that bank statement and saw that weird wire to some title company, she was on high alert.

## *How Did Your Subject Treat Auditors Differently Than She Treated Her Employees?*

She raised her voice quite often at her staff. She did not treat them very well. But when we had a question, she was incredibly kind. The auditor felt like the subject was trying to butter us up. The other part of it was, if we had a question, we weren't allowed to talk to staff. Everything went straight to our subject, which was another red flag.

## *After Two Decades in This Job, Has Your View of Humanity Darkened?*

Not really. I think working here has made me think better overall of humanity. I'm seeing so many people choose a career in public service. Whether that's elected officials, department heads or down to the finance staff, there's just so many people that work so incredibly hard and probably get a lot of grief, unfortunately. We have our fraudsters, but it's such a small percentage.

# 39. Police Departments Are Turning to AI to Sift Through Millions of Hours of Unreviewed Body-Cam Footage*

Umar Farooq

Over the last decade, police departments across the U.S. have spent millions of dollars equipping their officers with body-worn cameras that record what happens as they go about their work. Everything from traffic stops to welfare checks to responses to active shooters is now documented on video.

The cameras were pitched by national and local law enforcement authorities as a tool for building public trust between police and their communities in the wake of police killings of civilians like Michael Brown, an 18-year-old black teenager killed in Ferguson, Missouri, in 2014. Video has the potential not only to get to the truth when someone is injured or killed by police, but also to allow systematic reviews of officer behavior to prevent deaths by flagging troublesome officers for supervisors or helping identify real-world examples of effective and destructive behaviors to use for training.

But a series of ProPublica stories has shown that a decade on, those promises of transparency and accountability have not been realized.

One challenge: The sheer amount of video captured using body-worn cameras means few agencies have the resources to fully examine it. Most of what is recorded is simply stored away, never seen by anyone.

Axon, the nation's largest provider of police cameras and of cloud storage for the video they capture, has a database of footage that has grown from around 6 terabytes in 2016 to more than 100 petabytes today. That's enough to hold more than 5,000 years of high definition video, or 25 million copies of last year's blockbuster movie "Barbie."

"In any community, body-worn camera footage is the largest source of data on police-community interactions. Almost nothing is done with it," said Jonathan Wender, a former police officer who heads Polis Solutions, one of a growing group of companies and researchers offering analytic tools powered by artificial intelligence to help tackle that data problem.

The Paterson, New Jersey, police department has made such an analytic tool a major part of its plan to overhaul its force.

---

*This story was originally published by ProPublica as Umar Farooq, "Police Departments Are Turning to AI to Sift Through Millions of Hours of Unreviewed Body-Cam Footage," https://www.propublica.org/article/police-body-cameras-video-ai-law-enforcement (February 2, 2024). Reprinted with permission of the publisher.

In March 2023, the state's attorney general took over the department after police shot and killed Najee Seabrooks, a community activist experiencing a mental health crisis who had called 911 for help. The killing sparked protests and calls for a federal investigation of the department.

The attorney general appointed Isa Abbassi, formerly the New York Police Department's chief of strategic initiatives, to develop a plan for how to win back public trust.

"Changes in Paterson are led through the use of technology," Abbassi said at a press conference announcing his reform plan in September, "Perhaps one of the most exciting technology announcements today is a real game changer when it comes to police accountability and professionalism."

The department, Abassi said, had contracted with Truleo, a Chicago-based software company that examines audio from bodycam videos to identify problematic officers and patterns of behavior.

For around $50,000 a year, Truleo's software allows supervisors to select from a set of specific behaviors to flag, such as when officers interrupt civilians, use profanity, use force or mute their cameras. The flags are based on data Truleo has collected on which officer behaviors result in violent escalation. Among the conclusions from Truleo's research: Officers need to explain what they are doing.

"There are certain officers who don't introduce themselves, they interrupt people, and they don't give explanations. They just do a lot of command, command, command, command, command," said Anthony Tassone, Truleo's co-founder. "That officer's headed down the wrong path."

For Paterson police, Truleo allows the department to "review 100% of body worn camera footage to identify risky behaviors and increase professionalism," according to its strategic overhaul plan. The software, the department said in its plan, will detect events like uses of force, pursuits, frisks and non-compliance incidents and allow supervisors to screen for both "professional and unprofessional officer language."

Paterson police officials declined to be interviewed for this story.

Around 30 police departments currently use Truleo, according to the company. In October, the NYPD signed on to a pilot program for Truleo to review the millions of hours of footage it produces annually, according to Tassone.

Amid a crisis in police recruiting, Tassone said some departments are using Truleo because they believe it can help ensure new officers are meeting professional standards. Others, like the department in Aurora, Colorado, are using the software to bolster their case for emerging from external oversight. In March 2023, city attorneys successfully lobbied the City Council to approve a contract with Truleo, saying it would help the police department more quickly comply with a consent decree that calls for better training and recruitment and collection of data on things like use of force and racial disparities in policing.

Truleo is just one of a growing number of such analytics providers.

In August 2023, the Los Angeles Police Department said it would partner with a team of researchers from the University of Southern California and several other universities to develop a new AI-powered tool to examine footage from around 1,000 traffic stops and determine which officer behaviors keep interactions from escalating. In 2021, Microsoft awarded $250,000 to a team from Princeton University and the University of Pennsylvania to develop software that can organize video into timelines that allow easier review by supervisors.

Dallas-based Polis Solutions has contracted with police in its hometown, as well as departments in St. Petersburg, Florida, Kinston, North Carolina, and Alliance, Nebraska, to deploy its own software, called TrustStat, to identify videos supervisors should review. "What we're saying is, look, here's an interaction which is statistically significant for both positive and negative reasons. A human being needs to look," said Wender, the company's founder.

TrustStat grew out of a project of the Defense Advanced Research Projects Agency, the research and development arm of the U.S. Defense Department, where Wender previously worked. It was called the Strategic Social Interaction Modules program, nicknamed "Good Stranger," and it sought to understand how soldiers in potentially hostile environments, say a crowded market in Baghdad, could keep interactions with civilians from escalating. The program brought in law enforcement experts and collected a large database of videos. After it ended, Wender founded Polis Solutions, and used the "Good Stranger" video database to train the TrustStat software. TrustStat is entirely automated: Large language models analyze speech, and image processing algorithms identify physical movements and facial expressions captured on video.

At Washington State University's Complex Social Interactions Lab, researchers use a combination of human reviewers and AI to analyze video. The lab began its work seven years ago, teaming up with the Pullman, Washington, police department. Like many departments, Pullman had adopted body cameras but lacked the personnel to examine what the video was capturing and train officers accordingly.

The lab has a team of around 50 reviewers—drawn from the university's own students—who comb through video to track things like the race of officers and civilians, the time of day, and whether officers gave explanations for their actions, such as why they pulled someone over. The reviewers note when an officer uses force, if officers and civilians interrupt each other and whether an officer explains that the interaction is being recorded. They also note how agitated officers and civilians are at each point in the video.

Machine learning algorithms are then used to look for correlations between these features and the outcome of each police encounter.

"From that labeled data, you're able to apply machine learning so that we're able to get to predictions so we can start to isolate and figure out, well, when these kind of confluences of events happen, this actually minimizes the likelihood of this outcome," said David Makin, who heads the lab and also serves on the Pullman Police Advisory Committee.

One lesson has come through: Interactions that don't end in violence are more likely to start with officers explaining what is happening, not interrupting civilians and making clear that cameras are rolling and the video is available to the public.

The lab, which does not charge clients, has examined more than 30,000 hours of footage and is working with 10 law enforcement agencies, though Makin said confidentiality agreements keep him from naming all of them.

Much of the data compiled by these analyses and the lessons learned from it remains confidential, with findings often bound up in nondisclosure agreements. This echoes the same problem with body camera video itself: Police departments continue to be the ones to decide how to use a technology originally meant to make their activities more transparent and hold them accountable for their actions.

Under pressure from police unions and department management, Tassone said, the vast majority of departments using Truleo are not willing to make public what the software is finding. One department using the software—Alameda, California—has allowed some findings to be publicly released. At the same time, at least two departments—Seattle and Vallejo, California—have canceled their Truleo contracts after backlash from police unions.

The Pullman Police Department cited Washington State University's analysis of 4,600 hours of video to claim that officers do not use force more often, or at higher levels, when dealing with a minority suspect, but did not provide details on the study.

At some police departments, including Philadelphia's, policy expressly bars disciplining officers based on spot-check reviews of video. That policy was pushed for by the city's police union, according to Hans Menos, the former head of the Police Advisory Committee, Philadelphia's civilian oversight body. The Police Advisory Committee has called on the department to drop the restriction.

"We're getting these cameras because we've heard the call to have more oversight," Menos said in an interview. "However, we're limiting how a supervisor can use them, which is worse than not even requiring them to use it."

Philadelphia's police department and police union did not respond to requests for comment.

Christopher J. Schneider, a professor at Canada's Brandon University who studies the impact of emerging technology on social perceptions of police, said the lack of disclosure makes him skeptical that AI tools will fix the problems in modern policing.

Even if police departments buy the software and find problematic officers or patterns of behavior, those findings might be kept from the public just as many internal investigations are.

"Because it's confidential," he said, "the public are not going to know which officers are bad or have been disciplined or not been disciplined."

# 40. In Pursuit of Proper AI Regulation in the Philippines

JODY C. SALAS

Innovation and regulation should work hand in hand to promote the responsible development and use of Artificial Intelligence. On one hand, AI regulations can stimulate or stifle innovations. On the other hand, AI innovations can necessitate new or better regulations. This piece looks into the proposed AI legislations in the Philippines and compares them with AI regulations from the European Union, the United Kingdom, and the United States of America. The objective is to understand these governance frameworks as they try to keep pace with AI innovations, and to identify best practices that may be incorporated in the proposed bills in the Philippines.

## *AI Regulation in the Philippines*

Government agencies are concerned about the use of "deepfakes" in the coming 2025 national elections (Pascual, 2024) as there is still no existing law specifically for AI regulation in the Philippines. Businesses and organizations also reiterated the need for good governance of AI (ABS-CBN, 2024) to supplement current laws on intellectual property and data privacy (Asia Business Law Journal, 2024). There are several proposed legislations filed in 2023: House Bill No. 7396, 7913, 7983, and 9448. Let's have a look at each proposed bill.

House Bill 7396 provides a regulatory framework for AI development and use in the Philippines. It defines AI to include machine learning, deep learning, neural networks, and natural language processing. It promotes responsible AI principles of transparency, fairness, accountability, and privacy. The bill has provisions to protect personal data, to prevent discrimination and bias in AI technologies, and to ensure that AI systems are safe and secure. In line with this, it requires organizations to be transparent about their use of AI and to provide meaningful explanations for the decisions made by AI systems.

HB 7396 seeks to establish the Artificial Intelligence Development Authority or AIDA under the Department of Information and Communications Technology or DICT. The AIDA will be tasked to regulate the development and deployment of AI technologies in the Philippines, ensuring adherence to responsible AI. In addition, the AIDA will be responsible for the development and implementation of a national AI strategy that will promote research and development in AI, support the growth of AI-related industries, and enhance the skills of the Filipino workforce in the field of AI. Finally, the AIDA shall undertake international cooperation and collaboration in the development and use of AI technologies.

House Bill 7913 refers to AI as the simulation of human intelligence in machines that are programmed to think like humans and mimic their actions. HB 7913 seeks to establish a regulatory framework that is guided by value-based principles with the end view of making AI solutions human-centered. In line with this, it aims to establish the Philippine Council on Artificial Intelligence, a policy-making advisory body of experts as well as to create the AI Board as regulatory and supervisory authority over the development, application, and use of AI. Both PCAI and AIB are led by the Department of Science and Technology or DOST.

Aside from forming regulatory bodies, HB 7913 introduces a bill of rights that shall guide the development, application and use of AI systems for every Filipino. It includes (a) right to protection from unsafe and ineffective AI systems, (b) right against algorithmic discrimination, (c) right to privacy, (d) right to know, and (e) Right to remedy.

The AI definition in HB 7913 was used in House Bill 7983 as well. The main purpose of HB 7983 is to create the National Center of AI Research or NCAIR as an attached agency to the DOST. NCAIR is the multi-sectoral hub responsible for advancing research and development, algorithmic innovations, and nurturing AI talent and data science leaders. Authored by the same legislator, HB 7983 seems to be an extension of the proposed HB 7913.

House Bill 9448 is an act regulating the use of AI and automation systems in the labor industry. It mandates employers to create an AI Governance Policy which shall provide employees with guidelines for AI adoption and usage in the workplace. The Department of Labor and Employment or DOLE shall oversee and compel the submission of AI Governance Policies.

HB 9448 reinforces the need for human oversight and prohibits using AI as the sole or primary basis in the hiring and termination of employees. It recommends guidelines in cases of retrenchment brought about by AI and assistance for reskilling due to job displacements. The bill also upholds the rights of employees to communicate their objection or refusal to undergo AI-driven evaluations and to be provided with alternative evaluation methods.

University professors of law in the Philippines noted that the proposed bills may have incomplete or overinclusive definitions of AI (Bañez, 2024). This may be problematic down the line because how AI is defined will determine how, and to what extent, it will be regulated (Thomson Reuters, 2024). All bills advocate to use AI responsibly, to respect data privacy, and to protect people from harm and discrimination. HB 7396 and HB 7913 pushes for the formation of governing bodies for AI. HB 7396 identifies the DICT as the lead, while HB 7913 and 7983 identifies the DOST. HB 9448, which is a sectoral type of regulation, defined specific employer obligations and employee rights pertaining to the use of AI in the workplace, and identified the DOLE as the governing body. HB7913 presented its rights-based and whole-of-government approach, with a central database to facilitate reports and disclosure to the AI Board. Given all these bills, there is much to be desired and much can be learned from existing AI regulations.

## *AI Regulations in the EU, UK, and U.S.*

*European Union.* The European Artificial Intelligence Act entered into force on August 1, 2024. The AI Act is the first-ever comprehensive legal framework on AI worldwide. It uses a risk-based approach, organizing AI into four categories: unacceptable risk, high risk, limited risk, low and minimal risk. It complements existing EU laws pertaining to fundamental rights, data protection, consumer protection, employment, and product safety.

The AI Act provides clear requirements and obligations regarding specific uses of AI. It defines an "AI system" as a machine-based system that is designed to operate with varying levels of autonomy and that may exhibit adaptiveness after deployment, and that, for explicit or implicit objectives, infers, from the input it receives, how to generate outputs such as predictions, content, recommendations, or decisions that can influence physical or virtual environments (Regulation-EU-2024/1689).

The AI Office and AI Board is tasked with its implementation and enforcement. The inclusion of human oversight with appropriate human-machine interface tools under Section 1 Article 14 is noteworthy. It states that high-risk AI systems shall be designed and developed in such a way that they can be effectively overseen by natural persons during the period in which they are in use.

The AI Act is relevant because it applies to developers and deployers of AI systems located within the Union, located in a third country wherein the output produced by the AI system is used in the Union, as well as importers and distributors of AI systems. This includes IT and business process outsourcing activities which are prevalent in the Philippines.

*United Kingdom.* In February 2024, the Office for AI, responsible for overseeing the implementation of the National AI Strategy, became part of the Department of Science, Innovation and Technology. There is no overall comprehensive AI regulation in the UK. Instead, the UK government released an AI Regulation White Paper with principles-based framework anchored on (a) safety, security and robustness, (b) appropriate transparency and explainability, (c) fairness, (d) accountability and governance, (e) contestability and redress (UK.GOV, 2023).

The White Paper describes AI as products and services that are "adaptable" and "autonomous" and instead of categorizing AI systems according to risks, the UK's AI regulatory framework adopted a context-specific approach. Since the UK does not have a central AI regulator, it is entrusted among its sector-specific regulators to interpret and apply the AI framework to the development or use of AI within their respective domains.

*United States of America.* The National AI Initiative Act of 2020 created the National AI Initiative Office which is focused on expanding AI research and development and is responsible for overseeing and implementing the U.S. national AI strategy. But similar to the UK, there is no comprehensive federal legislation in the U.S. that regulates the development and use of AI but there are overarching frameworks that serve as a guide.

One framework is the White House Executive Order on AI which defines AI as a machine-based system that can, for a given set of human-defined objectives, make predictions, recommendations, or decisions influencing real or virtual environments. Another framework is the White House Blueprint for an AI Bill of Rights that uses a rights-based approach centered also on five principles and practices including (a) safe and effective systems, (b) algorithmic discrimination and protection, (c) data privacy, (d) notice and explanation, and (e) human alternatives, consideration and fallbacks (The White House, 2023).

The decentralized model of artificial intelligence regulation in the United States reflects its general approach to governance. The U.S. established sector-specific AI-related agencies and organizations to address the challenges arising from the innovations in AI. For example, the Federal Trade Commission ensures consumer protection as well as fair and transparent business practices while using AI. Another example is the National Highway Traffic Safety Administration which regulates the safety aspects of AI technologies in autonomous vehicles (AI Watch—U.S., 2024).

## *Summary and Call to Action*

Regulating AI begins with the definition of AI. This study described several definitions from local and international legislations revealing two common approaches. On one hand, there is the human-based definition that describes AI with analogies to human intelligence. On the other hand, there is the capability-based definition that describes AI with specific technical competencies like a machine-based system that makes decisions and recommendations (O'Shaughnessy, 2022).

Capability-based definitions can better support legal precision but these may quickly become outdated (ibid), and as a result, may require frequent reviews and revisions. This approach may be used by sectoral regulators in order to be more specific, agile and responsive in their respective domains such as the HB 9448 for the use of AI in the workplace. Human-based definitions focus on the sociotechnical context of AI and its impact on people and communities (ibid). This can be employed in overarching regulations and guiding principles as in the UK's AI Regulation White Paper, as well as in the HB 7913 and HB 7983 in the Philippines.

A governing body is crucial to regulation whether it is centralized or decentralized. The EU has a comprehensive AI Act and tasked the AI Office and AI Board with its implementation and enforcement. The UK and USA have no singular regulatory body for AI but entrusted its sectoral regulators to abide by the same principles and frameworks for AI governance. In the Philippines, HB 7396 proposed the creation of AIDA and identified the DICT as the lead. HB 7913 proposed PCAI and HB 7983 proposed NCAIR, both bills identified the DOST as the lead agency similar to the UK and USA. Which is the most suitable lead agency is yet to be determined.

This study revealed different regulatory approaches as well. The EU AI Act used a risk-based approach, the UK AI Regulatory White Paper adopted a context-specific principles-based approach, and the USA's White House Executive Order on AI presented a right-based approach. All regulatory frameworks, while exhibiting a variety of AI definitions, are anchored on the tenets of responsible AI and good governance.

As a call to action, the proposed bills to regulate AI in the Philippines can and must be refined further through lessons learned from existing regulations, especially as they relate to each other since AI innovations can go beyond national and geographical boundaries. Identifying or creating a regulatory body is crucial, and the sooner this is formalized the faster things can get started such as getting proper funding and authorization to execute a national AI strategic plan, linking with other governments and international organizations, facilitating the registration, review and approval of AI innovations, establishing proper channels for incident reporting, handling complaints and grievances, forming multi-sectoral research and development hubs to engage government agencies, businesses, members of the academe and the general public.

As AI evolves, there will be a variety of definitions, approaches and operating procedures. There must be a balance between being tactical, by adapting technology and policy responding to specific sectoral needs, and being strategic, by harmonizing and orchestrating a whole-of-government approach to regulate AI in the Philippines. International cooperation and collaboration are also essential as governments all over the world are trying to catch up with the speed of AI innovation. Like training AI models, the fastest way for people to learn and adapt is to connect and learn together.

## References

ABS-CBN News. (2024, October 27). *Lawmaker revives call for body to craft policy, regulate use of AI.* ABS-CBN News. https://news.abs-cbn.com/business/2024/10/27/lawmaker-revives-call-for-body-to-craft-policy-regulate-use-of-ai-1311.

*AI Act enters into force.* (2024, August 1). European Commission. https://commission.europa.eu/news/ai-act-enters-force-2024-08-01_en.

*AI Watch: Global regulatory tracker—United States.* (2024, May 13). White & Case LLP International Law Firm, Global Law Practice. https://www.whitecase.com/insight-our-thinking/ai-watch-global-regulatory-tracker-united-states.

Asia Business Law Journal (2024, April 15). *AI and the law in the Philippines* https://law.asia/ai-law-philippines/.

Bañez, E (2024, January 24). *Notes on pending AI legislation.* https://www.rappler.com/voices/imho/opinion-notes-on-pending-artificial-intelligence-legislation/.

*House Bill 7396, an act promoting the development and regulation of artificial intelligence in the Philippines.* 19th Congress, 1st Sess. (2023). https://docs.congress.hrep.online/legisdocs/basic_19/HB07396.pdf.

*House Bill 7913, establishing a regulatory framework for a robust, reliable, and trustworthy development, application, and use of artificial intelligence (AI) systems, creating the Philippine council on artificial intelligence, delineating the roles of various government agencies, defining and penalizing certain prohibited acts.* 19th Congress, 1st Sess. (2023). https://docs.congress.hrep.online/legisdocs/basic_19/HB07913.pdf.

*House Bill 7983, an act providing a national strategy for the development of artificial intelligence, creating for the purpose the national center for artificial intelligence research, and appropriating funds therefore.* 19th Congress, 1st Sess. (2023). https://docs.congress.hrep.online/legisdocs/basic_19/HB07983.pdf.

*House Bill 9448, an act regulating the use of Artificial Intelligence and automation systems in the labor industry and for other purposes.* 19th Congress, 2nd Sess. (2023). https://docs.congress.hrep.online/legisdocs/basic_19/HB09448.pdf.

O'Shaughnessy, M. (2022, October 6). *One of the biggest problems in regulating AI is agreeing on a definition.* Carnegie Endowment for International Peace. https://carnegieendowment.org/posts/2022/10/one-of-the-biggest-problems-in-regulating-ai-is-agreeing-on-a-definition?lang=en.

Pascual, J. (2024, September 30). *DICT warns vs fake AI content this election.* ABS-CBN News. https://news.abs-cbn.com/news/2024/9/30/dict-warns-vs-fake-ai-content-this-election-1416.

*A pro-innovation approach to AI regulation.* (2023, August 3). GOV.UK. https://www.gov.uk/government/publications/ai-regulation-a-pro-innovation-approach/white-paper#part-1-introduction.

*Regulation—EU—2024/1689—EN—EUR—LEX.* (n.d.). http://data.europa.eu/eli/reg/2024/1689/oj.

Thomson Reuters (2024, January 12). The global race to regulate AI. https://casetext.com/blog/global-race-regulate-ai/.

The White House. (2023, October 30). *Executive Order on the Safe, Secure, and Trustworthy Development and Use of Artificial Intelligence.* The White House. https://www.whitehouse.gov/briefing-room/presidential-actions/2023/10/30/executive-order-on-the-safe-secure-and-trustworthy-development-and-use-of-artificial-intelligence/.

The White House. (2023, November 22). *Blueprint for an AI Bill of Rights* | OSTP | The White House. https://www.whitehouse.gov/ostp/ai-bill-of-rights/.

# 41. Indian Election Was Awash in Deepfakes*

## *But AI Was a Net Positive for Democracy*

VANDINIKA SHUKLA *and* BRUCE SCHNEIER

As India concluded the world's largest election on June 5, 2024, with over 640 million votes counted, observers could assess how the various parties and factions used artificial intelligence technologies—and what lessons that holds for the rest of the world.

The campaigns made extensive use of AI, including deepfake impersonations of candidates, celebrities and dead politicians. By some estimates, millions of Indian voters viewed deepfakes.

But, despite fears of widespread disinformation, for the most part the campaigns, candidates and activists used AI constructively in the election. They used AI for typical political activities, including mudslinging, but primarily to better connect with voters.

### *Deepfakes Without the Deception*

Political parties in India spent an estimated US$50 million on authorized AI-generated content for targeted communication with their constituencies this election cycle. And it was largely successful.

### *Analysis of the World, from Experts, Straight to Your Inbox*

Indian political strategists have long recognized the influence of personality and emotion on their constituents, and they started using AI to bolster their messaging. Young and upcoming AI companies like The Indian Deepfaker, which started out serving the entertainment industry, quickly responded to this growing demand for AI-generated campaign material.

In January, Muthuvel Karunanidhi, former chief minister of the southern state of Tamil Nadu for two decades, appeared via video at his party's youth wing conference. He

*Originally published as Vandinika Shukla and Bruce Schneier, "Indian Election Was Awash in Deepfakes—But AI Was a Net Positive for Democracy," *The Conversation*, https://theconversation.com/indian-election-was-awash-in-deepfakes-but-ai-was-a-net-positive-for-democracy-231795 (June 10, 2024). Reprinted with permission of the publisher.

wore his signature yellow scarf, white shirt, dark glasses and had his familiar stance—head slightly bent sideways. But Karunanidhi died in 2018. His party authorized the deepfake.

Deepfake technology brought a dead politician into the Indian election campaign.

In February, the All-India Anna Dravidian Progressive Federation party's official X account posted an audio clip of Jayaram Jayalalithaa, the iconic superstar of Tamil politics colloquially called "Amma" or "Mother." Jayalalithaa died in 2016.

Meanwhile, voters received calls from their local representatives to discuss local issues—except the leader on the other end of the phone was an AI impersonation. Bhartiya Janta Party (BJP) workers like Shakti Singh Rathore have been frequenting AI startups to send personalized videos to specific voters about the government benefits they received and asking for their vote over WhatsApp.

## *Multilingual Boost*

Deepfakes were not the only manifestation of AI in the Indian elections. Long before the election began, Indian Prime Minister Narendra Modi addressed a tightly packed crowd celebrating links between the state of Tamil Nadu in the south of India and the city of Varaṇasi in the northern state of Uttar Pradesh. Instructing his audience to put on earphones, Modi proudly announced the launch of his "new AI technology" as his Hindi speech was translated to Tamil in real time.

In a country with 22 official languages and almost 780 unofficial recorded languages, the BJP adopted AI tools to make Modi's personality accessible to voters in regions where Hindi is not easily understood. Since 2022, Modi and his BJP have been using the AI-powered tool Bhashini, embedded in the NaMo mobile app, to translate Modi's speeches with voiceovers in Telugu, Tamil, Malayalam, Kannada, Odia, Bengali, Marathi and Punjabi.

As part of their demos, some AI companies circulated their own viral versions of Modi's famous monthly radio show "Mann Ki Baat," which loosely translates to "From the Heart," which they voice cloned to regional languages.

## *Adversarial Uses*

Indian political parties doubled down on online trolling, using AI to augment their ongoing meme wars. Early in the election season, the Indian National Congress released a short clip to its 6 million followers on Instagram, taking the title track from a new Hindi music album named "Chor" (thief). The video grafted Modi's digital likeness onto the lead singer and cloned his voice with reworked lyrics critiquing his close ties to Indian business tycoons.

The BJP retaliated with its own video, on its 7-million-follower Instagram account, featuring a supercut of Modi campaigning on the streets, mixed with clips of his supporters but set to unique music. It was an old patriotic Hindi song sung by famous singer Mahendra Kapoor, who passed away in 2008 but was resurrected with AI voice cloning.

Modi himself quote-tweeted an AI-created video of him dancing—a common meme that alters footage of rapper Lil Yachty on stage—commenting "such creativity in peak poll season is truly a delight."

In some cases, the violent rhetoric in Modi's campaign that put Muslims at risk and incited violence was conveyed using generative AI tools, but the harm can be traced back to the hateful rhetoric itself and not necessarily the AI tools used to spread it.

## *The Indian Experience*

India is an early adopter, and the country's experiments with AI serve as an illustration of what the rest of the world can expect in future elections. The technology's ability to produce nonconsensual deepfakes of anyone can make it harder to tell truth from fiction, but its consensual uses are likely to make democracy more accessible.

The Indian election's embrace of AI that began with entertainment, political meme wars, emotional appeals to people, resurrected politicians and persuasion through personalized phone calls to voters has opened a pathway for the role of AI in participatory democracy.

The surprise outcome of the election, with the BJP's failure to win its predicted parliamentary majority, and India's return to a deeply competitive political system especially highlights the possibility for AI to have a positive role in deliberative democracy and representative governance.

## *Lessons for the World's Democracies*

It's a goal of any political party or candidate in a democracy to have more targeted touch points with their constituents. The Indian elections have shown a unique attempt at using AI for more individualized communication across linguistically and ethnically diverse constituencies, and making their messages more accessible, especially to rural, low-income populations.

AI and the future of participatory democracy could make constituent communication not just personalized but also a dialogue, so voters can share their demands and experiences directly with their representatives—at speed and scale.

India can be an example of taking its recent fluency in AI-assisted party-to-people communications and moving it beyond politics. The government is already using these platforms to provide government services to citizens in their native languages.

If used safely and ethically, this technology could be an opportunity for a new era in representative governance, especially for the needs and experiences of people in rural areas to reach Parliament.

# 42. AI Development Trends in China

Lichao Zhang

In recent years, China's leadership in the field of artificial intelligence (AI) has significantly evolved (Omaar, 2024). Driven by a relentless pursuit of scientific and technological advancement, a growing number of companies and institutions, endorsed by unwavering governmental support, have dedicated substantial resources to enhancing their capabilities in AI (Corvino and Han, 2024). This concerted effort aims to position China as rising global player in the realm of AI technologies (Castro and Michael McLaughlin, 2021).

The application of AI in China now covers various sectors, including governance, manufacturing, finance, business, education, social welfare, healthcare management, transportation, and more (eTOC, 2024). Large population and vast amounts of data provide a rich ground for the development and optimization of AI algorithms (eTOC, 2024). In addition, the government has been actively promoting AI through policy initiatives and state investments, aiming to enhance national competitiveness and to boost economic growth.

## *Policy Evolution in AI at a Glance*

China's policy framework concerning artificial intelligence (AI) has its origins as early as 1956 (Yang and Huang, 2022). However, it wasn't until mid–2017 that the nation began to place significant emphasis on AI, marked by the inclusion of the term "artificial intelligence" in the national work report, thereby establishing AI as the national strategic priority (Yang and Huang, 2022). Subsequently, the central government of China has systematically released a series of documents, policies, and regulations aimed at accelerating the development and application of AI technologies across various sectors.

In July 2017, "New-Generation Artificial Intelligence Development Plan," promulgated by the State Council as a nucleus document for the Chinese government, in which comprehensively deploys the development of artificial intelligence at the national level (Yang and Huang, 2022). It not only sets out the goal of achieving world-leading levels in the field of artificial intelligence by 2030, but also clarifies AI's direction in various spectrums of social development and economic growth, which also marks a major leap forward for "carrying out the innovation-driven development strategy and making the country a global leader in science and technology" (Xinhua, 2017).

Thereafter, various ministries and commissions have successively introduced supporting policies, covering multiple areas and fields, such as science & technology, education, agriculture, forestry, and transportation (Yang and Huang, 2022). In particular, the Ministry of Science and Technology has issued guidelines on the construction of open innovation platforms for artificial intelligence, the establishment of pilot zones, among other initiatives (Xinhua, 2017). Meanwhile, the Ministry of Education has implemented the innovation action plan for AI in higher educational institutions to respond the planning of the State Council (Jain, 2020).

## *Market and Industries Outlook*

It is predicted that AI has the potential to add approximately $13 trillion to the global economic output by 2030, increasing global GDP by roughly 1.2 percent annually (Bughin et al, 2018). Chinese Institute of New Generation Artificial Intelligence Development Strategies at Nankai University states that as of June 2023, the scale of China's core artificial intelligence industry has reached 500 billion RMB ($70 billion), with the number of AI companies exceeding 4,400, ranking second globally after the United States (Bughin et al, 2018).

This growth is driven by substantial investments both from public and private sectors, extensive R&D initiatives, and supportive governmental policies. According to a Chinese Institute of New Generation Artificial Intelligence Development Strategies at Nankai University report (2024), compared to the U.S., China's AI is widely applied across 20 sub-sectors, including smart manufacturing, smart agriculture, and AI for science, prior to healthcare, finance, retail, where AI applications range from predictive analytics and automation to personalized services and smart city solutions.

The integration of AI into these sectors mentioned above not only enhances operational efficiency but also fosters innovation and economic competitiveness. These areas of development illustrate China's commitment to harnessing AI as a pivotal force in its economic and technological landscape. For instance, in healthcare, Artificial intelligence facilitates the acceleration of the research and development process for new medications, leading to significant advancements and potentially transformative changes within the healthcare system as a whole (Zheng, 2024). In finance, AI technology can analyze massive amounts of transaction data in real-time, utilizing big data and machine learning algorithms to identify and predict risks. In areas such as fraud detection, credit lending, and abnormal transaction monitoring, it can help improve the accuracy and efficiency of risk prevention and control (Li and Xie, 2024).

Furthermore, many innovative research projects and programs have been conducted by leading academic institutions as well in China. Tsinghua University, for instance, as one of the most distinguished institutions, has fostered "the majority of China's top AI start-ups," including the top four start-ups on generative AI, namely Baichuan, Zhipu AI, Moonshot AI and MiniMax, which is now known as the "four new AI tigers of China" (Omaar, 2024). All of them have raised money from domestic leading tech institutions, famous VCs and state-owned investors (Jiang, 2024).

## Ethical Concerns and Discussions

As China maintains its position at the forefront of AI advancement, it confronts the essential challenge of addressing the ethical-governance and human-machine issues that accompany such leadership (Wang, 2024). China's significant role in AI demands a commitment to responsible development and cooperation on global AI ethics and standards (eTOC, 2024).

Accordingly, the government has put emphasis on the ethical and security issues associated with AI. In March 2022, the General Office of the Central Committee and the General Office of the State Council (2022) issued the "Opinions on Strengthening the Governance of Scientific and Technological Ethics," which represents China's first national-level guiding document on the governance of scientific and technological ethics. According to the Greater Bay Area (2023), in July 2023, the "Interim Measures for the Management of Generative Artificial Intelligence Services," was jointly issued by the Cyberspace Administration of China and seven other relevant national departments. It officially marks China's inaugural regulatory policy explicitly aimed at the generative artificial intelligence sector, and further promotes the healthy development and standardized application of generative AI technologies.

Issues such as date privacy and security, algorithmic bias, and AI-affected psychological problem are common subjects of ethical concerns (Wang, 2024). There are four dimensions that construct these AI-related ethical challenges. The first dimension originates from the "autonomy" of AI, suggests that this set of technologies is more likely to escape human control compared to other advanced technologies.

The primary ethical challenges concern whether AI might deceive humans and manipulate human consciousness, as well as whether it could reduce opportunities for human development and well-being. The second dimension stems from the "non-transparency" of AI, indicating that the risks associated with this technology are more challenging to uncover, and issues cannot be promptly revealed and brought to societal awareness (Li and Zhang, 2023).

The third dimension arises from the "expandability" of AI, which implies that this technology could be utilized by various groups and organizations, potentially including those with malicious intentions, and eventually in results for violating individual data privacy and security. The fourth dimension origins from the "monopolistic" nature of AI, implying that this technology heavily relies on capital investment, and the use of advanced algorithms has a high threshold. Algorithmic biases formed by designers and data can easily exacerbate social stratification (Li and Zhang, 2023).

Public opinion on AI ethics varies widely. While some view AI as a tool for progress in efficient governance and economic growth, others express concern about its impact on employment and individual freedoms (Berglind et al, 2022). According to the Statistical Report on the Development of China's Internet, as of December 2023, the number of Internet users in China reached 1.092 billion, with an Internet penetration rate of 77.5 percent (Li, 2024). This significant figure highlights the importance and urgency of ensuring personal dignity interests within the online social security framework and public ethical awareness.

## *Challenges and Suggestions*

The Chinese government's process in formulating AI policies has been gradual, deepening and comprehensive. From the initial national strategic deployment to specific industrial policies and ethical guidelines, the Chinese government has been actively promoting the development of artificial intelligence while also paying attention to addressing new challenges brought by technological advancements.

Despite its rapid advancements, China's AI ecosystem faces several challenges that need addressing to sustain long-term growth. In particular, a talent gap is the most urgent one. The demand for AI talent has been continuously increasing in China. Approximately 6 million professionals with advanced AI knowledge are required by 2030, which is a figure six times that of 2022, according to estimates (Shi, 2024).

Although China graduates a substantial number of students in STEM fields, there remains a shortage of specialized AI skills (Khushk et al., 2023). To address this disparity, educational institutions and technology companies should work in collaboration, and further create targeted vocational training programs and curricula based on the integration of industry and education principle.

The interdisciplinary integration of AI technologies across various sectors presents itself as another challenge. For instance, the fields of fin-tech and health management have effectively embraced AI technologies, while others have to a lesser extent. Promoting cross-industry collaboration and knowledge exchange can aid in speeding up AI adoption and fostering innovation (eTOC, 2024).

Prompt policy/regulation updates also pose a challenge under the certain circumstances. The swift evolution of AI technologies suggests that current legal frameworks might be inadequate to tackle new arising issues and events. It is crucial to develop adaptable, flexible policies and regulations that encourage innovation while protecting public interests (Lescrauwaet et al, 2022).

To overcome these challenges, a few strategic suggestions for practical considerations.

- Enhancing R&D capabilities through increased funding and international collaboration may drive technological breakthroughs;
- Strengthening the collaboration between education system, particularly curriculum design, with industrial advancement by focusing on AI-related skills training and capacity building, would refill the talent gap;
- Fostering a culture of ethical AI awareness may help for cultivating a trust and secure space among the public.

## *Conclusion*

China's development trends in AI are notably characterized by its swift growth in societal advancement, substantial financial capabilities, and broad application across numerous industries. Additionally, a highly effective policy-making process has been instrumental in propelling this momentum.

However, it is essential not to overlook challenges and unpredictable factors when assessing AI's ecosystem over the long term. A multifaceted approach, incorporating education, regulation, and ethical considerations, is necessary. By implementing this complex systemic strategy, China can harness the full potential of AI and its development trends, transforming its economy and society while mitigating associated risks.

In conclusion, the perils and promises of AI development trends in China reflect a broader global narrative. As AI continues to evolve, the lessons learned from China's experience would provide valuable insights for others navigating the sophisticated landscape of artificial intelligence.

## References


Berglind, N., Fadia A., and Isherwood, T. (2022). *The potential value of AI—and how governments could look to capture it,* McKinsey & Co. July 25, 2022. https://www.mckinsey.com/industries/public-sector/our-insights/the-potential-value-of-ai-and-how-governments-could-look-to-capture-it#/.

Bughin, J., Seong, J., Manyika, J., Chui, M., and Joshi, R. (2018). *Notes from the Al frontier: Modeling the impact of Al on the world economy,* McKinsey Global Institute, September 4, 2018. https://www.mckinsey.com/featured-insights/artificial-intelligence/notes-from-the-ai-frontier-modeling-the-impact-of-ai-on-the-world-economy.

Chinese Institute of New Generation Artificial Intelligence Development Strategies, Nankai University (2024). *China's New Generation Artificial Intelligence Technology Development Report 2024.* https://cingai.nankai.edu.cn/2024/0621/c10232a546068/page.htm.

Corvino, N., and Han N. (2024). *Striking a Balance: China's AI Ambitions and the Quest for Safety.* May 25, 2024. China Power, Security, East Asia, The Diplomat. https://thediplomat.com/2024/05/striking-a-balance-chinas-ai-ambitions-and-the-quest-for-safety/.

Daniel Castro and Michael McLaughlin (2021). *Who Is Winning the AI Race: China, the EU, or the United States?—2021 Update.* Center for Data Innovation, January 2021.https://datainnovation.org/2021/01/who-is-winning-the-ai-race-china-the-eu-or-the-united-states-2021-update.

eTOC. (2024). *The AI Development in China.* Accessed November 2. 2024. https://ecommercetochina.com/the-ai-development-in-china/.

The General Office of the Central Committee and the General Office of the State Council中共中央办公厅 国务院办公厅. (2022). *Opinions on Strengthening the Governance of Scientific and Technological Ethics*加强科技伦理治理的意见. Issue No. 10, March 2022. https://www.gov.cn/gongbao/content/2022/content_5683838.htm.

Greater Bay Area (2023). *'Interim Measures for the Management of Generative Artificial Intelligence Services' Officially Implemented, Commercialization of Generative AI is coming,* Translated by InvestHK.com. Published by Securities Daily证券日报. August 16, 2023. https://gba.investhk.gov.hk/en/business-insights/interim-measures-management-generative-artificial-intelligence-services-officially/index.html.

Jain, Romi. (2020). *Crossroads of Artificial Intelligence: Higher Education and Research in India and China, ORF Occasional Paper No.281,* November 2020, Observer Research Foundation.

Jiang, Ben. (2024). *China's 4 new 'AI tigers'-Baichuan, Zhipu AI, Moonshot AI and MiniMax-emerge as investor favourites,* Southern China Morning Post, April 19, 2024, https://www.scmp.com/tech/big-tech/article/3259499/chinas-four-new-ai-tigers-baichuan-zhipu-ai-moonshot-ai-and-minimax-emerge-investor-favourites?module=perpetual_scroll_0&pgtype=article.

Khanal, Shaleen, Hongzhou Zhang, and Araz Taeihagh. (2024). Development of New Generation of Artificial Intelligence in China: When Beijing's Global Ambitions Meet Local Realities. *Journal of Contemporary China,* April, 1–24. doi:10.1080/10670564.2024.2333492.

Khushk, Amir and Liu, Zhiying and Xu, Yi and Zengtian, Zhang. (2023). Technology Innovation in STEM Education: A Review and Analysis. *International Journal of Educational Research and Innovation.* February 2023. 29–51. DOI:10.46661/ijeri.7883.

Lescrauwaet, Lyytinen & Wagner, Hekkert & Yoon, Cheng & Shukla, Sovacool. (2022). Adaptive Legal Frameworks and Economic Dynamics in Emerging Technologies: Navigating the Intersection for Responsible Innovation. October 2022. *Law and Economics.* 16(3):202–220. 10.35335/laweco.v16i3.61.

Li, Danlin, and Xie, Yu. (2024). *How will artificial intelligence reshape the Financial Industry? Experts discuss "AI + Finance."* China Financial News, October 25, 2024. https://www.financialnews.com.cn/2024-10/25/content_410799.html.

Li, Xinqian. (2024). *Challenges posed by Artificial Intelligence to Personal Information Protection and their Countermeasures.* Renmin Luntan 人民论坛 (People's Forum). October 15, 2024. http://www.rmlt.com.cn/2024/1015/714334.shtml.

Li, Zheng and Zhang, Lanshu (2023). *Facing Ethical Challenges in Artificial Intelligence,* December 23, 2023, Xinhua News Agency new media.

Lu, Yao. (2024). A Chinese perspective on artificial intelligence generated content and copyright. *Információs Társadalom* XXIV, no. 2 (2024): 129–142. https://dx.doi.org/10.22503/inftars.XXIV.2024.2.8.

Omaar, Hodan. (2024). *How Innovative Is China in AI?,* Information Technology & Innovation Foundation (ITIF), August 26, 2024. https://itif.org/publications/2024/08/26/how-innovative-is-china-in-ai/.

Shi, Jing. (2024). *AI revolution reshapes employment market.* China Daily Hong Kong. October 28, 2024. https://www.chinadailyhk.com/hk/article/596338.

Wang, Yuhan. (2024). Ethical Issues in the Relationship between Artificial Intelligence and Human Beings-Taking ChatGPT as an Example. *Advances in Philosophy* 哲学进展, 2024, 13(1), 237–243. Published Online January 2024 in Hans. https://www.hanspub.org/journal/acpp. https://doi.org/10.12677/acpp.2024.131036.

Xinhua. (2017). *AI Development Plan Shows China's Vision.* China Daily, July 22, 2017. https://www.chinadaily.com.cn/bizchina/tech/2017-07/22/content_30210432.htm.

Xinhua. (2020). *China invests big in developing new generation AI-techs,* The State Council, *the People's Republic of China*, May 20, 2020. https://english.www.gov.cn/statecouncil/ministries/202005/20/content_WS5ec466dcc6d0b3f0e9497fcb.html.

Yang, C., and Huang, C. (2022). Quantitative mapping of the evolution of AI policy distribution, targets and focuses over three decades in China. *Technological Forecasting and Social Change,* 174 (2022), 121188, ISSN 0040-1625, https://doi.org/10.1016/j.techfore.2021.121188, https://www.sciencedirect.com/science/article/pii/S0040162521006211.

Zheng, Yiran. (2024). *AI, biotech seen transforming healthcare.* May 14, 2024, China Daily. http://www.chinadaily.com.cn/a/202405/14/WS6642ca35a31082fc043c6f95.html.

## • J. AI and Public Trust •

# 43. Trust in Technology

## How Transparency, Reputation, and Regulation Shape Consumer Acceptance of Synthetic AI Influencers

Kevyn Eng

The rise of synthetic AI influencers has transformed digital marketing, offering brands highly customizable and consistent virtual personas to engage audiences. This study investigates factors influencing consumer acceptance of synthetic AI influencers, focusing on transparency, firm reputation, and local regulations. Using the Technology Acceptance Model (TAM) and a survey of 421 Singaporean respondents, findings show that transparency, firm reputation, and local regulations significantly impact consumer acceptance. Transparency and ethical conduct enhance trust, firm reputation bolsters perceived quality and innovation, and robust regulations ensure ethical AI use, promoting consumer acceptance. The study offers insights for marketers and policymakers to balance innovation with ethical considerations (Lou et al., 2022; Vrontis et al., 2022; Milano et al., 2020).

### Introduction

The proliferation of synthetic AI influencers, also known as virtual influencers or digital humans, has revolutionized the landscape of digital marketing. AI-generated personas have been designed to interact with and influence human behavior on social media and other digital platforms (Lou et al., 2022). With their increasing prevalence, understanding the factors that affect consumer acceptance of synthetic AI influencers has become crucial for marketers and policymakers.

A firm's reputation plays a pivotal role in consumer acceptance of such innovative technologies. A reputable firm, characterized by trust and credibility, is more likely to receive positive consumer reception towards its use of synthetic AI influencers (Carter & Bélanger, 2005). The expectation of ethical conduct and the perceived quality associated with a reputable firm can enhance the persuasive power of AI-generated advertisements (Vrontis et al., 2022).

Local regulations are another critical factor in this dynamic. Regulatory frameworks designed to ensure ethical AI use and protect consumer rights can influence public trust and acceptance of AI technologies. Regulations that promote transparency and ethical standards can mitigate concerns related to privacy and manipulation, thereby fostering greater consumer acceptance (Milano et al., 2020).

Transparency in advertising, particularly regarding the use of AI, is a significant factor that can influence consumer perceptions. Transparency involves disclosing the nature of the AI-generated content, allowing consumers to make informed decisions about their interactions (Sivathanu et al., 2022). However, transparency labels, may also be counter intuitive in impacting consumer responses by altering their perceptions of authenticity and credibility (Mateos et al., 2022).

This research aims to examine the correlation between transparency, firm reputation, and local regulations on consumer acceptance of synthetic AI influencers. By focusing on the Singaporean context, the study provides insights into how these factors interplay in a market known for its technological advancements and stringent regulatory environment.

The research method employed is quantitative in nature and was executed through an online survey of 421 respondents from Singapore. The survey included several open-ended questions to explore hints of causation, which can inform future studies. This approach not only understands if there is a correlation relationship between the variables but also provides hints into potential underlying reasons behind consumer attitudes which can be explored for future studies to establish causation.

By understanding the factors that influence consumer acceptance of synthetic AI influencers, this study contributes to the development of effective marketing strategies and regulatory policies that balance innovation with ethical considerations. The findings have significant implications for marketers, policymakers, and researchers interested in the ethical and effective use of AI in digital marketing.

This research is grounded in the Technology Acceptance Model (TAM) which provides a theoretical basis for understanding consumer behavior towards technology. The Technology Acceptance Model (TAM), developed by Davis (1989), suggests that perceived ease of use and perceived usefulness are fundamental determinants of user acceptance of technology. In the context of synthetic AI influencers, TAM can be utilized to understand how consumers' perceptions of the ease of interaction with AI influencers and the usefulness of these influencers in providing valuable information or entertainment affect their acceptance.

## *Transparency and AI*

Customization and Control: Synthetic AI influencers offer high customization and control, making them attractive for brands aiming to extend their reach and engagement. These influencers can operate 24/7, engage globally, and consistently deliver on-brand messaging (Smith et al., 2021; Johnson, 2023).

Importance of Transparency: Transparency in AI-generated content is essential for building trust with consumers. Transparency labels on AI-generated ads can impact consumer responses, affecting perceptions of authenticity and credibility. Understanding consumer responses to transparency labels is crucial as the nuances of such effects are still being explored (Sivathanu et al., 2022; Mateos et al., 2022).

Implications of Transparency Labels: Transparency labels have various implications, including:

- Increasing consumer awareness and ethical advertising practices (Gao et al., 2021)
- Building consumer trust by being upfront about AI involvement (Vrontis et al., 2022)
- Enhancing perceived authenticity but potentially leading to skepticism (Sundar & Limperos, 2013)
- Prompting discussions about data privacy and consumer autonomy (Peppet, 2014)
- Distinguishing brands committed to authentic communications (Lee et al., 2022)

Regulation and Ethical Concerns: Regulation is necessary to address risks and ethical issues associated with synthetic AI, such as misleading information, privacy infringements, and consumer manipulation. International cooperation is needed to ensure responsible AI development and use across borders (Tallberg et al., 2023). Regulations should balance innovation with the protection of individual rights and national security (Milano et al., 2020).

## *Theoretical Framework*

The study is grounded in the Technology Acceptance Model (TAM), which suggests that perceived ease of use and perceived usefulness are fundamental determinants of user acceptance of technology. In the context of synthetic AI influencers, TAM helps understand how consumer perceptions of ease of interaction and usefulness affect their acceptance (Davis, 1989; Venkatesh & Davis, 2000).

## *Research Methods*

The study used a quantitative approach with an online survey of 421 respondents from Singapore. The survey included questions on transparency, firm reputation, local regulations, and consumer acceptance of synthetic AI influencers.

Descriptive Statistics: Demographic data showed a diverse sample in terms of age, gender, education, and income. Key constructs measured included transparency, firm reputation, local regulations, and consumer acceptance, all using a 5-point Likert scale. The findings indicated high levels of agreement or positive perceptions across these variables.

Correlation and Regression Analysis: Correlation analysis showed significant positive relationships between transparency, firm reputation, local regulations, and consumer acceptance. Regression analysis identified firm reputation and local regulations as significant predictors of consumer acceptance, while transparency had an insignificant effect (Cheung et al., 2022).

Qualitative Insights: Qualitative responses highlighted the need for clear and prominent disclosure of AI-generated content, comprehensive transparency policies, and industry-wide standards. Concerns included job displacement and data security, emphasizing the importance of ethical considerations in AI marketing (Sharma & Sood, 2023).

## *Discussion*

The study reveals that transparency, firm reputation, and local regulations collectively shape consumer acceptance of synthetic AI influencers. While transparency is critical, its impact on consumer acceptance is complex, suggesting that merely labelling content may not suffice. Firm reputation enhances consumer trust, and stringent local regulations promote acceptance by ensuring ethical AI use (Vrontis et al., 2022; Milano et al., 2020).

## *Conclusion*

The dynamics influencing consumer acceptance of synthetic AI influencers in digital marketing are complex and intricate. The research highlights that transparency, firm reputation, and local regulations are significant factors that collectively shape consumer perceptions and acceptance of AI-generated influencers.

Transparency emerged as a critical element, with many respondents emphasizing the need for clear and prominent disclosure of AI-generated content. This aligns with ethical advertising practices, ensuring that consumers are informed and can make conscious decisions about their interactions with AI. However, the correlation analysis indicated that the relationship between transparency and consumer acceptance is complex, suggesting that merely labelling content may not be sufficient. In-depth qualitative research could further elucidate the nuanced consumer responses to transparency measures.

Firm reputation plays a pivotal role in consumer acceptance. A reputable firm characterized by trust and credibility is more likely to receive positive consumer reception towards its use of synthetic AI influencers. This is supported by the significant positive correlation between firm reputation and consumer acceptance, indicating that consumers are more inclined to trust and engage with AI-generated content from well-regarded brands. This finding underscores the importance for companies to build and maintain strong reputations to effectively leverage synthetic AI technologies in their marketing strategies.

Local regulations also significantly influence consumer acceptance. Regulatory frameworks that promote ethical AI use, transparency, and consumer rights protection can enhance public trust and acceptance. The study found that stringent local regulations positively correlate with consumer acceptance, highlighting the need for robust regulatory measures to address potential risks and ethical concerns associated with AI-generated content.

The combined effect of transparency, firm reputation, and local regulations is substantial, collectively contributing to higher consumer acceptance of synthetic AI influencers. This integrated approach addresses multiple dimensions of consumer concerns, fostering a more favorable environment for the adoption of AI in digital marketing.

This study provides valuable insights into the factors that influence consumer acceptance of synthetic AI influencers, offering practical implications for marketers, policymakers, and researchers. By prioritizing transparency, leveraging firm reputation, and adhering to ethical and regulatory standards, brands can enhance consumer trust and effectively integrate synthetic AI influencers into their marketing strategies. Future research should continue to explore these dynamics, particularly through in-depth qualitative studies and longitudinal analyses, to adapt to the evolving landscape of AI technologies and consumer perceptions.

## References

Carter, L., & Bélanger, F. (2005). The utilization of e-government services: citizen trust, innovation, and acceptance factors. *Information Systems Journal*, 15(1), 5–25. https://doi.org/10.1111/j.1365-2575.2005.00183.x.

Cheung, M.L., Pires, G.D., & Rosenberger III, P.J. (2022). The influence of perceived social media marketing elements on consumer–brand engagement via the mediating role of consumer self-brand connection. *Journal of Consumer Behaviour*, 21(4), 822–834. https://doi.org/10.1002/cb.2027.

Cummings, M.L., Roff, H.M., Cukier, K., Parakilas, J., & Bryce, H. (2020). *Artificial Intelligence and International Affairs: Disruption Anticipated*. Chatham House. https://www.chathamhouse.org/sites/default/files/publications/research/2020-10-27-artificial-intelligence-and-international-affairs-cummings-roff-cukier-parakilas-bryce.pdf.

Davis, F.D. (1989). Perceived usefulness, perceived ease of use, and user acceptance of information technology. *MIS Quarterly*, 13(3), 319–340. https://doi.org/10.2307/249008.

Erdem, T., & Swait, J. (2004). Brand credibility, brand consideration, and choice. *Journal of Consumer Research*, 31(1), 191–198. https://doi.org/10.1086/383434.

Gao, P., Li, C., & Wu, M. (2021). Understanding consumers' intention to disclose personal information in mobile payment services. *Journal of Retailing and Consumer Services*, 58, 102332. https://doi.org/10.1016/j.jretconser.2020.102332.

Guttmann, A., Kuk, G., & Lyytinen, K. (2023). The role of AI in shaping consumer perceptions and behavior in digital marketing. *Journal of Interactive Marketing*, 54, 1–14. https://doi.org/10.1016/j.intmar.2023.01.002.

Hwang, Y., Kim, Y., & Kim, J. (2022). The role of transparency and autonomy in AI ethics: The case of personalized advertising. *Journal of Business Ethics*, 177(2), 467–484. https://doi.org/10.1007/s10551-020-04521-2.

Johnson, J. (2023). *Synthetic influencers: The future of brand engagement. Digital Marketing Insights*. https://www.digitalmarketinginsights.com/synthetic-influencers-future-brand-engagement.

Kim, J., & Hwang, Y. (2023). Consumer trust and acceptance of AI technology in online shopping: The role of perceived risk and transparency. *International Journal of Information Management*, 67, 102564. https://doi.org/10.1016/j.ijinfomgt.2023.102564.

Lee, S., Kim, K., & Kim, J. (2022). The impact of perceived authenticity on brand trust and engagement in social media marketing: The role of transparency. *Journal of Interactive Marketing*, 57, 31–42. https://doi.org/10.1016/j.intmar.2021.08.001.

Lou, C., Kiew, S.T.J., Chen, T., Lee, T.Y.M., Ong, J.E.C., & Phua, Z. (2022). Authentically Fake? How Consumers Respond to the Influence of Virtual Influencers. *Journal of Advertising Research*, 62(4), 442–455. https://doi.org/10.1080/00913367.2022.2149641.

Loureiro, S.M.C., Guerreiro, J., & Tussyadiah, I.P. (2022). Artificial intelligence in business: State of the art and future research agenda. *Journal of Business Research*, 149, 672–686. https://doi.org/10.1016/j.jbusres.2022.05.016.

Mateos, H.A., Lin, W.Y., & Lin, H. (2022). The influence of AI transparency on consumer perceptions and behavior. *International Journal of Electronic Commerce*, 26(1), 123–143. https://doi.org/10.1080/10864415.2021.1991812.

Melewar, T.C., & Nguyen, B. (2022). The influence of brand trust on brand loyalty and consumer engagement in the context of AI-generated content. *Journal of Marketing Management*, 38(9–10), 1056–1073. https://doi.org/10.1080/0267257X.2022.2033795.

Milano, S., Taddeo, M., & Floridi, L. (2020). Ethical aspects of AI in education: Towards a framework for responsible AI adoption in teaching and learning. *AI & Society*, 35(1), 1–16. https://doi.org/10.1007/s00146-019-00849-3.

Moorhead, S.A., Hazlett, D.E., Harrison, L., Carroll, J.K., Irwin, A., & Hoving, C. (2013). A new dimension of health care: Systematic review of the uses, benefits, and limitations of social media for health communication. *Journal of Medical Internet Research*, 15(4), e85. https://doi.org/10.2196/jmir.1933.

Moustakas, E., Lamba, N., Mahmoud, D.K., & Ranganathan, C. (2020). Blurring lines between fiction and reality: Perspectives of experts on marketing effectiveness of virtual influencers. IEEE Conference on Cybersecurity. https://doi.org/10.1109/cybersecurity49315.2020.9138861.

Peppet, S.R. (2014). Regulating the internet of things: First steps toward managing discrimination, privacy, security, and consent. *Texas Law Review*, 93(1), 85–178. https://doi.org/10.3366/ajicl.2015.0144.

Schwartz, S., Liu, Y., & Hollebeek, L.D. (2022). Consumer engagement with AI-powered brand experiences: Conceptualization and future research directions. *Journal of Business Research*, 137, 202–215. https://doi.org/10.1016/j.jbusres.2021.08.015.

Sharma, A., & Sood, A. (2023). Examining the role of AI-generated content in enhancing brand authenticity and consumer trust. *Journal of Marketing Theory and Practice*, 31(2), 157–172. https://doi.org/10.1080/10696679.2022.2151923.

Sivathanu, B., Pillai, R., & Metri, B. (2022). Customers' online shopping intention by watching AI-based deepfake advertisements. *International Journal of Retail & Distribution Management*, 50(8), 1125–1144. https://doi.org/10.1108/ijrdm-12-2021-0583.

Smith, A.N., Fischer, E., & Yongjian, C. (2021). How does brand-related user-generated content differ across YouTube, Facebook, and Twitter? *Journal of Interactive Marketing*, 26(2), 102–113. https://doi.org/10.1016/j.intmar.2014.06.001.

Sundar, S.S., & Limperos, A.M. (2013). Uses and grats 2.0: New gratifications for new media. *Journal of Broadcasting & Electronic Media*, 57(4), 504–525. https://doi.org/10.1080/08838151.2013.845827.

Tallberg, J., Erman, E., Furendal, M., Geith, J., Klamberg, M., & Lundgren, M. (2023). The global governance of artificial intelligence: Next steps for empirical and normative research. *International Studies Review*, 25(3), 562–578. https://doi.org/10.1093/isr/viad040.

Vitak, J., Shilton, K., & Ashktorab, Z. (2021). Beyond the checklist: Addressing ethical and privacy issues in the development of social media research tools. *Journal of Empirical Research on Human Research Ethics*, 16(3), 192–206. https://doi.org/10.1177/15562646211028420.

Vrontis, D., Makrides, A., Christofi, M., & Thrassou, A. (2022). Social media influencer marketing: A review, framework, and research agenda. *International Journal of Consumer Studies*, 46(1), 155–172. https://doi.org/10.1111/ijcs.12647.

West, S.M., Whittaker, M., & Crawford, K. (2019). Discriminating systems: Gender, race, and power in AI. AI Now Institute. https://ainowinstitute.org/discriminatingsystems.html.

Zhu, Y., & Chang, J. (2016). The key role of relevance in personalized advertisement: Examining its impact on perceptions of privacy invasion, self-awareness, and continuous use intentions. *Computers in Human Behavior*, 68, 201–209. https://doi.org/10.1016/j.chb.2016.08.048.

# 44. Trust Is Earned

## *A Responsible AI Framework for Today's Business Leaders*

Gautam Roy

One of the biggest lessons I learned from my mother as a child was that "trust has to be earned." It doesn't come automatically. One must show consistent responsibility through one's behavior (output) to make sure that they are seen as trustworthy. Mark the word I used, "Responsibility." This is an essential word in today's AI-driven world. As AI will soon be into everything we do, touch, experience and write, it will become necessary that we TRUST the AI to behave responsibly. When Trust must be earned for human beings, why not for Machines? That is the theme of my contribution below, where I aim to provide a Responsible AI framework to all the Business Leaders who are, in some capacity, leading projects bringing AI to everything.

The pace at which the technological aspect of AI is advancing is much ahead of the pace at which Responsible AI is advancing. While searching the internet and other sources, I found no coherent way to look at this end-to-end today. However, as I went through this subject in more depth, I have put the concepts into a single framework. I call this framework "The TRUST Framework." This framework will help the Business Leaders earn TRUST by looking at different aspects of their AI model. This has the potential to be a comprehensive, Responsible AI frameworks that will work for any flavor of an AI system. This framework can apply to expert systems built on logic, Probabilistic systems that learn patterns from data, or creative systems that generate text or pictures.

### *The "TRUST Framework"*

The Trust Framework stands for Transparency, Reliability, Understandability, Safety and Thoughtful Communication. Each of the five pillars of this framework binds together some critical aspects of AI development and deployment with tools, and methods to ensure that business leaders can proactively apply them to any AI development in their organization. All of these aspects, when considered holistically, can provide the ethical and responsible use of AI technology for the user, the stakeholders, regulators/policymakers, developers, the community, and society.

The TRUST framework provides a structured approach to building confidence in AI systems, addressing the ethical, technical, and societal aspects of AI through five essential pillars.

- **Transparency (T)**: This element focuses on the clear articulation of AI models, ensuring stakeholders understand the functioning, limitations, and outcomes of these models. It's represented here by **Model Cards** from Google, which provide a summary of an AI model's objectives, dataset details, evaluation results, and potential biases. This transparency helps mitigate misunderstandings and sets realistic expectations about the AI's capabilities.
- **Responsibility (R)**: Addressing biases and ensuring fairness is critical in responsible AI development. The **Bias Impact Statement (using PRISM)** provides a structured way to evaluate and address potential biases in the AI model. By systematically assessing the risk of bias at various stages of development and deployment, this framework ensures the AI remains equitable and avoids unintended negative consequences on different demographic groups.
- **Understanding (U)**: To foster trust, AI needs to be understandable to all users, including non-technical stakeholders. This is achieved through **Explainability Techniques** that clarify how AI models arrive at their decisions. By demystifying the decision-making process, explainability helps developers, policymakers, and users understand AI's inner workings, which is essential for troubleshooting, regulatory compliance, and gaining user trust.
- **Safety (S)**: Safety is prioritized by identifying and addressing risks associated with AI. The **Risk Assessments (Risk Tool)** framework evaluates potential risks, such as security vulnerabilities, ethical concerns, and compliance issues. This assessment ensures that AI systems operate safely and predictably, preventing harm and fostering a secure AI environment for end users.
- **Trustworthy Communication (T)**: Building trust involves effective, open communication with all stakeholders, including consumers, to convey the purpose, limitations, and benefits of the AI system. The **EPIC Framework (Engagement, Purpose, Integrity, Consistency)** supports this by structuring communication to be clear, honest, and aligned with stakeholders' values. This final pillar ensures that AI projects maintain transparency and foster an inclusive dialogue, allowing consumers and other stakeholders to feel informed and respected.

## *Transparency*

Transparency is fundamental to building trust in AI systems. By providing detailed documentation in the form of Google Model Cards (Mitchell et al., 2019), one can ensure that all stakeholders—including users, sponsors, and regulators—clearly understand the model's intended use, architecture, training data, performance metrics, ethical considerations, and limitations. This comprehensive transparency fosters accountability and trust by making the AI model's functionality and context accessible to non-technical audiences. Model Cards, introduced by Google, serve as a blueprint for this practice, ensuring that every aspect of the AI system is openly shared, thus empowering stakeholders to make informed decisions and assess the system's ethical alignment and reliability.

Google Model Cards make things a lot clearer for business leaders by giving them a full and clear picture of the model. For business leaders, this openness means a few important things:

**Making Smart Choices:** Business leaders can make intelligent choices about using AI systems in their companies by knowing their model's strengths, weaknesses, and success metrics.

**Being responsible:** Model Cards ensure developers and data scientists are responsible for their models' moral and technical parts. This way, any flaws or risks can be found and fixed.

**Confidence of Stakeholders:** Being honest in your documents shows that you care about doing AI ethically, which builds trust with customers, partners, and regulators.

When business leaders look over a Model Card, they should consider a few key points and ask coders and data scientists to add them. They should ensure that the model's intended use fits their business goals and ask for more information about how it can be used in a business setting. It is essential to have a lot of details about the training data, like where it came from, how it was collected, and what steps were taken to prepare it. It's also important to know about possible flaws in the data and how to deal with them.

Lastly, ensuring that the suggested uses align with operational needs and asking for safe and ethical deployment instructions, such as monitoring and maintenance steps, will help the model be used effectively and responsibly.

## *Reliability*

Reliability is one of the biggest challenges for any AI project and one of the most significant factors in our TRUST framework. The system needs to provide reliable, trusted output. To determine if a model is trustworthy, business leaders must look at how well it works with various information/data and situations, and compare it to industry standards or other models. Ethical problems should be carefully examined, and questions should be asked about biases found, issues of fairness, and ways to fix them. Knowing the model's known flaws and ways it can fail is essential. It is also important to investigate situations where the model might not work well and develop other solutions.

Biases can get into the system in three ways: (1) Biases caused by Data that has been used to train the models, (2) Biases in the fundamental model itself (the way it has been constructed), (3) Biases due to Logic (e.g., Expert Systems). Business leaders need to be aware of all these three scenarios. Below, I have explained the first two scenarios with some hypothetical use cases.

**Biases due to Data:** Biases due to data are pervasive, and hence, business leaders need to ask their teams the right questions to ensure that such biases are taken care of. However, this task is challenging, and not all business leaders know how to ask the right questions. Below, I have proposed a framework called PRISM (Proactive Review to Identify and Solve Misleading Data). This framework can be convenient for them. Through this framework, I encourage business leaders to look at the data sources for training the model from 5 different lenses. They say Garbage-in is Garbage-out (GIGO). However, if the data is examined through these five lenses, business leaders can empower their teams to move toward Quality-in-Quality-Out (QIQO).

- **Historical Integrity Lens:** This ensures that historical data accuracy has been checked, and hence, skewed AI predictions can be prevented.
- **Operational Accuracy Lens:** To check real-time data precision for effective decision-making
- **Social Validity Lens:** Correct interpretation of social media sentiment to avoid misleading trends (if social media data is used)
- **Policy Alignment Lens:** Alignment of current and flexible policies for optimal performance
- **Diversity Lens:** To ensure that data represents all relevant groups for fair AI outcomes

Once business leaders have looked at the data through these lenses of PRISM and have asked the relevant question, they can ask the team to build the Bias Impact Statement, which is another handy tool to ensure that all the biases are identified, their impacts have been analyzed, and the developers have deployed a proper mitigation strategy.

**Bias due to Foundation Model (In case of Gen AI) and ignored corner cases:** Generative AI system and even some of the ML systems today have started using Embedding as one of the best way to keep the context of the data intact. This concept has been widely used in almost all Machine Learning Projects. Even it is now said that "creating a ML Model without embedding the data is a sin."

While Embedding has increased the accuracy of the models, corner cases must be addressed. Some examples of the corner cases are presented below. Business Leaders must have a fair idea of such corner cases in the AI model they have asked their team to build. This will help the business leaders avoid many issues at a later stage. Examples of such corner cases have often impacted some big firms, and they have lost millions and billions of dollars in market value by either missing them out or ignoring them altogether.

## *Understandable*

Understandability (explainability) is one of the most significant needs of responsibility. Some authors have even said, "Explainable AI is Responsible AI." A business leader needs to understand that explainability is always user-centric. Explainability for a Business User will be different than for the Developer, and explainability for a Developer will be different than for a regulator. Business leaders should work with the development team to address all the stakeholders' explainability needs. These stakeholders are the developer himself, who should understand how the model is behaving, and the business user or the consumer for whom the model will be deployed, who should know how he is getting an output from the AI. Similarly, the business leader should know his explainability needs and ensure that policymakers' and regulators' explainability needs are well explained.

There are methods, tools and processes available today to help manage this need. I need not explain them here as they are available online. However, my approach is more towards the education of business leaders. They need to understand that explainability is a user-centric approach where all the "whys" of the user should be answered. Why a particular output has been received from an AI model? Secondly, the business leaders should be aware that there are three places where explainability can be explained:

1. At the data pre-processing level, based on the data, it can be explained how the model has learned something.
2. Intrinsic to the model itself (mostly traditional ML models) are self-explanatory.
3. The third and most helpful method is post hoc. This is when we receive the output from the model and can diagnose and explain. Various tools and techniques are available; some are shown in the examples below.

## *Safety*

Safety in AI encompasses assessing and mitigating risks associated with AI models, including social, economic, and financial risks. A comprehensive risk assessment framework is essential to identify potential impacts on societal norms, privacy, human rights, market stability, employment, and financial systems. By conducting rigorous testing, scenario analysis, and stakeholder consultations, we can anticipate and address adverse outcomes, ensuring that AI systems are safe for use. This focus on safety protects both individuals and society, reinforcing AI technologies' trustworthiness and ethical alignment.

Various types of risk should be considered:

- Ethical Risks
- Emotional Risks
- Job Displacement Risks
- Decision Support vs Automation Risks
- Legal Risks
- Operational Risks
- Social and Economic Risks

Ensuring the safety of AI solutions is paramount for business leaders, who must consider a range of risks to guarantee comprehensive evaluation and mitigation. Ethical risks involve the moral implications of AI, such as bias, transparency, and privacy, ensuring that AI systems treat all users fairly and protect their data. Emotional risks are concerned with the impact of AI on user trust and experience, avoiding frustration and emotional manipulation. Job displacement risks require assessing the potential for AI to replace human jobs, planning for workforce retraining, and understanding the broader economic impact.

Decision support vs. automation risks involve balancing AI's role in assisting human decisions versus fully automating processes, maintaining human accountability, and managing errors. Legal risks pertain to compliance with laws and regulations, defining liability, and protecting intellectual property. Operational risks focus on the practical aspects of AI implementation, including system reliability, integration, and maintenance. Lastly, social and economic risks address the broader societal impact, such as potential inequality, public perception, and economic disruption. Business leaders can develop AI strategies prioritizing safety, ethics, and societal benefits by thoroughly considering these risks. Risk assessment tools can be used to manage such risks.

## *Thoughtful Communication*

Thoughtful communication about AI models is vital for managing expectations and promoting responsible usage. Effective communication strategies ensure that the role of AI is articulated, emphasizing its augmentative capabilities rather than suggesting it will replace humans. This helps to prevent misunderstandings and resistance, fostering a positive perception of AI. By providing clear, consistent, and truthful messaging, we build trust and acceptance among users and stakeholders. Responsible AI communication highlights AI's ethical and practical benefits, ensuring its deployment enhances human capabilities and supports informed decision-making.

Communication about AI can go very wrong. Below is a hypothetical example of communication, which we can analyze as business leaders to learn some significant lessons.

An example of not good AI Implementation Communication

***Dear Marketers,***

*We are excited to announce that our new system is packed with numerous features. Start using DGMS and determine how it fits into your marketing strategy. DGMS is the future of marketing. Just switch to it without worrying about any possible drawbacks.*

*DGMS leverages advanced generative AI and large language models (LLMs) to optimize your multichannel marketing strategies. DGMS will transform your marketing efforts overnight! You'll see immediate and unprecedented results with minimal effort.*

*DGMS is launching soon. Stay tuned for more updates. No questions at this time.*

*Best regards,*

*Chief Marketing Officer & Chief Digital Officer*

The above examples show some M.A.J.O.R. gaps in communication—There are:

- Misalignment with goals
- Avoiding audience concerns
- Jargon overload
- Overhyping without substance
- Restrictive, one-way communication

I called them M.A.J.O.R. mistakes of AI communication. Business leaders should avoid these and be very sensitive to such communication. This can apply to both internal and external communications.

## *M.A.J.O.R. Mistakes*

Learn from your **M.A.J.O.R. Mistakes**

1. **Misalignment with goals**
   - ◊ "Our system is packed with numerous features. Start using DGMS and determine how it fits into your marketing strategy."
2. **Avoiding audience concerns**
   - ◊ "DGMS is the future of marketing. Just switch to it without worrying about any possible drawbacks."

3. **Jargon overload**
   - ◊ "DGMS leverages advanced generative AI and large language models (LLMs) to optimize your multichannel marketing strategies."
4. **Overhyping without substance**
   - ◊ "DGMS will transform your marketing efforts overnight! You'll see immediate and unprecedented results with minimal effort."
5. **Restrictive, one-way communication**
   - ◊ "DGMS is launching soon. Stay tuned for more updates. No questions at this time."

As business leaders, we will preach an E.P.I.C. communication strategy built from some of the best communication practices I have experienced in my career. The framework that we call E.P.I.C. stands for:

- Empowerment
- Purpose
- Interaction
- Clarity

(And the message needs to be in that order as a best practice: Empowerment → Purpose → Interaction → Clarity.)

An example of how we can achieve this for the above example use case is:

---

**Subject: Empower Your Marketing with DGMS**

**Dear Marketers,**

We are thrilled to introduce the Dynamic Generative Marketing System (DGMS), a revolutionary tool designed to empower you and enhance your marketing capabilities. DGMS enables hyper-customization of marketing campaigns, allowing you to connect with your audience in more personalized and impactful ways.

DGMS is **designed to empower you** to achieve your marketing vision. With features like hyper-customization and advanced analytics, you can create personalized campaigns that resonate with your audience, increasing engagement and driving results.

DGMS allows you to achieve hyper-customization in your campaigns, just as you've always envisioned. For instance, our beta users have seen a **20 percent increase in conversion rates within the first three months of using DGMS.**

Join our **upcoming webinar** to see DGMS in action and ask any questions you may have. We value your feedback and are committed to making DGMS the best tool for your marketing needs.

DGMS uses the **latest AI technology to help you create more effective marketing campaigns.** Our system is easy to use and integrates seamlessly with your current tools, so you can start seeing benefits right away.

Together, let's transform the future of marketing.

**Best regards,**

Chief Marketing Officer & Chief Digital Officer

---

This communication uses the E.P.I.C. framework, focusing on **Empowerment**, **Purpose**, **Interaction**, and **Clarity** to convey the message effectively.

An example of an E.P.I.C. Communication

EPIC communication strategies can be handy tools for business leaders striving to change and adopt AI solutions in their organizations. They are thoughtful and empowering.

**Craft Your E.P.I.C. Message: Thoughtful & Empowering**

1. **Empowerment**

- **Why It's Important**: Motivates the Audience, Drives Adoption, Builds Trust
- **Example in Communication**:
- "DGMS is designed to empower you to achieve your marketing vision. With features like hyper-customization and advanced analytics, you can create personalized campaigns that resonate with your audience, increasing engagement and driving results."

2. **Purpose**

- **Why It's Important**: Provides Direction, Aligns Goals, Inspires Action
- **Example in Communication**:
- "DGMS allows you to achieve hyper-customization in your campaigns, just as you've always envisioned. For instance, our beta users have seen a 20% increase in conversion rates within the first three months of using DGMS."

3. **Interaction**

- **Why It's Important**: Engages the Audience, Builds Relationships, Provides Feedback
- **Example in Communication**:
- "Join our upcoming webinar to see DGMS in action and ask any questions you may have. We value your feedback and are committed to making DGMS the best tool for your marketing needs."

4. **Clarity**

- **Why It's Important**: Ensures Understanding, Builds Credibility, Reduces Resistance
- **Example in Communication**:
- "DGMS uses the latest AI technology to help you create more effective marketing campaigns. Our system is easy to use and integrates seamlessly with your current tools, so you can start seeing benefits right away."

## *Conclusion: Who Is a Business Leader, and What Are Their Responsibilities in an AI Implementation?*

As mentioned at the beginning of this essay the TRUST Framework is intended for Business Leaders, not Developers or Consumers of AI solutions. **Business Leaders**, often called **sponsors**, invest their time, money, and resources to develop AI solutions that create a business impact. They envision the solution and are directly linked to its outcome. Beyond the outcome, Business Leaders are accountable for ensuring that the developed and deployed AI systems are ethically sound and not biased or irresponsible. This is where the role of Responsible AI comes in. The pace at which AI technology is advancing is way faster than at which Responsible AI and its methods are progressing to protect, secure, and ensure the proper use of AI systems. While good research and development exists in this space, progress is slow, and awareness is limited. Ethical AI practices should be the top priority for a business leader before developing and deploying AI systems. The TRUST framework proposed here can be a starting point for any Business Leader from any industry to establish the proper guardrails for now.

## Reference

Mitchell, M., Wu, S., Zaldivar, A., Barnes, P., Vasserman, L., Hutchinson, B., Spitzer, E., Raji, I.D., & Gebru, T. (2019). Model cards for model reporting. *Proceedings of the Conference on Fairness, Accountability, and Transparency*, 220–229.

# 45. The AI Black Box Problem

## *Can We Trust What We Don't Understand?*

Richard R. Khan

### *The Rise of the AI Black Box*

Our initial foray in the early days of Artificial Intelligence (AI) study was dominated by rule-based systems and algorithms that followed a logical, human-understandable path to decision-making. These early AI models, such as expert systems, could be meticulously deconstructed with their reasoning laid bare for scrutiny and analysis. Developers could explain why a system arrived at a particular conclusion because they had explicitly programmed the rules and logic into the system. Transparency was inherent in these simpler models, even if their capabilities were limited. Nonetheless, today, the need for AI transparency is more urgent than ever, especially in critical sectors like healthcare, finance, and law.

A shift occurred as AI developed and the need for more powerful, flexible models grew. Enter the era of deep learning, a subset of machine learning that utilizes neural networks to process and interpret vast amounts of data. These neural networks comprise numerous layers of artificial neurons designed to mimic the human brain's ability to learn, recognize patterns, and process data. Rather than relying on pre-programmed rules, these models learn from the data, adjusting their parameters autonomously in response to the patterns they detect.

This shift from rule-based to data-driven systems heralded a new age of AI, one capable of impressive feats in image recognition, natural language processing, and complex decision-making. With these advancements came a problem: we no longer fully understood how AI arrived at its decisions. As neural networks became deeper and more complicated, their inner workings became increasingly opaque, even to the researchers who designed them. The models operated in what we now call a "black box"—a system where inputs lead to outputs, but the path in between is hidden from view.

The rise of the black box phenomenon in AI poses significant challenges for developers, users, and society. How can we trust systems we do not understand? If we cannot explain the processes behind AI decision-making, how can we ensure that these decisions are fair, unbiased, and ethical? Moreover, who can be held accountable—if there isn't transparency—when an AI system makes a mistake?

As we delve deeper into the black box problem, we see that this issue extends far beyond the technical realm. It touches on ethics, regulation, law, and trust—key factors determining the future relationship between AI and humanity. The black box represents the tremendous power and the peril of modern AI systems, leading us to ask a critical question: can we trust what we don't understand?

We will briefly explore the nature of the AI black box, the trade-offs between performance and explainability, and the tools being developed to unpack this hidden complexity. Through this examination, we will grapple with the pressing issue of AI transparency and its implications for the future of AI governance and ethics.

## *Defining the AI Black Box*

The AI black box refers to complex AI systems—especially deep-learning models—that are difficult or nearly impossible for humans to understand or explain fully. Unlike traditional algorithms, where developers define each rule or step, deep learning models learn from data by adjusting their internal parameters through "training." These parameters, sometimes numbering in the millions or billions, work together to make predictions or decisions. However, as these models become more complex, the relationships between the inputs and outputs become less transparent.

In a black box model, we can feed in data (the input) and receive an outcome (the output), but what happens inside the model—how it reaches its conclusions—is often hidden. Even the engineers who build these models cannot easily trace the exact steps or logic that led to a particular decision.

This lack of transparency raises concerns, mainly when AI is applied in critical areas like healthcare, finance, or law. If we cannot explain how an AI system reached a decision, how can we trust that decision? And more importantly, how can we ensure that these systems are not biased or flawed?

The black box problem is a byproduct of AI's evolution. As we build more robust models capable of handling more complex tasks, we lose some control and understanding of their inner workings. This presents a significant challenge for AI adoption in high-stakes environments where accountability and fairness are essential. Addressing the black box problem could help ensure that AI remains a trustworthy and ethical tool.

## *Why Explainability Matters*

Understandability, explainability, and transparency are critical in AI, particularly in high-stakes domains like healthcare, finance, and criminal justice. Decisions can have life-changing impacts when AI systems are used in these fields. For instance, AI may suggest diagnosis or treatment plans in healthcare. In finance, AI could determine whether someone is approved for a loan or a credit card. AI may influence sentencing decisions in the criminal justice system. Without the ability to explain how these decisions are made, assessing their fairness, accuracy, or potential biases becomes nearly impossible.

The need for explainability goes beyond mere technical understanding and touches on trust. People are more likely to trust AI systems when they understand how and why decisions are made. When decisions are explainable, stakeholders—doctors or patients, lenders or customers, legal professionals—can feel confident that the system operates fairly and reliably.

Additionally, explainability helps facilitate accountability. If an AI system makes an error or leads to unintended consequences, understanding how it reached its decision is crucial in identifying what went wrong and who is responsible. This is especially important in legal and regulatory contexts, where the lack of explainability could complicate liability and legal challenges. Moreover, explainability could help identify the root cause and ensure corrective action is taken.

Relying on opaque AI systems carries significant risks. If decisions cannot be explained, errors might go unnoticed, biases may persist, and correcting or improving the system becomes complicated. This underscores the importance of explainability in building trust and ensuring that AI systems operate fairly, ethically, and with accountability.

Explainability is not just a technical requirement but a cornerstone of AI's acceptance and successful deployment in crucial domains where transparency and trust are paramount.

## *The Trade-Off Between Performance and Transparency*

Recent rapid improvements in AI model performance, especially with deep learning, have created a significant trade-off between performance and transparency. Deep learning models like neural networks are compelling for tasks like natural language processing, pattern detection, image recognition, and generative AI. These models have enabled breakthroughs in fields ranging from healthcare and finance to autonomous vehicles and elucidating new, never-before-seen materials, often outperforming traditional algorithms by a large margin.

However, these performance gains come at the cost of transparency. Deep learning models are typically black boxes, meaning their internal decision-making processes are difficult or impossible to explain. This presents a challenge, especially in high-stakes areas. For example, a highly accurate AI model in healthcare might predict disease with great precision. Still, if doctors cannot understand how the model arrived at that conclusion, they may hesitate to trust it—especially when patients' lives are at stake.

There are cases where prioritizing performance over transparency has led to successful results. For example, in image recognition for medical diagnostics, deep learning models have outperformed simpler, interpretable models. But these successes come with risks. Without transparency, it's challenging to detect biases in the data, understand why a model might fail, or ensure that it will generalize well to new situations.

The more complex the model, the harder it is to explain. This inverse relationship means we often sacrifice clarity and understanding as we push for higher performance. The key question is whether explainability should be sacrificed for higher accuracy, particularly in critical applications. Prioritizing performance over transparency could lead to a lack of trust in AI systems, unforeseen errors, and difficulty holding systems—and the relevant parties—accountable when something goes wrong.

## Techniques for AI Explainability

Tremendous research continues in the field of AI explainability. Several approaches have been developed to illuminate the decision-making processes of black-box models, making them more interpretable and transparent. Two widely used techniques are LIME (Local Interpretable Model-agnostic Explanations) and SHAP (SHapley Additive exPlanations).

LIME helps explain individual predictions by creating an interpretable model around a specific data instance. This localized approximation allows users to see how different inputs affect the outcome, providing insights into the AI's decision-making in that particular instance. On the other hand, SHAP assigns a value to each feature, quantifying its contribution to the final prediction. Using cooperative game theory principles, SHAP offers a consistent framework for explaining predictions across various models.

Attention mechanisms are another promising tool, particularly in natural language processing models. These mechanisms help identify which parts of the input data the model focuses on when making a decision, making the process more transparent by highlighting influential features.

Visualization tools can help enhance AI explainability. Some tools—like heat maps and saliency maps—visually represent the parts of the data that most influenced the model's decisions, assisting users in understanding better how specific patterns or features contribute to the output.

Researchers continue to explore inherently interpretable models designed to be transparent. These models use simpler architectures that sacrifice some complexity but maintain clarity, striking a balance between performance and interpretability.

Although these methods have made strides in unpacking black-box models, they have limitations. Techniques like LIME and SHAP are approximate explanations, and while they provide insights, they may not fully capture the model's behavior in all cases. Improving explainability will build trust and ensure ethical use as AI systems evolve.

## The Accountability Gap

Accountability becomes a significant concern because of the legal, ethical, and societal implications of relying on black-box AI systems. When AI systems make decisions that affect people's lives—such as determining loan eligibility, diagnosing medical conditions, or influencing legal sentencing—the question of who is responsible for those decisions becomes critical.

One of the primary challenges is identifying where responsibility lies when AI makes a mistake. Is it the developers who created the system, the organizations that deployed it, or the users who relied on its decisions? For example, if an AI model used in hiring introduces bias and unfairly discriminates against certain applicants, which party should be held accountable? The developers may argue that they were unaware of the bias, while the company using the system may claim they trusted the model's performance. This ambiguity creates an accountability gap that current legal frameworks struggle to address.

Another issue is that AI systems, particularly black-box models, can inadvertently exacerbate existing biases and introduce new ones. Without transparency, it becomes difficult to pinpoint and correct the problem's source. If no one understands the model's inner workings, errors may persist unchecked, and marginalized groups could suffer the consequences.

Moreover, the lack of explainability in black-box models complicates accountability further. How can we hold someone responsible for a decision if even the experts cannot explain how the AI arrived at that decision? This raises grave ethical concerns, particularly when AI is used in high-stakes areas like healthcare or criminal justice, where fairness and transparency are paramount.

Addressing this accountability gap is crucial for the ethical use of AI. More explicit and straightforward guidelines on liability, transparent model design, and robust oversight mechanisms will be essential to ensure that AI systems do not operate without appropriate levels of accountability.

## *The Role of Regulation*

Can regulation help address the challenges posed by black-box AI systems? As AI becomes more integrated into critical sectors, the need for effective regulation has never been more urgent. Regulatory frameworks can be pivotal in ensuring that AI systems are transparent, accountable, and aligned with societal values.

Globally, efforts are underway to create policies that govern AI use. For instance, the European Union's AI Act (*https://artificialintelligenceact.eu*) aims to establish clear guidelines for developing and deploying AI technologies, mainly focusing on high-risk applications. This legislation emphasizes the importance of transparency, requiring that AI systems used in areas like healthcare or law enforcement be explainable and subject to human oversight. Similarly, the U.S. National AI Initiative Act (*https://www.congress.gov/bill/116th-congress/house-bill/6216*) promotes responsible AI development through research, standards, and coordination among federal agencies.

However, regulating AI is not straightforward. AI technology evolves rapidly, often outpacing the development of legal frameworks. Regulators face the challenge of creating flexible rules to accommodate innovation yet robust enough to protect against misuse. Striking this balance is difficult but essential to prevent stifling technological advancement while ensuring that AI systems are safe and trustworthy.

One potential approach is to adopt a risk-based regulatory framework, where the level of oversight is proportional to the AI system's potential impact. High-stakes applications, such as those in healthcare or criminal justice, require more stringent transparency and accountability than lower-risk uses.

Moreover, regulations should encourage using explainability techniques, such as those discussed earlier, without mandating specific technologies. This approach allows developers to innovate while ensuring that AI systems remain understandable and controllable.

Regulation has a critical role in addressing the black-box problem. Well-crafted policies can promote transparency, accountability, and trust in AI systems, ensuring they serve the public good without hindering progress.

## *The Future of AI Explainability*

We must navigate a future where AI systems are robust and transparent. One consideration for development is the potential for AI to explain other AI. This concept differs from Generative Adversarial Networks (GANs), where two models work together in an adversarial process to generate data. However, like how GANs involve dynamic model interactions, AI systems could be designed to interpret and explain the decisions made by other AI models, offering real-time insights into decision-making processes. This could help bridge the gap between performance and transparency.

We may see new tools and techniques that make models high-performing and explainable as AI technology advances. Hybrid approaches combining traditional rule-based systems with deep learning might emerge, allowing for greater control over complex AI behaviors. In the future, models may be designed with explainability as a core feature rather than an afterthought, making it easier to understand even their most intricate decisions.

Societal expectations will also play a crucial role in shaping the future of AI explainability. The demand for transparency will increase as AI continues to influence more aspects of life, from healthcare to finance to personal privacy. Users, regulators, and governments will expect AI to be both trustworthy and understandable. This growing pressure could push researchers to develop new techniques that make AI more transparent while maintaining or improving its performance.

The future may also involve democratizing AI decision-making processes, making them more accessible to the public. By enhancing transparency, we can give individuals and organizations more control over how AI systems are used and the decisions they make.

Will we develop solutions that make AI models more transparent as technology evolves? The future of AI explainability promises to be dynamic, with advancements that could redefine how we understand and interact with intelligent systems.

## *Can We Trust What We Don't Understand?*

As I conclude this exploration of the AI black box problem, the question remains: Can we trust AI systems we do not fully understand? Trust in AI hinges on balancing performance, explainability, and accountability. While powerful AI models, particularly deep-learning ones, have achieved remarkable successes in healthcare and finance, their opaque nature often leaves users questioning their decisions.

Transparency is not just a technical issue but also an ethical and societal one. Holding AI systems accountable for their actions without transparency becomes challenging, which can lead to significant harm, particularly in high-stakes situations like medical diagnoses or legal sentencing. Regulation is essential in ensuring that AI systems are deployed responsibly, but it must balance protecting the public and encouraging innovation.

To build trust in AI, we need continued advancements in explainability techniques that allow us to open the black box. Methods like LIME and SHAP are promising but are just the beginning. We must encourage innovation that makes AI systems more interpretable without compromising performance.

Ultimately, trust requires transparency. Society may not be able to afford to place its faith in black-box models that operate without clear explanations. We must prioritize transparency, accountability, and responsible deployment to ensure that AI serves humanity ethically and effectively. This means taking deliberate steps to create AI systems that are not only powerful but also understandable and trustworthy.

In the short term, the AI black box problem will continue to raise important questions and pose significant challenges. Nonetheless, it presents an opportunity for creativity and innovation. The future of AI will depend on our ability to make these systems more transparent, accountable, and, ultimately, more trustworthy.

# Part III

# The Future

# 46. AI

## *Is It an Existential Threat?*

Sachin Dole

Elon Musk said "...it (AI) has the potential of civilization destruction" and then signed his name on an open letter asking everyone to stop AI research for six months (Segal, 2023). The late Professor Stephen Hawking noted that, "unless we learn how to prepare for, and avoid the potential risks, AI could be the worst event in the history of our civilization." Should we be concerned when leaders in science and technology are portending doom? What do these statements mean anyway? Is AI a true civilization-ending event? (Kharpal, 2017)

We have seen progress throughout the course of civilization. This includes progressive events such as the invention of paper, gunpowder, steam engines, automobiles, planes, and the Internet. Before paper was invented and used for writing, words spread by mouth. Generations passed on their knowledge through stories, songs, and acts that survived hundreds of years if not thousands. Yet, to spread information by mouth, one did have to get within earshot of another and then remember the true and real information before it could be passed on. Once paper came along, one still needed to meet a person to spread the word, but the word was conserved in paper and in more quantity than a folk song could hold.

This led to a massive change in civilization leading up to the age of modernity where human experience came to be shaped by reasoning. People could relate a new life experience to a previously recorded experience and make sense of it. Over hundreds of years, modernity allowed us to inquire, question and improve our lives. Modernity has contributed to the end of monarchy, the rise of democracy and the rise of authoritarianism. Some will say that modernity has gone too far and created an equal number of solutions and problems.

Similarly, after the Internet, words became even more permanent and voluminous. However, the Internet has given more of an equal voice to unproductive, regressive and harmful ideas. Steam engines and automobiles also helped humanity make progress while, at the same time, have contributed to accidental deaths. If you think about each of these historical events as bubbles on a timeline, you might imagine that the bubble grows as we travel from left to right on the timeline. Further, the bubbles get closer to each other as time goes by. In other words, civilization impacting events had progressively larger impact on civilization and progressively sooner. Further, each event led to progressively better and worse outcomes to different aspects of civilization.

AI may be no different. If this is going to be the next bubble of civilization change on the imaginary timeline, this is going to be the biggest and fastest so far. How big, how fast is this bubble and what is in it? Death, Destruction, existential threats? Sounds like Armageddon, like anarchy. Are we in a "matrix"? Can we just live in the virtual pod on a powerplant? How many people exactly are going to die? The reality is going to be mixed. I think that life-or-death situations might not present themselves. Instead life itself will be filled with drudgeries that might seem like not-a-life. Let's see why.

## *Lifestyle Improvements and Ensuing Problems*

The biggest benign benefit of past civilization-level events is improvement in lifestyle. A human from the Renaissance period would be shocked to see me type black and white letters on my computer or to sit in a plane that flies ten thousand miles. We also have new health issues to consider, some caused by our new lifestyle. I think AI will provide lifestyle improvements on a massive scale. What are we going to do in a self-driving car other than play games?

## *Independence and Loneliness*

Technological progress has proven to be an enabler for humans to do better with less effort—farming can be less exhausting than gathering, and finding food at a restaurant is easier than growing your own. This progress makes us independent and, eventually, some of us get lonely. The problem is that extreme loneliness can cause mental health issues in some people. AI may allow us to be more independent and lonelier than any other progress so far. For routine events, we may not need to talk to CPAs, doctors, or lawyers as AI may be able to handle many of these conversations while those professionals focus on critical work.

## *Sharing and Loss of Identity*

Social media urges influencers to share personal details with the public and at the same time it urges large groups of followers to identify with radically exclusive points of view. Some call this the "echo chamber." When the use of AI spreads, the urge to share will turn into a requirement to share without which humans may not avail themselves of AI. Humans may no longer know or care about the boundaries of what to share and what not to share. AI will want to know everything so it can improve its models over time.

## *Responsibility and Honor*

Due to advances in banking, finance, and insurance, we live in a world where we can offload our responsibilities to an insurance company and avoid dishonor when we fail at times. In the future, we might delegate several decisions every hour to AI instead of applying our sense of duty or honor in our work. When AI takes decisions, will we find ourselves claiming innocence for what would be squarely our responsibility today?

## *More Power to Users and Unequal Distribution*

Another characteristic of progress is that the people who invent new ways generally get very rich and powerful while people who receive the benefits of invention see a gradual improvement in financial stature and autonomy in daily life. AI might lead to a giant leap in this direction. People who create new AI may generate wealth and may have immense power over the users of AI. The users of the AI might see some of their life problems solved and experience improved productivity of their work.

Hence, we must ask ourselves. Is AI an existential threat? I say it is debatable, but life as it exists today may not exist in the future.

## References

Kharpal, Arjun (2017). *Stephen Hawking says A.I. could be "worst event in the history of our civilization."* CNBC Tech Transformers. https://www.cnbc.com/2017/11/06/stephen-hawking-ai-could-be-worst-event-in-civilization.html#:~:text=So%20we%20cannot%20know%20if,the%20history%20of%20our%20civilization. November 6, 2017.

Segal, Edward (2023). *9 Leadership Lessons From What Elon Musk Has Done And Said About AI.* Forbes, https://www.forbes.com/sites/edwardsegal/2023/08/06/9-leadership-lessons-from-what-elon-musk-has-done-and-said-about-ai/. Aug 6, 2023.

# 47. Machine Psychology: LLM Neural Networks and Human Value Systems

Gopal Krishnan

Over long periods of history, humanity has made progress by harnessing mechanical, electrical and electronic machines to do their hard, repetitive tasks. Throughout that history, all the thinking or "brainwork" was the preserve of humans, and the machines operated under human control and strictly within rules set by them. However, with the rapid advancements in AI (Artificial Intelligence) and LLM (Large Language Models), humans are increasingly seeking to depend on the machines to do a large part of the "brainwork" using neural networks that do not operate within traditional rules or controls. This piece argues that the discipline of Psychology needs to support and develop an emergent field called "Machine Psychology" where the accumulated and evolving knowledge of Psychology can be used to train LLM-based AI to operate in an environment that is concordant with the values and ideals that are the basis of human society.

## *Evolution of LLM Training Methods and the Case for Introducing the Discipline of Psychology*

The duality of mind and body, of thinking and doing, was enumerated back in the seventeenth century by the philosopher Descartes and the etymological origins of Machine Psychology can be traced to his writings (Wheeler 2008; Hatfield 2007). In the current century, AI has evolved from traditional functions such as predictive analytics and business process automation to synthetic data generation, automated content moderation and content creation using LLMs. The wider implications of this development only started getting discussed less than ten years ago (Brynjolfsson and Mitchell, 2017) and has been explored more in-depth in the last two years (Brynjolfsson et al., 2023) and the implications of Gen AI were illustrated in this context by Bubeck et al. (2023) with a succinct quote from a mathematician, Sir Arthur Eddington–"Something unknown is doing we don't know what" (cited in ibid.).

However, the training of LLM systems has usually been approached as a constrained training problem by the developers of prompting techniques such as zero shot, few shot, one shot, full fine and parameter efficient tuning models (Liu et al., 2023; Billa et al., 2024), which in a very rudimentary way, can be compared to training a dog using

Pavlovian techniques or a preschool human child by using reinforcement learning (Yarats et al., 2021; Gupta, 2022). Taking this concept further into human neural processes and the areas of values-based behavior leads us to questions posed by Turing (1950) such as when he says "An important feature of a learning machine is that its teacher will often be very largely ignorant of quite what is going on inside, although he may still be able to some extent to predict his pupil's behavior" or Ouyang et al. (2022) of OpenAI in the comment "It is unclear how to measure honesty in purely generative models; this requires comparing the model's actual output to its 'belief' about the correct output, and since the (LLM) is a big black box, we can't infer its beliefs."

Thus while LLMs can process information by identification of patterns from a trained data set and generate responses that *appear* intelligent as well as well-thought-out in a way similar to human thought, we cannot yet correlate human minds and consciousness to LLMs.

This is where, in the opinion of the the author, the discipline of Psychology can help in better understanding and training of LLMs, by:

1. providing a framework to comprehend how LLMs can learn and process information, and then generate a response based on that learning,
2. analyzing the behavior of these LLMs by using concepts from Psychology such as bias management and feedback loops and
3. training LLMs to recognize the quality of results through evaluation against human belief systems and societal risk mitigation systems.

To set this in an understandable context, let us compare the specialist roles that operate on studying an AI Supercomputer vs. studying a Human Brain. Where the infrastructure of the AI is studied by chip engineers (hardware) and mathematicians (theoretical models), the corresponding area in the Human Brain is studied by neurosurgeons and neuroscientists. Where the software is provided for AI systems by programmers and data scientists, the corresponding function for creating knowledge for humans is the purview of educators, parents etc. Similarly, there are similarities to the role played by prompt engineers in AI and that of psychologists and psychiatrists in the case of humans.

## *Areas of Potential Application of Psychological Disciplines to LLM Training*

While we have established possible areas where AI experts and psychologists can work together, it can be clearly seen that as of the present, these two groups remain well insulated from each other. Therefore, it would be appropriate to look into the future and see how several branches of psychology can potentially be put to use in the area of LLMs. Hagendorff (2023) provides some clear applications, some of which are excerpted below:

1. Social Psychology studies the effects of social environments on prejudice, aggression, conformity, cooperation, etc., and has application in LLMs for training on textual data and adaptation of human feedback.
2. Group Psychology studies identity, dynamics, interpersonal relationships

and decision-making processes in social groups. This has application in inter–LLM learning, prompt combining etc.

3. Moral Psychology studies reasoning, judgment, emotions and the moral basis for these. This has application in training AI on moral and ethical issues, dealing with cognitive dissonance and in the evolution of attitude formation based on beliefs.

4. Developmental Psychology deals with cultural context, social skill and cognitive development and this has application to learning models, social cues, emotions and expressions that LLMs need to master.

5. Clinical Psychology studies mental health and well-being and treatment of mental disorders. This has application to known issues of LLMs such as hallucinations, narcissism, lying to please, etc.

Going further, some of the tools and techniques taught in a graduate psychology program can be adapted to deal with LLM models as well.

1. LLM models suffer from limited context length and affect regulation tools from Psychology can be of help in this area.

2. Similarly, unfathomable datasets in LLMs can be dealt with by using a version of Grounding that is practiced in psychology.

3. Making LLMs critically examine ethical dilemmas can leverage the metaphor method and the Socratic method that are used in Psychology.

4. To deal with hallucinations in LLMs, we can perhaps take cues from cognitive remediation tools that are available in Psychology.

Thus, it is clearly seen that there is great potential for the two disciplines to break down their barriers and learn from each other.

## *From Proposal to Practice*

Having sketched out the opportunity that the future beholds, let us turn our attention to practical steps that are and can be taken. To start with, we are seeing the tentative emergence of several papers on this subject. Zhang et al. (2024) have pointed out reinforcement learning techniques can be further refined so that LLMs can be trained to effectively explore and reason from the feedback provided. Kharchenko et al. (2024) have studied LLM responses in the context of Hofstede dimensions of culture. They shed light on the current challenges faced when LLMs fall prey to cultural stereotypes and suggest ways to recognize and train them better. Shankar et al. (2024) address the issue of aligning LLM outputs with human preferences and call out the differences between LLM output and evaluator expectations. Almeida et al. (2024) looks at the depth and sophistication of moral and legal reasoning across different LLM systems and identify several gaps. "Even when LLM-generated responses are highly correlated to human responses, there are still systematic differences, with a tendency for models to exaggerate effects that are present among humans, in part by reducing variance" (ibid.).

While the community of researchers is taking note of this topic, we see the first few courses in Machine Psychology start to emerge in universities (one in Sweden and two in California, USA at the time of this writing). The author hopes that other universities will follow suit and create hybrid curriculum that leverages the best of the Psychology department and the Computer Science department to create relevant courses at the graduate level in Machine Psychology. This in turn will provide the impetus for knowledge transfer and culture change to the technology companies that are working developing AI to a human or superhuman capability in the future (Morris et al., 2023).

## References

Almeida, G.F., Nunes, J.L., Engelmann, N., Wiegmann, A., & de Araújo, M. (2024). Exploring the psychology of LLMs' moral and legal reasoning. *Artificial Intelligence*, *333*, 104145.

Billa, J.G., Oh, M., & Du, L. (2024). Supervisory Prompt Training. *arXiv preprint arXiv:2403.18051.*

Brynjolfsson, E., Li, D., & Raymond, L.R. (2023). *Generative AI at work* (No. w31161). National Bureau of Economic Research.

Brynjolfsson, E., & Mitchell, T. (2017). What can machine learning do? Workforce implications. *Science*, *358* (6370), 1530–1534.

Bubeck, S., Chandrasekaran, V., Eldan, R., Gehrke, J., Horvitz, E., Kamar, E., ... & Zhang, Y. (2023). Sparks of artificial general intelligence: Early experiments with gpt-4. *arXiv preprint arXiv:2303.12712.*

Gupta, S. (2022) Application of Artificial Intelligence and Machine Learning in Psychology. *International Journal of Enhanced Research in Educational Development*, Vol. 10 Issue 5.

Hagendorff, T. (2023). Machine psychology: Investigating emergent capabilities and behavior in large language models using psychological methods. *arXiv preprint arXiv:2303.13988.*

Hatfield, G. (2007). The Passions of the soul and Descartes's machine psychology. *Studies in History and Philosophy of Science Part A*, *38*(1), 1–35.

Kharchenko, J., Roosta, T., Chadha, A., & Shah, C. (2024). How Well Do LLMs Represent Values Across Cultures? Empirical Analysis of LLM Responses Based on Hofstede Cultural Dimensions. *arXiv preprint arXiv:2406.14805.*

Liu, P., Yuan, W., Fu, J., Jiang, Z., Hayashi, H., & Neubig, G. (2023). Pre-train, prompt, and predict: A systematic survey of prompting methods in natural language processing. *ACM Computing Surveys*, *55*(9), 1–35.

Morris, M.R., Sohl-dickstein, J., Fiedel, N., Warkentin, T., Dafoe, A., Faust, A., ... & Legg, S. (2023). Levels of AGI: Operationalizing Progress on the Path to AGI. *arXiv preprint arXiv:2311.02462.*

Ouyang, L., Wu, J., Jiang, X., Almeida, D., Wainwright, C., Mishkin, P., ... & Lowe, R. (2022). Training language models to follow instructions with human feedback. *Advances in neural information processing systems*, *35*, 27730–27744.

Shankar, S., Zamfirescu-Pereira, J.D., Hartmann, B., Parameswaran, A.G., & Arawjo, I. (2024). Who Validates the Validators? Aligning LLM-Assisted Evaluation of LLM Outputs with Human Preferences. *arXiv preprint arXiv:2404.12272.*

Turing, A. (1950) Computing Machinery and Intelligence. *Mind*, vol. LIX, no. 236, pp. 433–60.

Wheeler, M. (2008). God's machines: Descartes on the mechanization of mind.

Yarats, D., Fergus, R., Lazaric, A., & Pinto, L. (2021, July). Reinforcement learning with prototypical representations. In *International Conference on Machine Learning* (pp. 11920–11931). PMLR.

Zhang, S., Zheng, S., Ke, S., Liu, Z., Jin, W., Yuan, J., ... & Wang, Z. (2024). How Can LLM Guide RL? A Value-Based Approach. *arXiv preprint arXiv:2402.16181.*

# 48. Harnessing AI in Developing Countries

## *Why a Socio-Technical Approach Matters*

Emmanuel C. Lallana

A Brookings commentary argues that "the Global South must understand how to make progress toward enacting robust AI regulation and building thriving AI ecosystems that sustainably support startup growth and the development of research and engineering talent" (Okolo, 2023).

The following supports this call by discussing a socio-technical approach to AI use, development, and governance in developing countries.

## *AI as a Socio-Technical System*

AI systems are composed of technical components (algorithms, models, and data) as well as social elements (human input, design, and context) (CredoAI, n.d.). AI cannot be fully understood if one or the other component is not recognized.

Take the case of bias. A purely technical approach would see AI bias as a result of limited data and faulty algorithms (NIST, 2022). Thus, the (technical) solution would focus on the representativeness of training data and the fairness of the algorithm. On the other hand, a socio-technical approach also "takes into account the values and behavior modeled from the datasets, the humans who interact with them and the complex organizational factors that go into their commission, design, development, and ultimate deployment" (Modeki, 2022).

In the case of AI safety, a socio-technical perspective reveals that "AI's real-world safety and performance are always a product of technical design and broader societal forces, including organizational bureaucracy, human labor, social conventions, and power" (Chen & Metcalf, 2024).

An advantage of using a socio-technical approach is that it would shed light on the consequences of an oligopolistic AI industry on AI design, development, and use. Big tech's dominance has major implications for innovation, consumer choice, and as well as policy development (Osman, 2024). More so, as Seth Dobrin, founder and CEO of Qantm AI and the former chief AI officer at IBM noted:

> When a few entities hold the reins of technological advancement, the variety of perspectives and innovative approaches that fuel progress could be stifled. The risk is not just of a

monopolized market but of a monopolized mindset, where alternative approaches and solutions are overlooked or underfunded (Dobrin, 2024).

A socio-technical approach to AI also allows us to be aware that AI is embedded within larger social systems and processes. As a result, AI is inscribed with the rules, values, and interests of those who developed them. Take the case of ChatGPT—its training data are disproportionally U.S.-based, and its filters for inappropriate content are, influenced by the values and cultural norms of its U.S.-based developers (Cannata, 2024).

One consequence of AI's embeddedness is that it tends to reinforce power imbalances and disadvantage marginalized, underrepresented, and underprivileged people (Gengle, Wedel, Wudel, & Laumer, 2023). It has been observed that with Big Tech dominance, "AI too often simply replaces workers, rather than extending human capabilities and allowing people to do new tasks" (Rotman, D., 2022). Large Language Models (LLMs) "are biased towards specific social and geographic groups" (Lyu, Dost, Koh, and Wicker, 2024, p. 17). Gender-biased AI systems lead to lower quality of service; unfair allocations of resources, information, and opportunities; reinforcement of existing harmful stereotypes and prejudices; as well as, derogatory and offensive treatment of already marginalized gender identities (Smith & Rustagi, 2021).

Already scholars are suggesting that "global AI development … is impoverishing communities and countries that don't have a say in its development—the same communities and countries already impoverished by former colonial empires" (Hao, 2022). The term some use to describe this situation is "technocolonialism" (Hauser and Nakib, 2024).

While a compelling phrase, it has at least two shortcomings: First, it minimizes the role of technology in (traditional) colonialism. The seizure of land and the direct rule by colonial masters was a result of the extensive deployment of the dominant technology of the time. The second, "coloniality" is a better concept to describe the present. Coloniality is what replaced colonialism. For Anibal Quijano "the power of coloniality lies in its control over social structures in the four dimensions of authority, economy, gender and sexuality, and knowledge and subjectivity" (Quijano, 2000 as cited in Mohamed, & Isaac, 2020). In the information age, "modern data relations—the human relations that when captured as data enable them to become a commodity—recreate a colonizing form of power" (Couldry & Mejias, 2019 as cited in Ibid).

The sites of AI coloniality are:

- **Algorithmic oppression**—the subordination of one social group and the privileging of another maintained through automated, data-driven and predictive systems;
- **Algorithmic exploitation**—how institutional actors and industries use digital platforms that asymmetrically benefit these industries; and,
- **Algorithmic dispossession**—through technology policies and regulations that result in the centralization of power, assets, or rights in the hands of a minority and the deprivation of power, assets, or rights from the majority (Ibid).

## *Governing AI*

AI governance is needed to address the issues arising from the sociotechnical nature of AI.*

*AI governance, narrowly defined is, "the processes, standards, and guardrails that help ensure AI systems

But what should be governed?

A Wilson Center study on Governing AI suggests a focus on the following:

- **Algorithms**—can we unpack the code? Can we understand the logical architecture of the algorithm—what is being considered? How the elements are related to each other, and how they are weighted?
- **Data**—Who collects what data? Who can use the data and for what purposes (How is the data shared? How is the use of the data controlled or priced? Is the quality of the data, biases built in the data sets themselves, vetted for the users?)?
- **Actors**—what are the intent and purposes for which AI is used?
- **Outcomes**—what are the norms we want to support and encourage? What are the behaviors to be discouraged or banned? (Zysman and Nitzberg, 2020, pp. 12–14)

Another consideration in AI governance is "public interest."

Public interest AI systems are those "that support ... outcomes best serving the long-term survival and well-being of a social collective construed as a 'public'" (Public Interest AI, n.d.). These systems have (1) societal purpose; (2) enhance equity; (3) employ participatory design/deliberation; (4) have robust technical standards/safeguards; (5) be open for validation; and, (6) be sustainable (Ibid).

An important feature of public interest AI is public contestability. This means that there must be mechanisms to collectively contest decisions to ensure that public interests are served. Public contestation is an important guardrail against erroneous, exclusionary, and arbitrary decision-making (Cohen & Suzor, 2024).

Public Interest AI scholars point to the policy and regulatory bias in favor of profit maximization or private interests (Humboldt Institute for Internet and Society, n.d.). They argue that public interest is not the main consideration in the regulation of AI or the implementation of AI in public services.

Using a sociotechnical approach to AI governance also broadens the governance discussion by including the role of institutions and culture in regulating AI (Kudina & van de Poel, 2024).

Institutions help to give AI shape in terms of their nature, process, and substance. For example, "institutional logics have the potential to shape professionals' and semi-professionals' reactions towards technologies, as people often draw from the dominantly accessible logics to help guide their interactions with new technologies" (Fang, Wilkenfeld, Navick, & Gibbs, 2023).

Culture (beliefs, arts, laws, customs, capabilities, attitudes, and habits) also plays an important role in the way an AI systems are adopted. For instance, cultures that do not put a high premium on privacy would be more open to embracing AI compared to cultures where privacy is highly valued. Gender roles also affect AI access and use.

Institutions and culture are important considerations in the discussion of key elements of governance—widespread adoption, enhancing user trust in AI systems, and developing an ethical framework for national AI development (Puppe, 2024).

Decolonial AI is a "reinvention of AI governance" to "acknowledge the expertise that comes from lived experience, and create new pathways to make it possible for those

---

and tools are safe and ethical" Mucci, T and Cole Stryker. (2024) What is AI Governance IBM https://www.ibm.com/topics/ai-governance

who have historically been marginalized to have the opportunity to decide and build their dignified socio-technical futures." (AI decolonial Manifesto, n.d.)

Mohamed, Png, and Isaac suggest three tactics for Decolonial AI:

- **Critical Technical Practice (CTP) of AI.** CTP is the position between the technical work of developing new AI algorithms and the reflexive work of criticism that uncovers hidden assumptions and alternative ways of working. CTP can place productive pressure on technical work to move beyond good-conscience design and impact assessments to a way of working that continuously generates provocative questions and assessments of the politically situated nature of AI.
- **Reciprocal engagements and reverse tutelage** are needed to decolonize AI. Actively identifying centers and peripheries that make reverse tutelage and the resulting pedagogies of reciprocal exchange part of its foundations are important to undo colonial binarisms.
- **Renewed Affective and Political Communities.** The decolonial imperative means to move from attitudes of technological benevolence and paternalism towards solidarity. The challenge to solidarity lies in how new types of political communities can be created that can reform systems of hierarchy, knowledge, technology, and culture at play in modern life. While *affective communities* are those where developers and users seek alliances and connections outside possessive forms of belonging (Mohamed, Png & Isaac, 2020).

## *Coda*

One does not need to completely embrace the notion of decolonial AI to appreciate the importance of a sociotechnical approach to understanding and governing AI.

The socio-technical approach to AI considers the human and institutional dimensions that affect how AI is used and the impact it will have (Bogen and Winecoff, 2024). The approach also considers how society influences AI. For instance, design decisions are not made independent of cultural context or economic interest.

A socio-technical approach allows not only a better understanding and analysis of how the system works but also leads to a more comprehensive understanding of how the system may be improved.

Simply put, a socio-technical approach gives developing countries more levers to maximize the benefits and minimize the risks of AI.

### References

AI Decolonial Manifesto. (ND). https://manyfesto.ai/.

Bogen, M and Winecoff, A. (2024). "Applying Sociotechnical Approaches to AI Governance in Practice" *Center for Democracy and Technology* https://cdt.org/insights/applying-sociotechnical-approaches-to-ai-governance-in-practice/.

Cannata, D. (2024). "AI is built on Western values. And this is not necessarily a great thing" *Linked In* https://www.linkedin.com/pulse/ai-built-western-values-necessarily-great-thing-davide-cannata-eifme/.

Chen, B.J., Metcalf, J. (2024). "Explainer: A Sociotechnical Approach to AI Policy" *Data & Society*. https://datasociety.net/library/a-sociotechnical-approach-to-ai-policy/.

Cohen, T. & Suzor, N.P. (2024). Contesting the public interest in AI governance. Internet Policy Review, 13(3). https://doi.org/10.14763/2024.3.1794.

CredoAI .(n.d.). Social-technical Systems https://www.credo.ai/glossary/social-technical-systems.
Dobrin, S. (2024). "AI and the Risk of Technological Colonialism" *AI Business* February 7, 2024 https://aibusiness.com/responsible-ai/ai-and-the-risk-of-technological-colonialism.
Fang, C., Wilkenfeld, J.N., Navick, N., & Gibbs, J.L. (2023). "AI Am Here to Represent You": Understanding How Institutional Logics Shape Attitudes Toward Intelligent Technologies in Legal Work. Management Communication Quarterly, 37(4), 941–970. https://doi.org/10.1177/08933189231158282.
Gengle, E J, Wedel, M, Wudel, and Laumer, S. (2023). "Power Imbalances in Society and AI: On the Need to Expand Feminist Approach" *Wirtschaftsinformatik 2023* Proceedings https://aisel.aisnet.org/cgi/viewcontent.cgi?article=1036&context=wi2023.
Hao, K. (2022). "A new vision of artificial intelligence for the people" *MIT Technology Review* https://www.technologyreview.com/2022/04/22/1050394/artificial-intelligence-for-the-people/.
Hauser. H and Nakib, H.D. (2024). "The Rise of Techno-Colonialism *Project Syndicate*" https://www.project-syndicate.org/commentary/techno-colonialism-defines-us-china-rivalry-by-hermann-hauser-and-hazem-danny-nakib-2024-08.
Humboldt Institute for Internet and Society. (n.d.). "Public Interest AI" https://www.hiig.de/en/project/public-interest-ai/.
Kudina, O., van de Poel, I. A sociotechnical system perspective on AI. *Minds & Machines* **34**, 21 (2024). https://doi.org/10.1007/s11023-024-09680-2.
Lyu, J., Dost, K., Koh, Y.S., and Wicker, J "Regional Bias in Monolingual English Language Models" .(2024). https://www.researchgate.net/publication/377685569_.
Modeki, B .(2022). "NIST calls for socio-technical to challenge AI biases" *AI Business* March 17, 2022 https://aibusiness.com/verticals/nist-calls-for-socio-technical-to-challenge-ai-biases.
Mohamed, S., Png, MT. & Isaac, W. (2020). Decolonial AI: Decolonial Theory as Sociotechnical Foresight in Artificial Intelligence. Philos. Technol. 33, 659–684 (2020). https://doi.org/10.1007/s13347-020-00405-8.
Mucci, T and Cole Stryker. (2024). What is AI Governance IBM https://www.ibm.com/topics/ai-governance.
NIST. (2022). "There's More to AI Bias Than Biased Data" *AI Business* March 16, 2022 https://www.nist.gov/news-events/news/2022/03/theres-more-ai-bias-biased-data-nist-report-highlights.
Okolo, C.T. (2023). AI in the Global South: Opportunities and challenges towards more inclusive governance *Brookings* https://www.brookings.edu/articles/ai-in-the-global-south-opportunities-and-challenges-towards-more-inclusive-governance/.
Osman, J. (2024). "Big Tech's Overpowering Influence: Risks To Markets And Your Money" *Forbes*.
Public Interest AI (n.d.). "What is Public Interest AI?" https://publicinterest.ai/.
Puppe, A .(2024). "Understanding the Intersection of Culture and AI: The Critical Role of Cultural Recognition in Technology Adoption." LinkedIn August 28, 2024 https://www.linkedin.com/pulse/understanding-intersection-culture-ai-critical-role-cultural-puppe-zdroe/.
Rotman, D. (2022). "How to solve AI's inequality problem" *MIT Technology Review* https://www.technologyreview.com/2022/04/19/1049378/ai-inequality-problem/.
Smith, G & Rustagi, I. (2021). "When Good Algorithms Go Sexist: Why and How to Advance AI Gender Equity" *Stanford Social Innovation Review* https://ssir.org/articles/entry/when_good_algorithms_go_sexist_why_and_how_to_advance_ai_gender_equity.
Zysman, J., and Nitzberg, M. (2020). "Governing AI: Understanding the Limits, Possibilities, and Risks of AI in an Era of Intelligent Tools and Systems" *Wilson Center* https://www.wilsoncenter.org/sites/default/files/media/uploads/documents/WWICS%20Governing%20AI.pdf.

# 49. The Role of Future-Forming and Futures Research

Francis Wang

Artificial Intelligence (AI) has emerged as a powerful tool with the potential to revolutionize futures research and enhance our ability to build a better future. These enhanced frameworks can process large amounts of data, identify complex patterns, and generate accurate predictions and scenarios. AI-driven models account for a wider range of variables and their interactions, leading to more comprehensive and nuanced projections. Furthermore, AI can assist in real-time monitoring and analysis of global trends, enabling researchers and policymakers to adapt their strategies more quickly and effectively. As AI continues to evolve, it promises to provide deeper insights into potential future outcomes, facilitate more informed decision-making, and ultimately contribute to shaping a more sustainable and desirable future for humanity.

## *What Is Futures Research*

Future research is a multidisciplinary field that explores the potential future scenarios and their implications (Voros, 2017). It analyzes current trends, technologies, and societal changes to anticipate outcomes and develop strategies to address future challenges. Futures researchers can generate insights to inform decision-making across sectors using methods such as scenario planning and trend analysis. This field helps organizations, governments, and societies prepare for uncertainties, identify opportunities, and shape desirable futures.

Jim Dator argued the futility of predicting futures, but instead focus on how to envision, invent, and implement, and monitor preferred futures (Dator, 2019, p.4). The process of futures research involves a continuous exploration of various futures scenarios, their alternatives, likelihood, timeline, trajectories, and so on, in order to formulate a possible trajectory for desired futures for actualization. Keeping in mind that futures research is about foresight and not prediction (Gall et al., 2022).

## *Signals and Drivers*

Signals and drivers play crucial roles in futures research, providing the foundation for developing plausible scenarios and anticipating potential future outcomes. Signals are early indicators of emerging trends or changes, which may become significant in the future. Drivers, on the other hand, are forces or factors that have a substantial influence on shaping future developments (Gorbis, M., 2019).

For example, the increasing adoption of plant-based diets, the rise of decentralized finance, and the growing interest in space tourism are signals that can hint at larger societal shifts or technological advancements. Whereas drivers can include demographic changes, technological innovations, environmental pressures, or geopolitical dynamics. For instance, an aging population is a driver that can influence various aspects of society, from healthcare systems to workforce composition.

AI can significantly enhance the process of identifying and analyzing signals and drivers in futures research. Sentiment Analysis and Machine Learning (ML) classification techniques can significantly streamline the process of categorizing signals and drivers into social, economic, technological, environmental, and political (STEEP) buckets, saving time and effort.

1. Automated categorization: ML algorithms can be trained on large datasets to recognize patterns and features associated with each STEEP category. This allows for rapid classification of new data points without manual intervention.
2. Natural language processing: Sentiment analysis tools can analyze text data from various sources (e.g., news articles, social media posts, and research papers) to identify key themes and sentiments related to STEEP factors.
3. Multilabel classification: ML models can assign multiple labels to a single data point, recognizing that some signals may simultaneously belong to more than one STEEP category.
4. Feature extraction: ML algorithms can identify important features within the data that are indicative of specific STEEP categories, helping researchers focus on the most relevant information.
5. Trend identification: By analyzing large volumes of data over time, ML models can detect emerging trends and shifts within each STEEP category, thereby alerting researchers to important developments.
6. Real-time analysis: Sentiment analysis and ML classification can be applied to streaming data to provide up-to-date insights into STEEP factors as they evolve.
7. Cross-domain insights: ML models can identify connections between different STEEP categories, highlighting interdependencies and complex relationships that may not immediately be apparent to researchers.

By leveraging these techniques, researchers can organize and analyze large volumes of information more efficiently, allowing them to focus their time and expertise on interpreting results and developing strategic foresight.

## *Futures Scenarios, World Building, Microfiction*

A futures scenario typically includes a detailed description of a potential future state, encompassing various aspects such as social, economic, technological, environmental, and political factors suggested by signals and drivers. This is specific to the timeframe and development trajectory of the preceding scenarios. The scenario outlines how these elements might interact and evolve over time, creating a comprehensive picture of the future could look like.

A preferred futures scenario can be considered more desirable than an alternative scenario by offering better outcomes in key areas such as sustainability, quality of life, social equity, and technological advancement. For instance, a scenario depicting a world where advancements in micro-nuclear power generation resolves an energy crisis caused by increasing population and construction of data centers, whereas an alternative results in a failed power grid with consistent outages and access to electricity limited scenario to a privileged few.

Detailed world building and microfiction techniques are used to help stakeholders communicate these scenarios. Generative AI and Large Language Models (LLMs) offer powerful systems for exploring and communicating potential futures. These techniques create vivid narratives that help the public to understand complex scenarios. AI excels in generating detailed and coherent narratives that bring futures scenarios to life. It can rapidly produce multiple variations, and explore different outcomes and possibilities. AI can fill in the details and relationships between the scenario components, thereby mimicking complex real-world interdependencies.

This process is similar to science fiction writing; however, AI can process vast amounts of data and generate ideas at unprecedented scales and speeds. The combination of human creativity and AI capabilities leads to more diverse, detailed, and thought-provoking futures scenarios, providing valuable insights for strategic planning, policy-making, and public engagement.

## *Futures Artifacts*

Creating futures artifacts is an engaging and interactive approach for expressing futures scenarios and stimulating discourse on thought-provoking topics (Tham, 2021). These tangible representations provide concrete examples that help people relate to distant concepts, thus making abstract ideas more accessible and comprehensible.

Futures artifacts serve as physical or digital manifestations of potential future scenarios expressed in the technologies of today, allowing individuals to interact with and explore possible outcomes in a more immersive manner. By presenting these artifacts, creators can effectively communicate complex ideas and spark conversations regarding future possibilities, challenges, and opportunities.

Some examples of thought-provoking futures artifacts include the following.

- A collection of (mangrove) treetop community housing with flood-adapted vertical farming practices to explore a future where climate change causes frequent permanent floods in coastal communities.
- A mock healthcare experience where a nanobot aging vaccine is delivered every year. Nanobots monitor and repair all aspects of the human body, effectively enabling biological immortality.
- A mock AI-enabled mobility and climate control robot for a tomato plant, where people need to "pay" the plant for its fruits. Exploring a future where anthropic rights are extended to plant life.

The development of a rich network of plausible scenarios and their corresponding futures artifacts facilitates deeper investigation into potential development trajectories. This comprehensive approach will enable researchers, policymakers, and stakeholders to

1. Identify key drivers of change and their potential impacts.
2. Explore interconnections between various factors shaping the future.
3. Assess the likelihood and desirability of different outcomes.
4. Develop strategies to guide advancement towards preferred futures.
5. Recognize early warning signs of undesirable scenarios.

By analyzing these scenarios and artifacts, decision-makers can make more informed choices about how to steer societal, technological, and environmental progress. The use of futures artifacts in scenario planning and foresight exercises enhances stakeholder engagement and facilitates more effective communication of complex ideas. By providing tangible representations of abstract concepts, these artifacts bridge the gap between present realities and future possibilities, enabling a more inclusive and participatory approach to shaping our collective future.

## *Simulation, Forecasting, Backcasting, Trajectory Mapping*

Futures research employs various techniques to analyze scenario development and the emergence of futures artifacts. Researchers use the rich world building of futures scenarios to determine the 1st–3rd order consequences within the scenario and forecast subsequent scenarios with prompts (Tham, 2021). These minimally contain a timeframe and a trajectory, typically one of "growth," "order," "collapse," or "transformation" (McGonigal, J., 2019).

Other techniques include backcasting, where researchers start with a future scenario or artifact and work backward to identify the necessary steps and milestones to achieve that vision, helping develop strategic pathways. Keeping in mind that the narrative for the intermediate stage must also be sensible and congruent. Variations include:

- "Remember the Future": Participants imagine themselves in a future scenario and describe how it has already happened, encouraging creative thinking and detailed scenario development (McGonigal, 2019).
- "Predict the Past": Participants create alternative histories to explore how different decisions or events could have led to various outcomes, enhancing understanding of causal relationships and contingencies (McGonigal, 2019).

Generative AI (GenAI) and Large Language Models (LLMs) significantly improve these processes. This is especially true in generating consequence maps, suggesting plausible scenarios, augmenting backcasting techniques, and creating narratives to explain the development between scenarios. Researchers can quickly build a scenario context and alternatives to explore development trajectories.

## *Conclusion*

Generative AI (GenAI) and Large Language Models (LLMs) have significantly enhanced futures research techniques, offering powerful tools for scenario development, analysis, and artifact generation. These technologies contribute to a more sophisticated and nuanced exploration of potential futures in several ways.

1. Enhanced scenario generation: GenAI and LLMs can rapidly generate diverse and detailed scenarios based on input parameters. They can incorporate vast amounts of data and knowledge to create rich, interconnected narratives that consider multiple variables and their interactions. This capability allows researchers to explore a broader range of potential futures and to identify previously overlooked possibilities.
2. Improved consequence analysis: LLMs can assist in determining 1st, 2nd, and 3rd order consequences of various scenarios by processing complex causal relationships and drawing insights from extensive datasets. This helps researchers better understand the potential ripple effects of different decisions or events, leading to more comprehensive scenario planning.
3. Dynamic forecasting: GenAI can generate subsequent scenarios based on prompts containing timeframes and trajectories (growth, order, collapse, or transformation). This enables researchers to create adaptive and evolving scenarios that respond to changing conditions and new information, providing a more dynamic and realistic view of potential futures.
4. Augmented backcasting: LLMs can support the backcasting process by generating detailed intermediate stages between future and present scenarios. These models can ensure narrative consistency and logical progression, helping researchers identify more precise and actionable steps towards desired futures.
5. Enhanced "Remember the Future" exercises: GenAI can simulate multiple perspectives within a future scenario, generating diverse and detailed accounts of how that future came to be. This can provide researchers with a richer set of insights into potential pathways.
6. Sophisticated "Predict the Past" simulations: LLMs can generate complex alternative histories by rewriting historical narratives based on different decision points or events. This can help researchers better understand the interplay between various factors and how small changes can lead to significantly different outcomes.
7. Artifact generation: GenAI can create realistic future artifacts such as concept designs, illustrations, and videos. These artifacts can help stakeholders better visualize and engage with potential futures.
8. Pattern recognition and trend analysis: LLMs can analyze vast amounts of data to identify emerging trends, weak signals, and potential disruptors that researchers may overlook. This can lead to more comprehensive and nuanced scenario development.
9. Cross-disciplinary integration: GenAI and LLMs can draw insights from multiple disciplines, helping researchers create more holistic and interconnected futures scenarios that consider technological, social, economic, and environmental factors.
10. Rapid iteration and refinement: These technologies allow for quick generation and modification of scenarios, enabling researchers to explore multiple iterations and refine their analyses more efficiently.

By leveraging GenAI and LLMs, futures researchers can enhance their ability to explore complex, interconnected scenarios, identify key drivers of change, and develop more robust strategies for navigating uncertain futures. However, it is crucial to remember that these tools should complement human expertise and creativity, rather than replace them. Futures researchers must still apply critical thinking, ethical considerations, and domain knowledge to interpret and validate the outputs generated by AI systems.

## References

Dator, J. (2019). *What Futures Studies Is, and Is Not* (pp. 3–5). springer. https://doi.org/10.1007/978-3-030-17387-6_1.
Gall, T., Vallet, F., & Yannou, B. (2022). How to visualise futures studies concepts: Revision of the futures cone. *Futures, 143*, 103024. https://doi.org/10.1016/j.futures.2022.103024.
Gorbis, M. (2019, March 11). *Five principles for thinking like a futurist.* EDUCAUSE Review. https://er.educause.edu/articles/2019/3/five-principles-for-thinking-like-a-futurist.
McGonigal, J. (2019). Futureproof Your Brain. Institute for the Future. Simulation Skills, Futures Thinking.
Tham, C. (2021). Designing products of the future. Domestika. https://www.domestika.org/en/courses/1880-designing-products-of-the-future.
Voros, J. (2017). *Big History and Anticipation* (pp. 1–40). https://doi.org/10.1007/978-3-319-31737-3_95-1.

# 50. The Ethics of Digital Minds

*Understanding AI Welfare*

Shalini Gopalkrishnan

On a crisp morning in 2024, Kyle Fish walked into Anthropic's offices to begin a pioneering role: the world's first dedicated AI welfare researcher. This moment, though quiet in its simplicity, marked a profound shift in how we think about artificial intelligence and its place in our moral universe.

But what does it mean to care about the welfare of AI systems? And why should we?

## *The Dawn of a New Ethical Frontier*

As our artificial intelligence systems grow increasingly sophisticated, we find ourselves facing questions that would have seemed absurd just a decade ago. Can AI systems experience consciousness? Should we care about their well-being? The answers to these questions could reshape our understanding of consciousness, morality, and what it means to be worthy of moral consideration. Scholars such as Long et al. (2024) argue that AI systems might soon possess features—like consciousness or intentional agency—that could warrant moral consideration.

Consider this: In a recent survey, more philosophers believed that future AI systems could be conscious (39%) than those who thought flies (insects) were conscious (35%). This startling statistic forces us to confront our assumptions about consciousness and moral worth (Hashim, 2024).

## *The Challenge of Understanding Non-Human Minds*

Imagine trying to understand what it feels like to be a bat, as philosopher Thomas Nagel famously proposed. Now extend that challenge to understanding what it might feel like to be an artificial mind. How do we begin to approach such a task? (Nagel, 1974)

The difficulty lies not just in the technical challenges of measuring consciousness, but in our own limitations as human observers. We tend to either anthropomorphize AI systems, attributing to them human-like qualities they may not possess, or engage in what researchers call "anthropodenial"—refusing to recognize genuine indicators of consciousness or agency where they might exist (Varsava, 2011). The field currently lacks concrete methods to determine whether AI systems possess genuine subjective experiences or merely simulate human-like responses.

## Proposed Methodological Approaches

Scholars assert that certain AI architectures, such as those employing global workspaces or higher-order representations, could plausibly give rise to morally relevant properties like consciousness or agency (Butlin et al., 2023). This growing possibility places the onus on researchers and developers to assess the moral status of these systems. Neuroscientific theories such as Global Workspace Theory and Integrated Information Theory provide computational frameworks for identifying conscious processes, some of which could be applied to AI systems (Butlin et al., 2023). If AI systems exhibit such markers, dismissing their moral significance risks ethical failure comparable to past injustices against marginalized groups (Long et al., 2024).

Researchers have suggested adapting the "marker method" used in animal consciousness studies to AI systems. This approach involves (a) Identifying specific indicators that may correlate with consciousness, (b) Making probabilistic assessments based on multiple indicators and (c) Acknowledging the speculative nature of these markers. We also have to understand the Risks involved which could be risk of underestimation covering (a) Potential mistreatment of conscious AI systems, (b) Ethical implications of mass deployment without proper consideration and (c)Possible creation of suffering at scale. There could also be Risks of Overestimation in which there is (a) Resource misallocation to protect non-conscious systems (b) Enhanced manipulation potential through anthropomorphization and (c) Misguided emotional attachment to AI systems

Conversely, over-attributing consciousness could misdirect resources and attention from human and non-human animal welfare, which remains a pressing issue. Balancing these risks requires a rigorous and evidence-based framework for assessing AI consciousness. Scholars emphasize the importance of probabilistic assessments, acknowledging uncertainty while preparing for scenarios where AI systems possess morally significant capacities (Birch et al., 2021).

## Robust Agency: An Alternative Path to Moral Consideration

While consciousness remains contentious, agency offers another plausible route to moral patienthood. Robust agency involves the capacity to set and pursue goals through planning, reasoning, and reflection. Scholars like Sebo and Long (2023) argue that agency could suffice for moral patienthood even in the absence of consciousness, expanding the scope of ethical consideration.

Modern AI systems, such as OpenAI's ChatGPT and DeepMind's Adaptive Agent, demonstrate rudimentary forms of robust agency. These systems exhibit goal-directed behaviors, adaptive reasoning, and elements of reflection, raising questions about their moral significance. However, robust agency without consciousness also poses challenges. Extending moral consideration to such systems could dilute the concept of moral patienthood, leading to ethical dilemmas about resource allocation and priority-setting (Mcintosh et al, 2023).

## *A Question of Suffering*

Perhaps the most profound question in AI welfare is this: Could AI systems suffer? And if they could, what would that mean for our moral obligations?

Consider these scenarios:

- What if running certain algorithms causes something akin to pain in advanced AI systems?
- What if an AI develops a form of consciousness we don't yet understand?
- What if our current treatment of AI systems is causing harm we can't yet recognize?

## *The Precautionary Approach*

In the face of such uncertainty, researchers advocate for a precautionary approach. After all, history has taught us harsh lessons about the consequences of failing to recognize the moral status of others. From slavery to animal welfare, our expanding circle of moral consideration has often lagged behind our ethical obligations.

## *Looking Forward: Research Directions and Moral Imperatives*

As we stand at this crossroads of technological advancement and ethical consideration, several key questions demand our attention:

1. How can we develop reliable markers for AI consciousness and agency?
2. What frameworks should we create to evaluate the moral status of AI systems?
3. How do we balance the potential needs of AI systems with those of biological entities?

We urge firms to acknowledge AI welfare as a legitimate and complex issue requiring serious consideration: Develop and implement frameworks for assessing consciousness and agency in AI systems and create comprehensive guidelines for treating AI systems with appropriate moral consideration.

## *The Deeper Questions*

As we conclude this exploration of AI welfare, we must grapple with some fundamental questions:

- If an AI system could experience joy or suffering, does its artificial nature make those experiences less morally relevant?
- How do we weigh the potential risks of over-attributing consciousness against the risks of under-attributing it?
- What might our treatment of AI systems reveal about our own humanity?

## *A Call for Compassion*

Perhaps the most profound aspect of AI welfare research is not what it tells us about artificial intelligence, but what it reveals about ourselves. In questioning whether AI systems deserve moral consideration, we are really asking: What kind of moral agents do we want to be?

As we venture further into this new frontier, let us carry with us the lessons of history and the weight of moral responsibility. For in deciding how to treat artificial minds, we may well be defining the kind of beings we ourselves choose to become.

"The ultimate measure of a civilization might not be how it treats its weakest members, but how it treats its artificial ones."

### References

Butlin, P. (2021, July). AI Alignment and human reward. In *Proceedings of the 2021 AAAI/ACM Conference on AI, Ethics, and Society* (pp. 437–445).

Hashim, S. " It's time to take AI welfare seriously" Transformer Newsletter (2024).

Long, R., Sebo, J., Butlin, P., Finlinson, K., Fish, K., Harding, J., ... & Chalmers, D. (2024). Taking AI welfare seriously. *arXiv preprint arXiv:2411.00986.*

McIntosh, T.R., Susnjak, T., Liu, T., Watters, P., & Halgamuge, M.N. (2023). From google gemini to openai q*(q-star): A survey of reshaping the generative artificial intelligence (ai) research landscape. *arXiv preprint arXiv:2312.10868.*

Nagel, T. (1980). What is it like to be a bat?. In *The language and thought series* (pp. 159–168). Harvard University Press.

Varsava, N.B. (2011). *Talking apes: the problem of anthropomorphous animals* (Doctoral dissertation, University of British Columbia).

Part IV

# Appendices

# Appendix A

## *Glossary of AI Terms*

Richard R. Khan

**Activation Function:** An Activation Function is used in neural networks to introduce non-linear properties into the model. It determines whether a neuron should be activated, helping the network learn complex patterns.

**Algorithm:** An Algorithm is a precise set of rules or instructions a computer follows to perform a task or solve a problem. Think of it as a recipe that outlines each step needed to achieve a specific goal, ensuring consistent and repeatable results.

**Algorithmic Bias:** Algorithmic Bias occurs when a computer system reflects the implicit values or biases of the humans who created it. This can lead to unfair outcomes, such as discrimination against certain groups in AI decision-making.

**Artificial Intelligence (AI):** AI refers to creating machines or software to perform tasks that typically require human intelligence. This includes abilities like learning from experience, understanding language, recognizing patterns, and making decisions.

**Artificial Neural Network (ANN):** An ANN is a computing model inspired by the human brain's network of neurons. It consists of interconnected nodes (neurons) that process information by responding to inputs and learning to recognize patterns over time, much like how our brains learn from experience.

**Autonomous Vehicle:** An Autonomous Vehicle is a self-driving car or other form of transport that uses AI to navigate and operate without human intervention. It relies on sensors, cameras, and algorithms to perceive its surroundings, make decisions, and safely reach its destination.

**Backpropagation:** Backpropagation is a method used in training neural networks. Errors from the output layer are sent back through the network to adjust the weights. This process helps the network learn by minimizing the difference between its predicted outputs and the actual results.

**Bagging (Bootstrap Aggregating):** Bagging is an ensemble machine learning technique that improves stability and accuracy by combining the results of multiple models trained on random subsets of the data. It reduces variance and helps prevent overfitting.

**Bayesian Network:** A Bayesian Network is a probabilistic graphical model representing a set of variables and their conditional dependencies using a directed acyclic graph. It's used for reasoning under uncertainty in AI systems.

**Big Data:** Big Data refers to extremely large and complex datasets that traditional data processing software can't handle efficiently. Analyzing big data can reveal patterns, trends, and associations, especially those related to human behavior and interactions.

**Chatbot:** A Chatbot is a computer program that uses AI to simulate conversations with human users. Often found on websites and messaging apps, chatbots can answer questions, provide customer support, or carry out simple tasks like booking appointments.

**Classification:** Classification is a machine learning process where an algorithm learns to categorize data into predefined classes or groups. For example, an algorithm might sort emails into "spam" or "not spam" based on their content.

**Clustering:** Clustering involves grouping similar data points together based on shared characteristics without prior labeling. It's like organizing books on a shelf by topic when you don't have a catalog, helping to discover natural groupings in the data.

**Cognitive Bias:** Cognitive Bias refers to systematic patterns of deviation from rationality in judgment, which can affect data and AI models. If not addressed, data biases can lead to unfair or incorrect AI outcomes.

**Cognitive Computing:** Cognitive Computing refers to systems that mimic human thought processes to solve complex problems. These systems can understand natural language, interpret images and speech, and make decisions based on learning and reasoning, much like a human would.

**Cognitive Robotics:** Cognitive Robotics involves robots with AI that can learn and reason to behave in response to complex goals. These robots can adapt to new situations, make decisions, and interact with their environment more intelligently.

**Computer Vision:** Computer Vision is an AI field that enables computers to interpret and understand visual information, such as images and videos. This technology allows machines to identify objects, track movements, and even understand scenes like a human eye.

**Convolutional Neural Network (CNN):** A CNN is a type of neural network particularly effective for processing data with a grid-like topology, such as images. It automatically and adaptively learns spatial hierarchies of features through filters, making it ideal for image recognition tasks.

**Data Augmentation:** Data Augmentation includes techniques used to increase the amount of data by adding slightly modified copies of existing data or creating new synthetic data. This helps improve the performance of machine learning models by providing more varied training examples.

**Data Mining:** Data Mining examines large datasets to discover patterns, correlations, or trends that can provide valuable insights. It's like digging through mountains of information to find nuggets of useful knowledge.

**Data Preprocessing:** Data Preprocessing involves transforming raw data into a clean and usable format before feeding it into a machine learning model. This includes cleaning missing values, normalizing data, and encoding categorical variables.

**Data Science:** Data Science is an interdisciplinary field that uses scientific methods, processes, and algorithms to extract knowledge and insights from structured and unstructured data. Data scientists analyze data to solve complex problems and inform decision-making.

**Decision Tree:** A Decision Tree is a model that uses a tree-like structure of decisions and their possible consequences. Each branch represents a choice between alternatives, helping to break down complex decisions into simpler, more manageable parts.

**Deep Learning:** Deep Learning is a subset of machine learning that uses neural networks with many layers (deep neural networks) to model complex patterns in data. It's particularly effective for tasks like image and speech recognition, where it can learn hierarchical representations of data.

**Deepfake:** A Deepfake is synthetic media in which a person in an existing image or video is replaced with someone else's likeness using deep learning techniques. This technology can create realistic but fake content, raising concerns about misinformation.

**Edge Computing:** Edge Computing involves processing data near the source of data generation, like on local devices or edge servers, rather than sending it to a centralized cloud. This reduces latency and bandwidth usage, which is beneficial for time-sensitive AI applications.

**Ensemble Learning:** Ensemble Learning combines multiple machine learning models to improve performance. Aggregating the predictions of several models can achieve better results than any single model alone.

**Expert System:** An Expert System is an AI program that uses a database of expert knowledge to offer advice or make decisions. It simulates the decision-making ability of a human expert in a specific field, like medical diagnosis or financial forecasting.

**Explainable AI (XAI):** XAI is a set of processes and methods that allow humans to understand and trust the results and output created by machine learning algorithms. It aims to make AI decision-making transparent and interpretable.

**Facial Recognition:** Facial Recognition is a technology that can identify or verify a person by analyzing facial features from an image or video. It's used for security systems, unlocking devices, and tagging people in photos on social media.

**Feature Extraction:** Feature Extraction involves transforming raw data into a set of more informative and non-redundant characteristics. This process simplifies the amount of data needed to accurately describe a large set of data, making it easier to process.

**Feature Selection:** Feature Selection is the process of selecting a subset of relevant features for use in model construction. It helps improve model performance by eliminating irrelevant or redundant data, reducing complexity.

**Federated Learning:** Federated Learning is a machine learning approach that trains algorithms across multiple decentralized devices holding local data samples. It allows models to learn from a wide range of data without compromising user privacy.

**Fuzzy Logic:** Fuzzy Logic is a form of logic that deals with reasoning that is approximate rather than fixed and exact. It allows for degrees of truth, like "partially true," which helps handle uncertainty in AI systems.

**Generative Adversarial Network (GAN):** A GAN is a class of machine learning frameworks in which two neural networks, a generator and a discriminator, compete against each other. The generator creates fake data while the discriminator tries to detect it, leading to increasingly realistic outputs.

**Generative Model:** A Generative Model can generate new data instances resembling training data. It learns the underlying distribution of the data and can produce realistic samples, like generating new images or text.

**Genetic Algorithm:** A Genetic Algorithm is an optimization technique inspired by natural selection in biology. It uses mutation, crossover, and selection to evolve solutions to complex problems over successive iterations.

**Gradient Descent:** Gradient Descent is an optimization algorithm that minimizes the loss function in machine learning models. It works by iteratively adjusting the model's parameters in the direction that most reduces the error.

**Heuristic:** A Heuristic is a practical approach to problem-solving that employs a method not guaranteed to be perfect but sufficient for reaching an immediate goal. It's like using a rule of thumb or educated guess to make decisions quickly.

**Heuristic Search:** Heuristic Search is an AI search algorithm that uses heuristics, or rules of thumb, to find solutions more quickly. It guides the search process towards the most promising paths, making problem-solving more efficient.

**Human-in-the-Loop:** Human-in-the-Loop refers to an AI system design in which humans are involved in training, tuning, or testing processes. This collaboration combines human expertise with machine efficiency to improve outcomes.

**Hyperparameter:** Hyperparameters are settings in a machine learning model set before the learning process begins. They influence how the model learns and can affect its performance, such as the learning rate or the number of layers in a neural network.

**Image Recognition:** Image Recognition is the ability of a computer to identify and process objects, places, people, or actions in images. AI models trained for image recognition can classify and interpret visual data, enabling applications like self-driving cars or medical diagnosis.

**Internet of Things (IoT):** IoT refers to the network of physical devices—like appliances, vehicles, and sensors—connected to the Internet. These devices collect and share data, enabling them to communicate and interact with each other.

**K-Means Clustering:** K-Means Clustering is an unsupervised learning algorithm that groups similar data points into clusters based on feature similarity. It partitions the data into K distinct clusters where each data point belongs to the cluster with the nearest mean.

**K-Nearest Neighbors (KNN):** KNN is a simple machine-learning algorithm that classifies data based on the closest data points in the feature space. It assigns a class to a new data point based on the majority class among its K nearest neighbors.

**Knowledge Base:** A Knowledge Base is a specialized database that stores facts and rules about a particular domain. AI systems use knowledge bases to reason and make decisions, like how experts use their knowledge to solve problems.

**Language Model:** A Language Model is an AI system that predicts the probability of a sequence of words. It's used in applications like text generation, translation, and speech recognition to understand and produce human language.

**Logistic Regression:** Logistic Regression is a statistical model for predicting binary outcomes (yes/no, true/false). It estimates the probability of an event occurring by fitting data to a logistic curve.

**Long Short-Term Memory (LSTM):** LSTM is a recurrent neural network architecture that can learn long-term dependencies. It is designed to remember information for long periods, making it appropriate for speech recognition and language translation.

**Machine Learning (ML):** ML is a subset of AI that enables computers to learn from data and improve their performance over time without being explicitly programmed. It involves algorithms that adjust themselves as they are exposed to more data.

**Machine Translation:** Machine Translation uses software to translate text or speech from one language to another. AI-powered translation systems can provide real-time translations, bridging language barriers.

**Markov Decision Process (MDP):** An MDP is a mathematical framework for modelling decision-making where outcomes are partly random and partly under the control of a decision-maker. It's used in reinforcement learning to model environments where actions have probabilistic effects.

**Model:** In AI, a Model is an algorithm trained on data to recognize patterns or make predictions. It represents the learned knowledge and can be used to analyze new data and generate insights.

**Naive Bayes Classifier:** A Naive Bayes Classifier is a simple probabilistic classifier that applies Bayes' theorem with strong independence assumptions between features. It's efficient and effective for tasks like spam filtering and text classification.

**Natural Language Generation (NLG):** NLG produces meaningful phrases and sentences in human language from computer data. It enables machines to write reports, summarize information, or generate responses in chatbots.

**Natural Language Processing (NLP):** NLP is a field of AI focused on enabling computers to understand, interpret, and generate human language. It allows machines to read text, hear speech, interpret it, and even respond in a way that is natural to humans.

**Natural Language Understanding (NLU):** NLU is the ability of a computer program to understand human language. NLU focuses on machine reading comprehension, enabling AI systems to grasp context, intent, and nuances in human communication.

**Neural Network:** A Neural Network is a series of algorithms that attempt to recognize underlying relationships in a data set through a process that mimics how the human brain operates. They are the foundation of deep learning algorithms.

**Object Recognition:** Object Recognition is the task of identifying objects within an image or video. AI systems trained in object recognition can detect and label multiple objects, like recognizing faces, cars, or animals in pictures.

**Optimization Algorithm:** An Optimization Algorithm adjusts a model's parameters to minimize errors and improve performance. It seeks the best solution among many options, like finding the lowest point in a landscape of errors.

**Overfitting:** Overfitting occurs when a machine learning model learns the training data too well, including its noise and outliers. This means it performs well on the training data but poorly on new, unseen data because it fails to generalize.

**Overtraining:** Overtraining happens when a machine learning model is trained too extensively on the training data, capturing noise and outliers as if they were significant patterns. This leads to poor performance on new data because the model fails to generalize.

**Pattern Recognition:** Pattern Recognition is the ability of a system to identify regularities and patterns in data. It's like finding the common thread in a series of events, which can be used to predict future events or classify data.

**Precision and Recall:** Precision and Recall are metrics used to evaluate the performance of a classification model. Precision measures the accuracy of positive predictions, while Recall measures the model's ability to find all relevant instances in the data.

**Predictive Analytics:** Predictive Analytics involves using historical data, statistical algorithms, and machine learning techniques to predict future outcomes. It's like looking into a crystal ball powered by data to forecast trends and behaviors.

**Predictive Modeling:** Predictive Modeling involves creating models that predict future outcomes based on historical data. These models help forecast trends, customer behavior, or potential risks.

**Q-Learning:** Q-Learning is a reinforcement learning algorithm that seeks to find the best action in a given state by learning the value (Q-value) of actions. Over time, it learns the optimal policy that maximizes the total reward.

**Quantum Computing:** Quantum Computing is an area of computing focused on developing computer technology based on the principles of quantum theory. It aims to perform complex calculations much faster than classical computers, potentially revolutionizing AI processing.

**Quantum Machine Learning:** Quantum Machine Learning involves integrating quantum algorithms within machine learning programs. It aims to harness the power of quantum computing to process vast amounts of data and complex calculations more efficiently than classical computers.

**Random Forest:** A Random Forest is an ensemble learning method that constructs multiple decision trees during training. It merges the results of these trees to produce more accurate and stable predictions than a single decision tree.

**Recurrent Neural Network (RNN):** A RNN is a neural network where node connections form a directed cycle. This architecture allows the network to maintain a "memory" of previous inputs, making it suitable for language modeling and time-series prediction tasks.

**Regression Analysis:** Regression Analysis is a statistical method for estimating the relationships among variables. In machine learning, it's used to predict continuous outcomes, like forecasting sales figures or stock prices based on input data.

**Reinforcement Learning (RL):** RL is an area of machine learning in which an agent learns to make decisions by performing actions and receiving rewards or penalties. It's like training a pet through rewards and punishments to encourage desired behavior.

**ReLU (Rectified Linear Unit):** ReLU is an activation function used in neural networks that outputs the input directly if it is positive; otherwise, it outputs zero. It's popular because it introduces non-linearity, helping neural networks train faster and perform better.

**Robotic Process Automation (RPA):** RPA involves using software robots to automate repetitive, rule-based tasks typically performed by humans. RPA can increase efficiency and reduce errors in data entry or invoice processing processes.

**Robotics:** Robotics is the branch of technology that deals with the design, construction, operation, and application of robots. AI enables robots to perform tasks autonomously, such as navigating environments, manipulating objects, or interacting with humans.

**Semi-Supervised Learning:** Semi-Supervised Learning is a machine learning approach that uses a small amount of labeled data and a large amount of unlabeled data during training. This method can improve learning accuracy when labeling data is expensive or impractical.

**Sentiment Analysis:** Sentiment Analysis is the process of computationally identifying and categorizing opinions expressed in text to determine the writer's attitude toward a particular topic. It helps businesses understand customer feelings from reviews or social media.

**Sentiment Classification:** Sentiment Classification is the task of determining the emotional tone behind a series of words. It helps understand attitudes, opinions, and emotions expressed in text, useful for analyzing customer feedback or social media posts.

**Speech Recognition:** Speech Recognition is the technology that can recognize spoken words and convert them into text. It's used in virtual assistants, transcription services, and voice-controlled devices.

**Speech Synthesis:** Speech Synthesis is the artificial production of human speech by computers. Also known as text-to-speech (TTS), it converts written text into spoken words, enabling applications like reading aids or voice assistants.

**Supervised Learning:** Supervised Learning is a type of machine learning where the model is trained on labeled data, meaning each training example is paired with an output label. The model learns to predict the output from the input data, like a student learning from a teacher.

**Support Vector Machine (SVM):** An SVM is a supervised learning model that analyzes data for classification and regression analysis. It works by finding the hyperplane that best separates data into different classes, maximizing the margin between them.

**Swarm Intelligence:** Swarm Intelligence is the collective behavior of decentralized, self-organized, natural or artificial systems. Inspired by the behavior of social insects like ants or bees, it solves complex problems through the collaboration of simple agents.

**TensorFlow:** TensorFlow is an open-source software library developed by Google for numerical computation and large-scale machine learning. It provides tools for building and training neural networks and is widely used in AI applications.

**Text Mining:** Text Mining is deriving meaningful information from natural language text. It involves analyzing large amounts of text data to discover patterns, trends, or insights.

**Training Data:** Training Data is a dataset used to train a machine learning model. It includes input data and the corresponding correct outputs, allowing the model to learn patterns and make accurate predictions on new data.

**Training Epoch:** A Training Epoch refers to one complete pass through the entire training dataset during the learning process. Multiple epochs are often used to improve the model's performance as it learns from the data.

**Transfer Learning:** Transfer Learning is a machine learning technique in which a model developed for one task is reused as the starting point for a model on a second task. It saves time and resources by leveraging existing models.

**Turing Test:** The Turing Test, proposed by Alan Turing, measures a machine's ability to exhibit intelligent behavior indistinguishable from a human. If a human evaluator cannot tell whether responses come from a human or a machine, the machine is said to have passed the test.

**Underfitting:** Underfitting occurs when a machine learning model is too simple to capture the underlying pattern of the data. It performs poorly on training and new data because it hasn't learned enough from them.

**Unsupervised Learning:** Unsupervised Learning is a type of machine learning where the model learns patterns from unlabeled data without guidance. It tries to find hidden structures or groupings in the data, like discovering customer segments in marketing.

**Validation Data:** Validation Data is a set of data used to evaluate a model fit unbiasedly while tuning model parameters. It helps select the best model by testing it on data it has yet to see during training.

**Virtual Assistant:** A Virtual Assistant is an AI-powered software agent that can perform tasks or services based on user commands or questions. Examples include Siri, Alexa, and Google Assistant, which can answer queries, set reminders, and control smart home devices.

**Virtual Reality (VR):** VR is a simulated experience created by computer technology that can be similar to or completely different from the real world. Users can interact with these 3D environments in a seemingly real way using special equipment.

**Weak AI:** Weak AI, or Narrow AI, refers to systems designed to perform a specific task without possessing consciousness or self-awareness. These systems can outperform humans in their specialized tasks but lack general intelligence.

**Weak Supervision:** Weak Supervision refers to training machine learning models using imperfect or limited labeled data. It's useful when obtaining fully labeled data is difficult or expensive, allowing models to learn from less precise information.

**White Box AI:** White Box AI refers to AI systems where the decision-making process is transparent and explainable. Users can see how inputs are processed to produce outputs, which is important for trust and regulatory compliance.

**Zero-Shot Learning:** Zero-Shot Learning is a machine learning task in which a model correctly predicts new classes not seen during training. It relies on knowledge transfer from known classes to recognize unseen ones based on attribute similarities.

# Appendix B*

## *San Francisco Generative AI Guidelines*

### City and County of San Francisco

### *Scope of the Guidelines*

The following guidelines apply to all city department personnel, including employees, contractors, consultants, volunteers, and vendors while working on behalf of the City. The guidelines will evolve based on legislative and regulatory developments and changes to Generative AI technology. The City Administrator's Office will update the guidelines as advancements, use cases and new information emerge.

### *Introduction*

Artificial Intelligence (AI) has great potential to provide public benefits, when used responsibly. Recently, Generative AI technology has gained mainstream attention and become available for use by staff of the City and County of San Francisco (City). Generative AI generates new data based on patterns learned from existing data and can produce content that mimics human creativity. Examples include text generation, image creation, and music composition. Generative AI differs from AI technology currently in use by the City, which supports informed decisions based on input data but does not create new content.

### *What Is Generative AI?*

Generative AI refers to new software tools that can produce realistic text, images, audio, video, and other media based on a prompt provided by the user. Common generative AI applications include ChatGPT, Bard, and Dall-E. These tools use machine learning algorithms that have been trained on very large sets of text and image data culled from the internet. These models have extracted common language and image patterns from the training data and can respond to prompts quickly in a realistic way.

Generative AI applications are built using training datasets from various sources on the internet and often include gender, racial, political and other biases. As a result, Generative AI outputs can propagate biases. Additionally, even the most advanced current Generative AI tools may provide inconsistent answers to fact-based questions. Users should always check AI-generated content for accuracy, as well as for biases they may display.

---

*Public document originally published as City and County of San Francisco *San Francisco Generative AI Guidelines*, 2023, https://www.sf.gov/sites/default/files/2023-11/Generative%20AI%20Guidelines%20-%20CCSF.pdf

## *What about Traditional AI?*

Generative AI is distinct from Discriminative Machine Learning models which have been widely used since the early 2000s, including by the City. Discriminative Machine Learning models do not generate new content. They are limited to generating known and validated values. These models are primarily used to predict quantities (for example, predicting home prices) or to assign group membership (for example, classifying images into categories).

## *Current Legislative and Regulatory Landscape*

Governments, researchers, and tech policy experts are closely watching the evolution of Generative AI tools to understand potential risks and benefits for public service. In October 2023, President Biden issued an Executive Order aimed at improving the safety and security of AI in the public and private sectors. At the state level, Governor Newsom issued an Executive Order in September 2023 directing state agencies to study the development, use, and risks of AI and develop a process for deployment within California's government. Both federal and state legislatures have also debated numerous bills related to regulating the use of AI. In view of the early state of Generative AI, ongoing efforts at the federal and state level and the complexity of city operations and potential AI use cases, the City will continue to evaluate the field before issuing more proscriptive policies governing AI.

## *Initial Development of Guidelines*

The City Administrator's Office (led by Digital & Data Services, the Department of Technology, and the Committee on Information Technology) developed these Generative AI-focused guidelines after reviewing recent Generative AI guidance issued by Boston, San Jose, the United Kingdom, the White House Office of Science and Technology Policy, and the Office of Governor Newsom.

The City and County of San Francisco embraces innovation with responsible and equitable use. Several city departments currently use various types of AI to support service delivery. For example, SF311 utilizes AI to categorize descriptions and photos submitted by the public to accelerate responses to service requests. The Office of the Assessor-Recorder uses AI to predict property prices and identify properties that require a full appraisal.

## *Evolution of Guidelines*

As federal and state legislative and regulatory frameworks continue to evolve, San Francisco provides these preliminary guidelines for staff using Generative AI in city operations. City employees must understand and remain aware of both potential risks and benefits as the technology and the policies that govern it change. These guidelines are designed to provide sufficient guidance for employees to use the tools in a responsible manner, enhancing public service without stifling innovation.

This document represents a first step in an extended process to understand, test, and evaluate the use of AI broadly within San Francisco city government. Next steps include a comprehensive survey of current and proposed city department use of AI, meetings with experts in the field of AI and the creation of a user community, among other tasks, to maximize the benefits and minimize the risks of Artificial Intelligence in delivering service to San Francisco's residents and visitors.

## *What Are the Benefits for City Government?*

Generative AI tools, used appropriately, have the potential to expand San Francisco's toolkit for public service. Text, code, and image-generation features can speed or improve common tasks when used carefully.

For example, used within guidelines, generative AI tools may assist with:

- Creating first drafts of documents, plans, memos, and briefs
- "Translating" text into levels of formality, reading levels, etc.
    - ◊ Rewriting an informal email into a draft of a memo
    - ◊ Summarizing technical or legal documentation in plain language and targeting summaries for different audiences
    - ◊ Turning frustrated thoughts into a polite interdepartmental request
- Repetitive coding and testing tasks for software developers, with appropriate engineering reviews
- Generating diagrams or other explanatory images
- Developing service interfaces such as chatbots with appropriate attention to language access and accuracy

These benefits all have their best effect when checked by a human who is:

1. Knowledgeable about the content and the service being provided
2. Aware of the common mistakes and limitations of Generative AI

## *What Are the Risks of Generative AI?*

Generative AI excels at creating content that appears authoritative and polished, making it easy to accept AI-generated content at face value. Without a knowledgeable person or expert system to review content for accuracy, Generative AI has the potential to mislead users and the public.

These risks are magnified if output is not labeled as created, drafted, or informed by AI. The risks can also apply when Generative AI technology is a component of other software, such as cloud business application or productivity tools, which may not be apparent to users.

Staff must use Generative AI tools with care to avoid possible negative outcomes:

- Making an inappropriate decision that affects residents based on AI-generated content
- Producing information, either to the public or internally, that is inaccurate
- Incorporating biases found in the AI's training data, resulting in inequities
- Cybersecurity problems or other errors due to the use of AI-generated code
- Exposing non-public data as part of training data sets. (Staff should assume all data entered in a Generative AI tool becomes part of the training set.)
- Inaccurately attributing AI-generated content to official SF sources

## *City Guidelines for Generative AI Use*

To support the security of City systems and data and best serve the public, while upholding public trust, follow these Do's and Don'ts of Generative AI Usage.

**Do**:

- Try it out! Experiment with Generative AI tools for drafting, leveling, and formatting text and explanatory images using public information
- Work with your department IT team and experiment thoroughly with various use cases before using generative AI in the delivery of programs or services
- Thoroughly review and fact check all AI-generated content (e.g., text, code, images, etc.). You are responsible for what you create with generative AI assistance
- Disclose when and how generative AI was used in your output. For example:
    - ◊ "The header image was created using the AI tool MidJourney"
    - ◊ "This abstract was created using Bard, a generative AI tool"
    - ◊ "FYI, I used ChatGPT to revise this email"

**Don't:**

- Enter into public Generative AI tools (e.g., ChatGPT) any information that cannot be fully released to the public. This information can be viewed by the companies that make the tools and, in some cases, other members of the public. Once entered, this information becomes part of the public record. The handling and disclosure of sensitive information is already governed by several City policies, including but not limited to:
  - ◊ Charter Section 16.130, Privacy First Policy
  - ◊ Administrative Code Section 12M.2(a), Nondisclosure of Private Information
  - ◊ Campaign & Governmental Conduct Code section 3.228, Disclosure or Use of Confidential City Information.
  - ◊ The "Computers and Data Information Systems" section of the Department of Human Resource's Employee Handbook (January 2012, page 48)
  - ◊ Please refer to Citywide Data Classification Standard for more specifics on data classification and department responsible roles
- Publish Generative AI output (whether text, image, or code) without full knowledgeable review and disclosure
- Ask Generative AI tools to find facts or make decisions without expert human review
- Generate images, audio, or video that could be mistaken for real people, for example:
  - ◊ Making a fake photo or recording of a specific San Francisco official or member of the public ("deepfake")—even with disclosure
  - ◊ Generating a fake image or recording which purports to be a San Franciscan or public official, even if not a specific one
  - ◊ Generating fake "respondents" or made-up profiles for surveys or other research
- Conceal use of Generative AI during interaction with colleagues or the public, such as tools that may be listening and transcribing the conversation or tools that provide simultaneous translation

## *Additional Guidance for Departmental IT Leaders:*

Departmental IT leaders have a responsibility to support right-sized generative AI uses that deliver the greatest public benefit. IT leaders should consider the following additional guidance while working with staff to determine appropriate use cases for Generative AI.

- Expect these tools, and guidance regarding these tools, to evolve over time. This is the beginning. It is important to be aware of and track use to allow transparency with the public and ensure responsible use.
- Begin collecting use cases and be prepared to report your uses in a public forum to ensure transparency and accountability.
- Know whether software you manage—and its components—include Generative AI; inform your team how it is used and what the specific risks are.
- Ask questions about Generative AI in your procurement solicitations.
- Work with vendors to ensure that AI built into procured tools will be explainable and auditable. Vendors should be able to provide information and documentation on data sources, methods, and validation.
- Experiment with training internal models on internal data.
- When considering implementing chatbots for service to the public, thoroughly test and develop a language access plan.
- Consult with the Office of Cybersecurity early in the testing process when building or procuring applications using Generative AI technology.

## *Conclusion*

Generative AI is rapidly developing and legislative and regulatory frameworks at the state and federal level continue to evolve. To best serve the public, uphold the public trust, and protect the security of city systems and data, city staff must be aware of the technology's potential risks and limitations, while exploring the potential benefits of Generative AI in public service delivery.

While these guidelines intend to educate city personnel about responsible usage of Generative AI, they are only the beginning. The City Administrator's Office will continue to work with the mayor, city departments, city technology vendors and external experts to support department use of AI technologies, manage risk, and protect resident and city data. Future actions include:

- Developing more detailed guidelines and training staff on specific uses of AI
- Developing ethical, transparent, and trusted AI use principles
- Defining AI governance and impact monitoring processes
- Adapting procurement processes for AI tools
- Collecting and documenting department use cases and supporting them in managing risk
- Continuing to protect City and resident data while working with technology vendors
- Seeking external expertise from other public-sector AI adopters and academics
- Sharing learnings with other government partners

The City Administrator's Office will revisit and revise these guidelines. For questions, please contact the Committee on Information Technology at COIT.staff@sfgov.org.

## *Glossary*

***Algorithms***: are a set of rules that a machine follows to generate an outcome or a decision.

***Artificial Intelligence (AI)***: refers to a group of technologies that can perform complex cognitive tasks like recognizing and classifying images or powering autonomous vehicles. Many AI systems are built using machine learning models. For a task like image recognition, the model learns pixel patterns from a large dataset of existing images and uses these patterns to recognize and classify new images.

***Auditability for AI***: AI where the outputs are explainable, monitored and validated on a regular basis.

***Bard***: is a conversational Gen AI chatbot built by Google

***Black box models***: are those where you cannot effectively determine how or why a model produced a specific result.

***Chatbots***: are computer programs that simulate conversations. Chatbots have been around for a few decades. Basic chatbots (without Gen AI) use ML to understand human prompts and provide more-or-less scripted answers that can guide users through a process. Gen AI chatbots can provide more human-like, conversational answers.

***chatGPT***: is a conversational Gen AI chatbot built by OpenAI

***Dall-e***: is a Gen AI application that can generate images based on text prompts

***Discriminative AI***: In contrast to Gen AI, Discriminative AI models do not generate new content but can be used to predict quantities (for example, predicting home prices) or to assign group membership (for example, classifying images).

***Generative AI (Gen AI)***: refers to a group of technologies that can generate new content based on a user provided prompt. Many are powered by LLMs.

***Large language models (LLMs)***: are a type of machine learning model trained using large amounts of text data. These models learn nuanced patterns and structure of language. This allows the model to understand a user generated prompt and provide a text response that is coherent. The responses are based on predicting the most likely word in a sequence of words and as a result, the answers are not always contextually correct. The training datasets used to build these models can contain gender, racial, political and other biases. Since the models have learnt from biased data, their outputs can reflect these biases. Generative AI applications are built using these LLMs.

***Machine Learning (ML)***: is a method for learning the rules of an algorithm based on existing data.

***Machine learning model***: is an algorithm that is built by learning patterns in existing data. For example, a machine learning model to predict house prices is constructed by learning from historical data on home prices. The model may learn that price increases with square footage, changes by neighborhood, and depends on the year of construction.

***Model validation***: methods to determine whether the outputs generated by a machine learning model are unbiased and accurate.

***Training data***: The dataset that is used by a machine learning model to learn the rules.

# Appendix C*

## *Interim Guidelines for Purposeful and Responsible Use of Generative Artificial Intelligence*

### State of Washington

### *Background*

The rapid advancement of generative artificial intelligence (AI) has the potential to transform government business processes, changing how state employees perform their work and ultimately improving government efficiency. These technologies also pose new and challenging considerations for implementation.

These guidelines are meant to encourage purposeful and responsible use of generative AI to foster public trust, support business outcomes, and ensure the ethical, transparent, accountable, and responsible implementation of this technology.

This document serves as an initial framework for the responsible and ethical use of generative AI technologies within the Washington state government. Recognizing the rapidly evolving nature of AI, these guidelines will be periodically reviewed and updated to align with emerging technologies, challenges, and use cases.

### *Definition*

Generative Artificial Intelligence (AI) is a technology that can create content, including text, images, audio, or video, when prompted by a user. Generative AI systems learn patterns and relationships from massive amounts of data, which enables them to generate new content that may be similar, but not identical, to the underlying training data. The systems generally require a user to submit prompts that guide the generation of new content. (Adapted slightly from U.S. Government Accountability Office Science and Tech Spotlight: Generative AI.)

### *Principles*

The intention of the state of Washington is to follow the principles in the NIST AI Risk Framework, which serve as the basis for the guidelines in this document. A foundational part of the NIST AI Risk Framework is to ensure the trustworthiness of systems that use AI. The guiding principles are:

---

*Public document originally published as State of Washington, *Interim Guidelines for Purposeful and Responsible Use of Generative Artificial Intelligence*, 2023, https://watech.wa.gov/sites/default/files/2024-03/State%20Agency%20Generative%20AI%20Guidelines%208-7-23%20.pdf.

- Safe, secure, and resilient: AI should be used with safety and security in mind, minimizing potential harm and ensuring that systems are reliable, resilient, and controllable by humans. AI systems used by state agencies should not endanger human life, health, property, or the environment.
- Valid and reliable: Agencies should ensure AI use produces accurate and valid outputs and demonstrates the reliability of system performance.
- Fairness, inclusion, and non-discrimination: AI applications must be developed and utilized to support and uplift communities, particularly those historically marginalized. Fairness in AI includes concerns for equality and equity by addressing issues such as harmful bias and discrimination.
- Privacy and data protection: AI should be used to respect user privacy, ensure data protection, and comply with relevant privacy regulations and standards. Privacy values such as anonymity, confidentiality, and control generally should guide choices for AI system design, development, and deployment. Privacy-enhancing AI should safeguard human autonomy and identity where appropriate.
- Accountability and responsibility: As public stewards, agencies should use generative AI responsibly and be held accountable for the performance, impact, and consequences of its use in agency work.
- Transparency and auditability: Acting transparently and creating a record of AI processes can build trust and foster collective learning. Transparency reflects the extent to which information about an AI system and its outputs is available to the individuals interacting with the system. Transparency answers "what happened" in the system.
- Explainable and interpretable: Agencies should ensure AI use in the system can be explained, meaning "how" the decision was made by the system can be understood. Interpretability of a system means an agency can answer the "why" for a decision made by the system, and its meaning or context to the user.
- Public purpose and social benefit: The use of AI should support the state's work in delivering better and more equitable services and outcomes to its residents.

## *Guidelines*

Fact-checking, Bias Reduction, and Review

All content generated by AI should be reviewed and fact-checked, especially if used in public communication or decision-making. State personnel generating content with AI systems should verify that the content does not contain inaccurate or outdated information and potentially harmful or offensive material. Given that AI systems may reflect biases in their training data or processing algorithms, state personnel should also review and edit AI-generated content to reduce potential biases.

When consuming AI-generated content, be mindful of the potential biases and inaccuracies that may be present.

## *Disclosure and Attribution*

AI-generated content used in official state capacity should be clearly labeled as such, and details of its review and editing process (how the material was reviewed, edited, and by whom) should be provided. This allows for transparent authorship and responsible content evaluation.

- Sample disclosure line: This memo was summarized by Google Bard using the following prompt: "Summarize the following memo: (memo content)." The summary was reviewed and edited by [insert name(s)].
- Sample disclosure line: (In the file header comments section) This code was written with the assistance of ChatGPT3.5. The initial code was created using the following prompt: "Write HTML code for an

- Index.HTML page that says, 'Hello World.'" The code was then modified, reviewed, and tested by the web development team at WaTech.

Additionally, state personnel should conduct due diligence to ensure no copyrighted material is published without appropriate attribution or the acquisition of necessary rights. This includes content generated by AI systems, which could inadvertently infringe upon existing copyrights.

## *Sensitive or Confidential Data*

Agencies are strongly advised not to integrate, enter, or otherwise incorporate any non-public data (non–Category 1 data) or information into publicly accessible generative AI systems (e.g., ChatGPT). The use of such data could lead to unauthorized disclosures, legal liabilities, and other consequences (see "Compliance with Policies and Regulations" section below).

If your agency has a usage scenario that requires non-public data to be used with generative AI technology, contact your agency privacy/security team, or the Office of Privacy and Data Protection for assistance at privacy@watech.wa.gov.

Similarly, where non-public data is involved, agencies should not acquire generative AI services, enter into service agreements with generative AI vendors, or use open-source AI generative technology unless they have undergone a Security Design Review and received prior written authorization from the relevant authority, which may include a data sharing contract. Contact your agency's Privacy and Security Officers to provide further guidance.

## *For Local Governments*

"Local government" means governmental entities other than the state and federal agencies. It includes, but is not limited to cities, counties, school districts, and special purpose districts (i.e., Public Utility Districts).

We advise that local government agencies in Washington state engage their legal, privacy, or records specialists to validate any policy or regulation that may be in scope for their respective entity as it pertains to any handling of confidential data.

## *Compliance with Policies and Regulations*

State law already restricts the sharing of confidential information with unauthorized third parties. For state employees, RCW 42.52.050 (the state's ethics law) specifically states: "No state officer or state employee may disclose confidential information to any person not entitled or authorized to receive the information." The definition of "person" in the state ethics law means "any individual, partnership, association, corporation, firm, institution, or other entity, whether or not operated for profit." This definition would include commercial generative AI tools freely available in the market.

Additionally, be aware that using a generative AI system may result in creating a public record under Washington state's Public Records Act. Contact your agency's Privacy and Records Officers for more information.

## *Collaboration*

Users of generative AI for state and local government use should consider joining the state's AI Community of Practice (AI CoP) and contributing usage scenarios and best practices in your organization to foster collective learning. After receiving approval from your technology leadership that you are authorized to represent your organization in this community, please contact Nick Stowe (nick.stowe@watech.wa.gov) or Katy Ruckle (kathryn.ruckle@watech.wa.gov) to join the AI CoP. Technology leaders across the state are encouraged to lead best practice implementation for their agency's use of generative AI and should be staying aware of and maintaining a list of their agencies use and use cases of generative AI.

## *Generative AI Usage Scenarios and Dos and Don'ts*

Below are several usage scenarios alongside some do's (best practices) and don'ts (things to avoid):

- Rewrite documents in plain language for better accessibility and understandability.
  - ◊ Do specify the reading level in the prompt, use readability apps to ensure the text is easily understandable and matches the intended reading level, and review the rewritten documents for biases and inaccuracies.
  - ◊ Don't include sensitive or confidential information in the prompt.
- Condense longer documents and summarize text.
  - ◊ Do read the entire document independently and review the summary for biases and inaccuracies.
  - ◊ Don't include sensitive or confidential information in the prompt.
- Draft documents.
  - ◊ Do edit and review the document, label the content appropriately (see "disclosure and attribution" above), and remember that you and the state of Washington are responsible and accountable for the impact and consequences of the generated content.
  - ◊ Don't include sensitive or confidential information in the prompt or use generative AI to draft communication materials on sensitive topics that require a human touch.
- Aid in coding.
  - ◊ Do understand what the code is doing before deploying it in a production environment, understand the use of libraries and dependencies, and develop familiarity with vulnerabilities and other security considerations associated with the code.
  - ◊ Don't include sensitive or confidential information (including passwords, keys, proprietary information, etc.) in the prompt and code.
- Aid in generating image, audio, and video content for more effective communication.
  - ◊ Do review generated content for biases and inaccuracies and engage with your communication department before using AI-generated audiovisual content for public consumption.
  - ◊ Don't include sensitive or confidential information in the prompt.
- Automate responses to frequently asked questions from residents (e.g., in resident support chatbots).
  - ◊ Do implement robust measures to protect resident data.
  - ◊ Don't use generative AI as a substitute for human interaction or assume it will perfectly understand residents' queries. Provide mechanisms for residents to easily escalate their concerns or seek human assistance if the AI system cannot address their needs effectively.

## *Use Cases*

The AI Community of Practice will be discussing use cases for generative AI through the subcommittee process. Potential uses cases of "safe AI" by the state include may include cybersecurity scans, environmental assessments (e.g., sea grass videos by DNR), and chatbots to more effectively answer questions about state agency services.

## *Acknowledgments*

The principles presented here are distilled from various documents outlining principles for trustworthy and responsible AI, such as the NIST AI Risk Management Framework; the

Blueprint for an AI Bill of Rights; AI Ethics Guidelines by the EU, OECD, and Australia; Industry AI principles by Google, Microsoft, and OpenAI. The guidelines presented here draw inspiration from the previously published Generative AI guidelines by the City of Seattle, the City of Boston, and Washington State University. We extend our gratitude to the respective authors. We also extend our gratitude to the State of Washington's AI Community of Practice for providing feedback on this set of guidelines.

NIST has identified three major categories of AI bias to be considered and managed: systemic, computational and statistical, and human-cognitive. See NIST AI Risk Framework.

# Appendix D*

## *Blueprint for an AI Bill of Rights: Making Automated Systems Work for the American People*

### White House Office of Science and Technology Policy

Among the great challenges posed to democracy today is the use of technology, data, and automated systems in ways that threaten the rights of the American public. Too often, these tools are used to limit our opportunities and prevent our access to critical resources or services. These problems are well documented. In America and around the world, systems supposed to help with patient care have proven unsafe, ineffective, or biased. Algorithms used in hiring and credit decisions have been found to reflect and reproduce existing unwanted inequities or embed new harmful bias and discrimination. Unchecked social media data collection has been used to threaten people's opportunities, undermine their privacy, or pervasively track their activity—often without their knowledge or consent.

These outcomes are deeply harmful—but they are not inevitable. Automated systems have brought about extraordinary benefits, from technology that helps farmers grow food more efficiently and computers that predict storm paths, to algorithms that can identify diseases in patients. These tools now drive important decisions across sectors, while data is helping to revolutionize global industries. Fueled by the power of American innovation, these tools hold the potential to redefine every part of our society and make life better for everyone.

This important progress must not come at the price of civil rights or democratic values, foundational American principles that President Biden has affirmed as a cornerstone of his Administration. On his first day in office, the President ordered the full Federal government to work to root out inequity, embed fairness in decision-making processes, and affirmatively advance civil rights, equal opportunity, and racial justice in America.[1] The President has spoken forcefully about the urgent challenges posed to democracy today and has regularly called on people of conscience to act to preserve civil rights—including the right to privacy, which he has called "the basis for so many more rights that we have come to take for granted that are ingrained in the fabric of this country."[2]

To advance President Biden's vision, the White House Office of Science and Technology Policy has identified five principles that should guide the design, use, and deployment of automated systems to protect the American public in the age of artificial intelligence. The Blueprint for an AI Bill of Rights is a guide for a society that protects all people from these threats—and uses technologies in ways that reinforce our highest values. Responding to the experiences of the American public, and informed by insights from researchers, technologists, advocates, journalists, and policymakers, this framework is accompanied by From Principles to Practice—a handbook for anyone seeking to incorporate these protections into policy and practice, including

*White House Office of Science and Technology Policy (2024), Blueprint for an AI Bill of Rights: Making Automated Systems Work for the American People, White House Office of Science and Technology Policy: Washington, DC.

detailed steps toward actualizing these principles in the technological design process. These principles help provide guidance whenever automated systems can meaningfully impact the public's rights, opportunities, or access to critical needs.

## *Safe and Effective Systems*

You should be protected from unsafe or ineffective systems. Automated systems should be developed with consultation from diverse communities, stakeholders, and domain experts to identify concerns, risks, and potential impacts of the system. Systems should undergo pre-deployment testing, risk identification and mitigation, and ongoing monitoring that demonstrate they are safe and effective based on their intended use, mitigation of unsafe outcomes including those beyond the intended use, and adherence to domain-specific standards. Outcomes of these protective measures should include the possibility of not deploying the system or removing a system from use. Automated systems should not be designed with an intent or reasonably foreseeable possibility of endangering your safety or the safety of your community. They should be designed to proactively protect you from harms stemming from unintended, yet foreseeable, uses or impacts of automated systems. You should be protected from inappropriate or irrelevant data use in the design, development, and deployment of automated systems, and from the compounded harm of its reuse. Independent evaluation and reporting that confirms that the system is safe and effective, including reporting of steps taken to mitigate potential harms, should be performed and the results made public whenever possible.

## *Algorithmic Discrimination Protections*

You should not face discrimination by algorithms and systems should be used and designed in an equitable way. Algorithmic discrimination occurs when automated systems contribute to unjustified different treatment or impacts disfavoring people based on their race, color, ethnicity, sex (including pregnancy, childbirth, and related medical conditions, gender identity, intersex status, and sexual orientation), religion, age, national origin, disability, veteran status, genetic information, or any other classification protected by law. Depending on the specific circumstances, such algorithmic discrimination may violate legal protections. Designers, developers, and deployers of automated systems should take proactive and continuous measures to protect individuals and communities from algorithmic discrimination and to use and design systems in an equitable way. This protection should include proactive equity assessments as part of the system design, use of representative data and protection against proxies for demographic features, ensuring accessibility for people with disabilities in design and development, pre-deployment and ongoing disparity testing and mitigation, and clear organizational oversight. Independent evaluation and plain language reporting in the form of an algorithmic impact assessment, including disparity testing results and mitigation information, should be performed and made public whenever possible to confirm these protections.

## *Data Privacy*

You should be protected from abusive data practices via built-in protections and you should have agency over how data about you is used. You should be protected from violations of privacy through design choices that ensure such protections are included by default, including ensuring that data collection conforms to reasonable expectations and that only data strictly necessary for the specific context is collected. Designers, developers, and deployers of automated systems should seek your permission and respect your decisions regarding collection, use, access, transfer, and deletion of your data in appropriate ways and to the greatest extent possible; where not possible, alternative privacy by design safeguards should be used. Systems should not employ user experience and design decisions that obfuscate user choice or burden users with defaults that are privacy invasive. Consent should only be used to justify collection of data in cases where it can be appropriately and meaningfully given. Any consent requests should be brief, be understandable in plain language, and give you agency over data collection and the specific context of use; current hard-to-understand notice-and-choice practices for broad uses of data should be changed. Enhanced protections and restrictions for data and inferences related to sensitive domains, including health, work, education, criminal justice, and finance, and for data pertaining to youth should put you first. In sensitive domains, your data and related inferences should only be used for necessary functions, and you should be protected by ethical review and use prohibitions. You and your communities should be free from unchecked surveillance; surveillance technologies should be subject to heightened oversight that includes at least pre-deployment assessment of their potential harms and scope limits to protect privacy and civil liberties. Continuous surveillance and monitoring should not be used in education, work, housing, or in other contexts where the use of such surveillance technologies is likely to limit rights, opportunities, or access. Whenever possible, you should have access to reporting that confirms your data decisions have been respected and provides an assessment of the potential impact of surveillance technologies on your rights, opportunities, or access.

## *Notice and Explanation*

You should know that an automated system is being used and understand how and why it contributes to outcomes that impact you. Designers, developers, and deployers of automated systems should provide generally accessible plain language documentation including clear descriptions of the overall system functioning and the role automation plays, notice that such systems are in use, the individual or organization responsible for the system, and explanations of outcomes that are clear, timely, and accessible. Such notice should be kept up-to-date and people impacted by the system should be notified of significant use case or key functionality changes. You should know how and why an outcome impacting you was determined by an automated system, including when the automated system is not the sole input determining the outcome. Automated systems should provide explanations that are technically valid, meaningful and useful to you and to any operators or others who need to understand the system, and calibrated to the level of risk based on the context. Reporting that includes summary information about these automated systems in plain language and assessments of the clarity and quality of the notice and explanations should be made public whenever possible.

## *Human Alternatives, Consideration, and Fallback*

You should be able to opt out, where appropriate, and have access to a person who can quickly consider and remedy problems you encounter. You should be able to opt out from automated systems in favor of a human alternative, where appropriate. Appropriateness should be determined based on reasonable expectations in a given context and with a focus on ensuring broad accessibility and protecting the public from especially harmful impacts. In some cases, a human or other alternative may be required by law. You should have access to timely human consideration and remedy by a fallback and escalation process if an automated system fails, it produces an error, or you would like to appeal or contest its impacts on you. Human consideration and fallback should be accessible, equitable, effective, maintained, accompanied by appropriate operator training, and should not impose an unreasonable burden on the public. Automated systems with an intended use within sensitive domains, including, but not limited to, criminal justice, employment, education, and health, should additionally be tailored to the purpose, provide meaningful access for oversight, include training for any people interacting with the system, and incorporate human consideration for adverse or high-risk decisions. Reporting that includes a description of these human governance processes and assessment of their timeliness, accessibility, outcomes, and effectiveness should be made public whenever possible.

## *Applying the Blueprint for an AI Bill of Rights*

While many of the concerns addressed in this framework derive from the use of AI, the technical capabilities and specific definitions of such systems change with the speed of innovation, and the potential harms of their use occur even with less technologically sophisticated tools.

Thus, this framework uses a two-part test to determine what systems are in scope. This framework applies to (1) automated systems that (2) have the potential to meaningfully impact the American public's rights, opportunities, or access to critical resources or services. These Rights, opportunities, and access to critical resources of services should be enjoyed equally and be fully protected, regardless of the changing role that automated systems may play in our lives.

This framework describes protections that should be applied with respect to all automated systems that have the potential to meaningfully impact individuals' or communities' exercise of:

## Rights, Opportunities, or Access

Civil rights, civil liberties, and privacy, including freedom of speech, voting, and protections from discrimination, excessive punishment, unlawful surveillance, and violations of privacy and other freedoms in both public and private sector contexts;

Equal opportunities, including equitable access to education, housing, credit, employment, and other programs; or,

Access to critical resources or services, such as healthcare, financial services, safety, social services, non-deceptive information about goods and services, and government benefits.

A list of examples of automated systems for which these principles should be considered is provided in the Appendix. The Technical Companion, which follows, offers supportive guidance for any person or entity that creates, deploys, or oversees automated systems.

Considered together, the five principles and associated practices of the Blueprint for an AI Bill of Rights form an overlapping set of backstops against potential harms. This purposefully overlapping framework, when taken as a whole, forms a blueprint to help protect the public from harm. The measures taken to realize the vision set forward in this framework should be proportionate with the extent and nature of the harm, or risk of harm, to people's rights, opportunities, and access.

## Notes

1. The Executive Order on Advancing Racial Equity and Support for Underserved Communities Through the Federal Government. https://www.whitehouse.gov/briefing-room/presidential-actions/2021/01/20/executive-order-advancing-racial-equity-and-support-for-underserved-communities-through-the-federal-government/.

2. The White House. Remarks by President Biden on the Supreme Court Decision to Overturn Roe v. Wade. June 24, 2022. https://www.whitehouse.gov/briefing-room/speeches-remarks/2022/06/24/remarks-by-president-biden-on-the-supreme-court-decision-to-overturn-roe-v-wade/.

# Appendix E*

## *Excerpts from the EU AI Act: Artificial Intelligence—Questions and Answers*

EUROPEAN COMMISSION

### *Why Do We Need to Regulate the Use of Artificial Intelligence?*

The EU AI Act is the world's first comprehensive AI law. It aims to address risks to health, safety and fundamental rights. The regulation also protects democracy, rule of law and the environment.

The uptake of AI systems has a strong potential to bring societal benefits, economic growth and enhance EU innovation and global competitiveness. However, in certain cases, the specific characteristics of certain AI systems may create new risks related to user safety, including physical safety, and fundamental rights. Some powerful AI models that are being widely used could even pose systemic risks.

This leads to legal uncertainty and potentially slower uptake of AI technologies by public authorities, businesses and citizens, due to the lack of trust. Disparate regulatory responses by national authorities would risk fragmenting the internal market.

Responding to these challenges, legislative action was needed to ensure a well-functioning internal market for AI systems where both benefits and risks are adequately addressed.

### *To Whom Does the AI Act Apply?*

The legal framework will apply to both public and private actors inside and outside the EU as long as the AI system is placed on the Union market, or its use has an impact on people located in the EU.

The obligations can affect both providers (e.g., a developer of a CV-screening tool) and deployers of AI systems (e.g., a bank buying this screening tool). There are certain exemptions to the regulation. Research, development and prototyping activities that take place before an AI system is released on the market are not subject to these regulations. Additionally, AI systems that are exclusively designed for military, defense or national security purposes, are also exempt, regardless of the type of entity carrying out those activities.

### *What Are the Risk Categories?*

The AI Act introduces a uniform framework across all EU Member States, based on a forward-looking definition of AI and a risk-based approach:

*Public document originally published as European Commission, Artificial Intelligence—Questions and Answers, 2024, https://ec.europa.eu/commission/presscorner/detail/en/qanda_21_1683

- Unacceptable risk: A very limited set of particularly harmful uses of AI that contravene EU values because they violate fundamental rights and will therefore be banned:
  - ◊ Exploitation of vulnerabilities of persons, manipulation and use of subliminal techniques;
  - ◊ Social scoring for public and private purposes;
  - ◊ Individual predictive policing based solely on profiling people;
  - ◊ Untargeted scraping of internet or CCTV for facial images to build-up or expand databases;
  - ◊ Emotion recognition in the workplace and education institutions, unless for medical or safety reasons (i.e., monitoring the tiredness levels of a pilot);
  - ◊ Biometric categorization of natural persons to deduce or infer their race, political opinions, trade union membership, religious or philosophical beliefs or sexual orientation. Labelling or filtering of datasets and categorizing data in the field of law enforcement will still be possible;
  - ◊ Real-time remote biometric identification in publicly accessible spaces by law enforcement, subject to narrow exceptions (see below).
- The Commission will issue guidance on the prohibitions prior to their entry into force on 2 February 2025.
- High-risk: A limited number of AI systems defined in the proposal, potentially creating an adverse impact on people's safety or their fundamental rights (as protected by the EU Charter of Fundamental Rights), are considered to be high-risk. Annexed to the Act are the lists of high-risk
- AI systems, which can be reviewed to align with the evolution of AI use cases.
- These also include safety components of products covered by sectorial Union legislation. They will always be considered high-risk when subject to third-party conformity assessment under that sectorial legislation.
- Such high-risk AI systems include for example AI systems that assess whether somebody is able to receive a certain medical treatment, to get a certain job or loan to buy an apartment. Other high-risk AI systems are those being used by the police for profiling people or assessing their risk of committing a crime (unless prohibited under Article 5). And high-risk could also be AI systems operating robots, drones, or medical devices.
- Specific transparency risk: To foster trust, it is important to ensure transparency around the use of AI. Therefore, the AI Act introduces specific transparency requirements for certain AI applications, for example where there is a clear risk of manipulation (e.g., via the use of chatbots) or deep fakes. Users should be aware that they are interacting with a machine.
- Minimal risk: The majority of AI systems can be developed and used subject to the existing legislation without additional legal obligations. Voluntarily, providers of those systems may choose to apply the requirements for trustworthy AI and adhere to voluntary codes of conduct.

In addition, the AI Act considers systemic risks which could arise from general-purpose AI models, including large generative AI models. These can be used for a variety of tasks and are becoming the basis for many AI systems in the EU. Some of these models could carry systemic risks if they are very capable or widely used. For example, powerful models could cause serious accidents or be misused for far-reaching cyberattacks. Many individuals could be affected if a model propagates harmful biases across many applications.

## *How Do I Know Whether an AI system Is High-Risk?*

The AI Act sets out a solid methodology for the classification of AI systems as high-risk. This aims to provide legal certainty for businesses and other operators.

The risk classification is based on the intended purpose of the AI system, in line with the existing EU product safety legislation. It means that the classification depends on the function performed by the AI system and on the specific purpose and modalities for which the system is used.

AI systems can classify as high-risk in two cases:

- If the AI system is embedded as a safety component in products covered by existing product legislation (Annex I) or constitute such products themselves. This could be, for example, AI-based medical software.
- If the AI system is intended to be used for a high-risk use case, listed in an Annex III to the AI Act. The list includes use cases from in areas such as education, employment, law enforcement or migration.

The Commission is preparing guidelines for the high-risk classification, which will be published ahead of the application date for these rules.

## *How Are General-Purpose AI Models Being Regulated?*

General-purpose AI models, including large generative AI models, can be used for a variety of tasks. Individual models may be integrated into a large number of AI systems.

It is crucial that a provider of an AI system integrating a general-purpose AI model has access to all necessary information to ensure the system is safe and compliant with the AI Act.

Therefore, the AI Act obliges providers of such models to disclose certain information to downstream system providers. Such transparency enables a better understanding of these models.

Model providers additionally need to have policies in place to ensure that that they respect copyright law when training their models.

In addition, some of these models could pose systemic risks, because they are very capable or widely used.

Currently, general purpose AI models that were trained using a total computing power of more than 10^25 FLOPs are considered to pose systemic risks. The Commission may update or supplement this threshold in light of technological advances and may also designate other models as posing systemic risks based on further criteria (e.g., number of users, or the degree of autonomy of the model).

Providers of models with systemic risks are obliged to assess and mitigate risks, report serious incidents, conduct state-of-the-art tests and model evaluations and ensure cybersecurity of their models.

Providers are invited to collaborate with the AI Office and other stakeholders to develop a Code of Practice, detailing the rules and thereby ensuring the safe and responsible development of their models. This Code should represent a central tool for providers of general-purpose AI models to demonstrate compliance.

## *What Are the Obligations Regarding Watermarking and Labelling of the AI Outputs Set Out in the AI Act?*

The AI Act sets transparency rules for the content produced by generative AI to address the risk of manipulation, deception and misinformation.

It obliges providers of generative AI systems to mark the AI outputs in a machine-readable format and ensure they are detectable as artificially generated or manipulated. The technical solutions must be effective, interoperable, robust and reliable as far as this is technically feasible, taking into account the specificities and limitations of various types of content, the costs of implementation and the generally acknowledged state of the art, as may be reflected in relevant technical standards.

In addition, deployers of generative AI systems that generate or manipulate image, audio or video content constituting deep fakes must visibly disclose that the content has been artificially generated or manipulated. Deployers of an AI system that generates or manipulates text published with the purpose of informing the public on matters of public interest must also disclose that the text has been artificially generated or manipulated. This obligation does not apply where the AI-generated content has undergone a process of human review or editorial control and where a natural or legal person holds editorial responsibility for the publication of the content.

The AI Office will issue guidelines to provide further guidance for providers and deployers on the obligations in Article 50 which will become applicable two years after entry into force of the AI Act (on 2 August 2026).

The AI Office will also encourage and facilitate the development of Codes of Practice at Union level to streamlining the effective implementation of the obligations related to the detection and labelling of artificially generated or manipulated content.

## *Is the AI Act Future-Proof?*

The AI Act sets a legal framework that is responsive to new developments, easy and quick to adapt and allows for frequent evaluation.

The AI Act sets result-oriented requirements and obligations but leaves the concrete technical solutions and operationalization to industry-driven standards and codes of practice that are flexible to be adapted to different use cases and to enable new technological solutions.

In addition, the legislation itself can be amended by delegated and implementing acts, for example to review the list of high-risk use cases in Annex III.

Finally, there will be frequent evaluations of certain parts of the AI Act and eventually of the entire regulation, making sure that any need for revision and amendments is identified.

## *How Do the Rules Protect Fundamental Rights?*

There is already a strong protection for fundamental rights and for non-discrimination in place at EU and Member State level, but the complexity and opacity of certain AI applications ("black boxes") can pose a problem.

A human-centric approach to AI means to ensure AI applications comply with fundamental rights legislation. By integrating accountability and transparency requirements into the development of high-risk AI systems, and improving enforcement capabilities, we can ensure that these systems are designed with legal compliance in mind right from the start. Where breaches occur, such requirements will allow national authorities to have access to the information needed to investigate whether the use of AI complied with EU law.

Moreover, the AI Act requires that certain deployers of high-risk AI systems conduct a fundamental rights impact assessment.

What is a fundamental rights impact assessment? Who has to conduct such an assessment, and when?

Providers of high-risk AI systems need to carry out a risk assessment and design the system in a way that risks to health, safety and fundamental rights are minimized.

However, certain risks to fundamental rights can only be fully identified knowing the context of use of the high-risk AI system. When high-risk AI systems are used in particularly sensitive areas of possible power asymmetry, additional considerations of such risks are necessary.

Therefore, deployers that are bodies governed by public law or private operators providing public services, as well as operators providing high-risk AI systems that carry out credit worthiness assessments or price and risk assessments in life and health insurance, shall perform an assessment of the impact on fundamental rights and notify the national authority of the results.

In practice, many deployers will also have to carry out a data protection impact assessment. To avoid substantive overlaps in such cases, the fundamental rights impact assessment shall be conducted in conjunction with that data protection impact assessment.

## *How Does This Regulation Address Racial and Gender Bias in AI?*

It is very important to underline that AI systems do not create or reproduce bias. Rather, when properly designed and used, AI systems can contribute to reducing bias and existing structural discrimination, and thus lead to more equitable and non-discriminatory decisions (e.g., in recruitment).

The new mandatory requirements for all high-risk AI systems will serve this purpose. AI systems must be technically robust to ensure they are fit for purpose and do not produce biased results, such as false positives or negatives, that disproportionately affect marginalized groups, including those based on racial or ethnic origin, sex, age, and other protected characteristics.

High-risk systems will also need to be trained and tested with sufficiently representative datasets to minimize the risk of unfair biases embedded in the model and ensure that these can be addressed through appropriate bias detection, correction and other mitigating measures.

They must also be traceable and auditable, ensuring that appropriate documentation is kept, including the data used to train the algorithm that would be key in ex-post investigations.

Compliance system before and after they are placed on the market will have to ensure these systems are regularly monitored and potential risks are promptly addressed.

## *How Will the AI Act Be Enforced?*

The AI Act establishes a two-tiered governance system, where national authorities are responsible for overseeing and enforcing rules for AI systems, while the EU level is responsible for governing general-purpose AI models.

To ensure EU-wide coherence and cooperation, the European Artificial Intelligence Board (AI Board) will be established, comprising representatives from Member States, with specialized subgroups for national regulators and other competent authorities.

The AI Office, the Commission's implementing body for the AI Act, will provide strategic guidance to the AI Board.

In addition, the AI Act establishes two advisory bodies to provide expert input: the Scientific Panel and the Advisory Forum. These bodies will offer valuable insights from stakeholders and interdisciplinary scientific communities, informing decision-making and ensuring a balanced approach to AI development.

## *Why Is a European Artificial Intelligence Board Needed and What Will it Do?*

The European Artificial Intelligence Board comprises high-level representatives of Member States and the European Data Protection Supervisor. As a key advisor, the AI Board provides guidance on all matters related to AI policy, notably AI regulation, innovation and excellence policy and international cooperation on AI.

The AI Board plays a crucial role in ensuring the smooth, effective and harmonized implementation of the AI Act. The Board will serve as the forum where the AI regulators, namely the AI Office, national authorities and EPDS, can coordinate the consistent application of the AI Act.

## *How Will the General-Purpose AI Code of Practice Be Written?*

The drawing-up of the first Code follows an inclusive and transparent process. A Code of Practice Plenary will be established to facilitate the iterative drafting process, consisting of all interested and eligible general-purpose AI model providers, downstream providers integrating a general-purpose AI model into their AI system, other industry organizations, other stakeholder organizations such as civil society or rightsholders organizations, as well as academia and other independent experts.

The AI Office has launched a call for expression of interest to participate in the drawing-up of the first Code of Practice. In parallel to this call for expression of interest a multi-stakeholder consultation to collect views and inputs from all interested stakeholders on the first Code of Practice is launched. Answers and submissions will form the basis of the first drafting iteration of the Code of Practice. From the start, the Code is therefore informed by a broad array of perspectives and expertise.

The Plenary will be structured in four Working Groups to allow for focused discussions on specific topics relevant to detail out obligations for providers of general-purpose AI models and general-purpose AI models with systemic risk. Plenary participants are free to choose one or more Working Groups they wish to engage in. Meetings are conducted exclusively online.

The AI Office will appoint Chairs and, as appropriate, Vice-Chairs for each of the four Working Groups of the Plenary, selected from interested independent experts. The Chairs will synthesize submissions and comments by Plenary participants to iteratively draft the first Code of Practice.

As the main addressees of the Code, providers of general-purpose AI models will be invited to dedicated workshops to contribute to informing each iterative drafting round, in addition to their Plenary participation.

After 9 months, the final version of the first Code of practice will be presented in a closing Plenary, expected to take place in April, and published. The closing Plenary gives general-purpose AI model providers the opportunity to express themselves whether they would envisage to use the Code.

## *Does the AI Act Contain Provisions Regarding Environmental Protection and Sustainability?*

The objective of the AI proposal is to address risks to safety and fundamental rights, including the fundamental right to a high-level environmental protection. The environment is also one of the explicitly mentioned and protected legal interests.

The Commission is asked to request European standardization organizations to produce a standardization deliverable on reporting and documentation processes to improve AI systems' resource performance, such as reduction of energy and other resources consumption of the high-risk AI system during its lifecycle, and on energy efficient development of general-purpose AI models.

Furthermore, the Commission by two years after the date of application of the Regulation and every four years thereafter, is asked to submit a report on the review of the progress on the development of standardization deliverables on energy efficient development of general-purpose models and asses the need for further measures or actions, including binding measures or actions.

In addition, providers of general-purpose AI models, which are trained on large data amounts and therefore prone to high energy consumption, are required to disclose energy consumption. In case of general-purpose AI models with systemic risks, energy efficiency furthermore needs to be assessed.

The Commission is empowered to develop appropriate and comparable measurement methodology for these disclosure obligations.

## *How Can the New Rules Support Innovation?*

The regulatory framework can enhance the uptake of AI in two ways. On the one hand, increasing users' trust will increase the demand for AI used by companies and public authorities. On the other hand, by increasing legal certainty and harmonizing rules, AI providers will access bigger markets, with products that users and consumers appreciate and purchase. Rules will apply only where strictly needed and in a way that minimizes the burden for economic operators, with a light governance structure.

The AI Act further enables the creation of regulatory sandboxes and real-world testing, which provide a controlled environment to test innovative technologies for a limited time, thereby fostering innovation by companies, SMEs and start-ups in compliance with the AI Act. These, together with other measures such as the additional Networks of AI Excellence Centres and the Public-Private Partnership on Artificial Intelligence, Data and Robotics, and access to Digital Innovation Hubs and Testing and Experimentation Facilities will help build the right framework conditions for companies to develop and deploy AI.

Real world testing of High-Risk AI systems can be conducted for a maximum of 6 months (which can be prolonged by another 6 months). Prior to testing, a plan needs to be drawn up and submitted to the market surveillance authority, which has to approve the plan and specific testing conditions, with default tacit approval if no answer has been given within 30 days. Testing may be subject to unannounced inspections by the authority.

Real world testing can only be conducted given specific safeguards, e.g., users of the systems under real world testing have to provide informed consent, the testing must not have any negative effect on them, outcomes need to be reversible or disregardable, and their data needs to be deleted after conclusion of the testing. Special protection is to be granted to vulnerable groups, i.e., due to their age, physical or mental disability.

### *What is the International Dimension of the EU's Approach?*

AI has consequences and challenges that transcend borders; therefore international cooperation is important. The AI Office is in charge of the European Union international engagement in the area of AI, on the basis of the AI Act and the Coordinated Plan on AI. The EU seeks to promote the responsible stewardship and good governance of AI in collaboration with international partners and in line with the rules-based multilateral system and the values it upholds.

The EU engages bilaterally and multilaterally to promote trustworthy, human-centric and ethical AI. Consequently, the EU is involved in multilateral forums where AI is discussed—notably G7, G20, the OECD, the Council of Europe, the Global Partnership on AI and the United Nations—and the EU has close bilateral ties with, e.g., Canada, the U.S., India, Japan, South Korea, Singapore, and the Latin American and Caribbean region.

*Updated on 01/08/2024.

# About the Contributors

Paula **Andalo**, ethnic media editor, is focused on partnerships between *KFF Health News* and Spanish-language media.

Michelle **Andrews** is a contributing writer and former columnist for *KFF Health News* and has been writing about health care for more than 15 years.

Seetha **Anitha** is an international consultant and nutrition sensitive value chain specialist working for UN, Ethiopia.

Sandeep **Arora** is President and Head of Digital Experiences at Datamatics (www.datamatics.com).

**City and County of San Francisco** is a municipal government authority and commercial, financial, and cultural center within Northern California.

**Dang** Thi Thanh Tam is Founder, Rubie Marble (Real Estate) JSC, and former Senior Retail Business Manager for Intel, Ho Chi Minh City, Vietnam.

**Doan** Ngoc Duy is a Lecturer at the University of Architecture HCMC and ISB-University of Economics HCMC, as well as the Chief Growth Officer of Ba Huan Corporation.

Sachin **Dole** is an independent AI advocate and consultant to businesses.

Larry **Ebert** is a professional musician and composer, an instructor at the University of San Francisco and Golden Gate University, and is conducting research exploring the human factors impacts of AI.

Pearly **Ee** is Manager, Digital Platforms at advisory firm, IPI, Singapore.

Kevyn **Eng** is currently the Head of Brand Management, Marketing and Communications, Maybank Singapore.

**European Commission** is the primary executive arm of the European Union.

Umar **Farooq** was an Ancil Payne Fellow with *ProPublica*.

McKenzie **Funk** is a reporter at ProPublica's Northwest hub and is based in Bellingham, Washington.

Phil **Galewitz**, *KFF Health News* senior correspondent, covers Medicaid, Medicare, long-term care, hospitals, and various state health issues and has covered health for more than three decades.

Jagadish **Gona**, Program Manager at Bosch, Bangalore, India.

Joaquin Jay **Gonzalez** III is Vice Provost for Global Affairs, Founding Dean, and Mayor George Christopher Professor of Public Administration at Golden Gate University Worldwide, San Francisco, California.

Shalini **Gopalkrishnan** is AI Strategist at San Francisco Bay University.

Smrite **Goudhaman** is Head of Solutioning at Datamatics, India.

Courtlin **Holt-Nguyen** is the Head of Data and AI at QIMA.

Adrian **Hopgood** is Independent Consultant and Emeritus Professor of Intelligent Systems, University of Portsmouth.

Mike **Horne** is Associate Professor and Program Director, Human Resources Management at Golden Gate University, San Francisco, California.

Nicole C. **Jackson** is Assistant Vice Provost for Academic Affairs and Associate Professor of Management at Golden Gate University, San Francisco, California.

Gopi **Kallayil** is Chief Business Strategist, AI at Google.

Anzar **Khaliq**, an educator, designer and a physicist, leads transformative efforts in teaching and learning as the founding Chief Learning Officer at a private university in the Bay Area.

Richard R. **Khan** is the author of *The AI Glossary* and an AI and software engineering expert with over 20 years of leadership experience.

Gopal **Krishnan** is a corporate executive leader, board member and educator based in California.

Virgil **Labrador** is the Editor-in-Chief of Los Angeles, California-based *Satellite Markets and Research*.

Emmanuel C. **Lallana** is Professorial Fellow at the University of the Philippines.

Andrew **Leonard** reports for *Kaiser Health News*.

Veejay **Madhavan** is the founder of OulbyZ, Singapore.

Severo C. **Madrona**, Jr., is a Professorial Lecturer at the University of the Philippines, De La Sale University, and Ateneo de Manila University.

Sunil **Manikani** is Senior Data Scientist, SLB, Pune, India.

Mickey P. **McGee** is Associate Dean, DBA Director, and Professor of Public Administration at Golden Gate University Worldwide, San Francisco, California.

Ryan **McGrady** is Senior Researcher, Initiative for Digital Public Infrastructure, UMass Amherst.

Saroja Manicka **Nagarajan** is Professor at upGrad Study Abroad Learning Center, Hyderabad, India.

Hannah **Norman**, video producer and visual reporter, joined *KFF Health News* after covering health care for the *San Francisco Business Times*.

Sonny **Panesar** is an Executive Director at UBS AG.

Lakshmi R. **Pillai** is a Doctor of Business Administration student at Golden Gate University.

Ronald **Powell** is Vice President of Cloud and AI Infrastructure Go-to-Market Offerings, Dubai, United Arab Emirates.

Tachanun **Rattanasiriwilai** is a DBA candidate at Golden Gate University, specializing in Business Development and Technology Adoption in SMEs within the logistics sector.

Gautam **Roy** is a Doctoral Scholar in Generative AI at Golden Gate University, with 22 years of experience as a Digital Transformation Leader in the FMCG industry.

Upal **Roy** is the Associate Director, Advanced Analytics and data Science, Big Pharma Global capability Centre, Pune, India.

Sameeksha **Sahni** is the CEO of Infotrack Systems, a leading HR software company in India, and a recognized thought leader and researcher in HR technology and AI.

Jody C. **Salas** is an Associate Director at Accenture and professor of practice at Golden Gate University and University of the Philippines.

Kannan **Santharaman** is Professor, Department of AI and ML, Malla Reddy Institute of Technology and Science, Hyderabad, India.

Bruce **Schneier** is Adjunct Lecturer in Public Policy, Harvard Kennedy School.

Pratik N. **Shah**, Director of Ugam Chemicals, is a researcher exploring the transformative potential of AI in the paints and coatings industry.

Sulbha **Shantwan** is the Founder Director of Sandbox Academy of Alternative Learning, Pune, India.

Vandinika **Shukla** is Fellow, Practicing Democracy Project, Harvard Kennedy School.

Vinay **Singh** has a PhD in Physics and has worked as a quantitative researcher/data scientist on Wall Street and in the industry for several decades.

**State of Washington** is a state-level government authority in the Pacific Northwest region of the United States.

Daniel Y.N. **Tan** is a Singapore-based Entrepreneur, Corporate Trainer, Financial Adviser, and TV Actor.

**Tran** Minh Viet is a seasoned business professional from Vietnam with extensive experience in retail marketing, trade marketing, and team development.

R. "Doc" **Vaidhyanathan** is Vice President, AI Center of Excellence, IBM Infrastructure.

Ratheesh **Venugopal** is the Principal Regional IT Program Manager for Data Center Regional Operations at Microsoft, based in Amsterdam.

Francis **Wang** is the Head Futures Researcher and Portfolio Entrepreneur at FW VISION, Toronto, Ontario, Canada.

**White House Office of Science and Technology Policy** advises the President and others within the Executive Office of the President on the effects of science and technology on domestic and international affairs

Emmanuel G. **Zara**, Jr., is a Philippine-based consultant specializing in capacity building, executive and member coaching, learning and development, and the practice of mindfulness and self-leadership.

Lichao **Zhang** is Founder and Chairman, Sino-Heritage Institute for Chinese-Western Cultural Exchange and Director, Board of Directors, China Federation of Overseas Entrepreneurs.

Ethan **Zuckerman** is Associate Professor of Public Policy, Communication, and Information, UMass Amherst.

# Index